Human Dignity in Bic

This volume brings together a collection of essays that rigorously examine the concept of human dignity from its metaphysical foundations to its polemical deployment in bioethical controversies. The volume falls into three parts, beginning with meta-level perspectives and moving to concrete applications.

Part I analyzes human dignity through a worldview lens, exploring the source and meaning of human dignity from naturalist, postmodernist, Protestant, and Catholic vantages, respectively, letting each side explain and defend its own conception. Part II moves from metaphysical moorings to key areas of macro-level influence: international politics, U.S. law, and biological science. These chapters examine the legitimacy of the concept of dignity in documents by international political bodies, the role of dignity in U.S. jurisprudence, and the implications—and challenges—for dignity posed by Darwinism. Part III shifts from macro-level topics to concrete applications by examining the rhetoric of human dignity in specific controversies: embryonic stem cell research, abortion, human–animal chimeras, euthanasia and palliative care, psychotropic drugs, and assisted reproductive technologies. Each chapter analyzes the rhetorical use of human dignity by opposing camps, assessing the utility of the concept and whether a different concept or approach can be a *more* productive means of framing or guiding the debate.

Stephen Dilley is associate professor of philosophy at St. Edward's University, USA.

Nathan J. Palpant is senior research fellow at the University of Washington's Institute for Stem Cell and Regenerative Medicine, USA.

Routledge Annals of Bioethics

Series Editors: Mark J. Cherry, St. Edward's University, USA
Ana Smith Iltis, Saint Louis University, USA

1 **Regional Perspectives in Bioethics**
Edited by Mark J. Cherry and John F. Peppin

2 **Religious Perspectives on Bioethics**
Edited by Mark J. Cherry, Ana Smith Iltis, and John F. Peppin

3 **Research Ethics**
Edited by Ana Smith Iltis

4 **Thomistic Principles and Bioethics**
Jason T. Eberl

5 **The Ethics of Genetic Engineering**
Roberta M. Berry

6 **Legal Perspectives in Bioethics**
Edited by Ana Smith Iltis, Sandra H. Johnson, and Barbara A. Hinze

7 **Biomedical Research and Beyond**
Expanding the Ethics of Inquiry
Christopher O. Tollefsen

8 **Practical Autonomy and Bioethics**
James Stacey Taylor

9 **The Ethics of Abortion**
Women's Rights, Human Life, and the Question of Justice
Christopher Kaczor

10 **Bioethics, Public Moral Argument, and Social Responsibility**
Edited by Nancy M. P. King and Michael J. Hyde

11 **The Ethics of Gender-Specific Disease**
Mary Ann Cutter

12 **Death, Posthumous Harm, and Bioethics**
James Stacey Taylor

13 **Human Dignity in Bioethics**
From Worldviews to the Public Square
Edited by Stephen Dilley and Nathan J. Palpant

Human Dignity in Bioethics

From Worldviews to the Public Square

**Edited by Stephen Dilley and
Nathan J. Palpant**

Routledge
Taylor & Francis Group

NEW YORK AND LONDON

First published 2013
by Routledge
711 Third Avenue, New York, NY 10017

Simultaneously published in the UK
by Routledge
2 Park Square, Milton Park, Abingdon, Oxon OX14 4RN

*Routledge is an imprint of the Taylor & Francis Group,
an informa business*

Library of Congress Cataloging in Publication Data

Human dignity in bioethics : from worldviews to the public square / edited
    by Stephen Dilley and Nathan J. Palpant.
       p. cm. — (Routledge annals of bioethics ; 13)
    Includes bibliographical references and index.
  1. Bioethics.   2. Dignity.   I. Dilley, Stephen, 1974–
II. Palpant, Nathan J., 1979–
    QH332.H846 2012
    174.2—dc23
    2012026308

ISBN: 978-0-415-65931-4 (hbk)
ISBN: 978-0-203-07500-5 (ebk)

Typeset in Sabon
by Apex CoVantage, LLC

Printed and bound in the United States of America by Publishers Graphics,
LLC on sustainably sourced paper.

From Nathan

To Darien

To the soul and the lips
Of the woman I love,
I dedicate
Every word, every
Whisper, all the earth,
All the fire in my song
      —Pablo Neruda

From Stephen

To Madeline Laine Dilley
My daughter,
in whom my soul delights

If ever any beauty I did see,
Which I desir'd, and got, 'twas but a dreame of thee.
      —John Donne

Contents

Foreword   ix
Acknowledgments   xiii

Setting the Stage

1 Human Dignity in the Throes? An Introduction
   to the Volume   3
   STEPHEN DILLEY AND NATHAN J. PALPANT

2 Human Exceptionalism and the *Imago Dei:*
   The Tradition of Human Dignity   19
   DAVID H. CALHOUN

Part I: The Source and Meaning of Human Dignity in Worldview Context

3 A Catholic Perspective on Human Dignity   49
   CHRISTOPHER TOLLEFSEN

4 A Protestant Perspective on Human Dignity   67
   PAUL COPAN

5 Postmodern Perspectives on Human Dignity   86
   MARK DIETRICH TSCHAEPE

6 Dignity for Skeptics: A Naturalistic View
   of Human Dignity   103
   RICHARD MCCLELLAND

Part II: The Politics, Law, and Science of Human Dignity

7 International Policy and a Universal Conception
of Human Dignity   127
ROBERTO ANDORNO

8 Human Dignity and the Law   142
O. CARTER SNEAD

9 Prospects for Human Dignity after Darwin   166
DAVID H. CALHOUN

Part III: The Rhetoric of Human Dignity in Bioethics

10 Human Dignity and the New Reproductive Technologies   201
AUDREY R. CHAPMAN

11 The Language of Human Dignity in the Abortion Debate   219
SCOTT RAE

12 Human Dignity and the Debate over Early Human Embryos   239
NATHAN J. PALPANT AND SUZANNE HOLLAND

13 Human Dignity in End-of-Life Issues: From Palliative
Care to Euthanasia   264
THOMAS R. MCCORMICK

14 The Evolving Bioethical Landscape of Human–Animal
Chimeras   282
JOHN LOIKE

15 Psychotropic Drugs and the Brain:
A Neurological Perspective on Human Dignity   300
WILLIAM P. CHESHIRE JR.

Bibliography   323
Contributors   367
Index   373

Foreword

Human dignity is a key concept invoked in the ethical and political discourse of national and international bodies, treaties, declarations of human rights, religious teachings, and all manner of ethical writings. The meaning of *human dignity*, the grounds for invoking the concept, and the value of human dignity with respect to other important values remain topics of heated debate, of which this book is the latest example.

Intuitively, *dignity* has evocative power in Western thought. To take one of many examples where the concept is invoked, consider the ethical assessment of displays of the human body in terms of dignity. The National Library of Medicine of the United States National Institutes of Health launched the Visible Human Project in 1986 to "create complete, anatomically detailed, three-dimensional representations of normal male and female human bodies." Fifteen years later, the project was completed. It is available online at www.nlm.nih.gov/research/visible/visible_human.html and on a set of 60 CDs.

The bodies of a 39-year-old man executed by lethal injection in Texas and a 59-year-old woman from Maryland who died of a heart attack were used in the Visible Human Project. Under the supervision of a team at the University of Colorado Health Sciences Center, the cadavers were sent to Colorado, where they were thoroughly X-rayed and imaged with CT and MRI scans. They were then encased in blue gelatin and frozen. While frozen, they were sliced very carefully into exceedingly thin cross-sections from head to toe. The man was carved into 1,871 one-millimeter slices, the woman into 5,189 slices, each a third of a millimeter thick.

If someone had done this to a body in their basement, it would seem offensive and repugnant. It would, in the words of some, be a violation of human dignity—disrespectful of a human body. But the Visible Human Project has elicited no criticism of that nature. The information from the micro-dissection is used in many classrooms, exhibits, and publications. The slicing of two human beings into tiny sections is celebrated as a valued tool of science rather than as a violation of human dignity.

On the other hand, somewhat ironically, recent exhibits of preserved human bodies at museums and other exhibition sites have generated a good deal of criticism and comments about violations of human dignity.

Two major exhibits have been touring around the world—Body Worlds and Bodies: The Exhibition. Both have generated a great deal of attention and controversy. These are commercial displays involving relatively new techniques in which water and fat in the body's tissues are replaced with plastic. Plastination allows the preservation of almost all human tissues and parts and allows much more realistic depictions of the body in action. These exhibits have drawn huge crowds to science museums and exhibition parks all over North America, Europe, and other parts of the world. These exhibits, unlike the Visible Human Project, have also drawn fierce criticism. Why does slicing a body into tiny bits and putting the results online evoke little criticism, whereas displaying a dead body dribbling a basketball where the nerves and muscles involved can be readily appreciated gives rise to controversy?

Some insist showing real human remains that have been plastinated in museums, malls, civic centers, science museums, and other settings violates the body's intrinsic dignity. A typical argument concerning dignity and exhibitions like Body Worlds is as follows:

> The relevant claim is that human bodies may be used in a manner that disrespects their intrinsic worth only where the need is sufficiently great, e.g., to develop therapies or save a life. Such great need imposes a substantial duty on society that outweighs the near universal imperative not to treat human corpses as objects of consumption. (Burns 2007: 13)

Is this argument correct? Do these kinds of exhibitions violate the intrinsic dignity of bodies? Would we permit treating dead bodies like pure objects if it meant saving lives, as seems to be true of the Visible Human Project, but not just for education or entertainment, as seems true of Body Worlds? And if violation of a human body is an affront to human dignity, should legislative bodies set any limits on such exhibitions and should there be restrictions on who can see such exhibitions?

The reader of the essays in this volume should be in a much better position to both pose these questions and articulate a response to them. My own view is that it is not obvious that exhibitions of the human body in a plastinated state for educational and artistic purposes violate or disrespect the intrinsic worth of the human body. It surely is possible to display human remains in ways that are disrespectful. And it is also possible, as is known from many horrific examples throughout human history, including the mythic story of a triumphant Achilles dragging the fallen hero Hector's lifeless body in front of the Gates of Troy again and again, to disrespect human remains. But using human bodies and parts, artfully and even whimsically displayed, to engage public audiences so that they appreciate the beauty, shape, function, anatomy, and integration of the components of the human body seems to be not only respectful of the human body but precisely the means to both engage and cultivate that respect.

Not everyone believes that human dignity exists. A prominent U.S. bio-ethicist, Ruth Macklin of Albert Einstein College of Medicine in New York, declared in a short editorial in the *British Medical Journal* that "dignity is a useless concept in bioethics" (Macklin 2003: 1419). She went on to argue that outside of being used as a placeholder when asking for respect for persons or individual autonomy, dignity was mere window dressing in ethical argumentation.

That shot across the bow did not slow down the invocation of the idea of human dignity in various pronouncements offered by the United Nations, the President's Council on Bioethics, the Council of Europe, the Vatican, the European Union, and various athletic and scholarly associations on a variety of bioethical and non-bioethical subjects in the years that have followed. But Macklin's dismissal of the whole idea of human dignity clearly stuck in the analytical craw of those relying on the concept in their various arguments and positions. Some of the contributors to this volume are responding to those like Macklin who see no value, meaning, or basis for invocations of dignity.

Macklin says that the invocation of human dignity only makes sense if interpreted as a claim about respect for persons or as a claim about respect for individual autonomy. Is that really so? Many of us feel revulsion at what happens to persons who are tortured, raped, or trafficked for organs. Part of the reason is that we know that torture often involves humiliation or degradation of the victim. Respect for persons hardly seems sufficient for capturing the evil involved. Torture plays on cultural and social views of propriety that go beyond simply respecting the individual as a person to the broader social realms of what makes that person feel safe, whole, and a respected member of society.

Similarly, human dignity can refer, and has since Cicero and the Stoics, to the virtues that are esteemed in a society. Making a claim of dignity about a person or a group is more than simply acknowledging the right of individuals to act autonomously. Even the Christian notion of dignity invoking both humanity's likeness to God and praising human creations such as art, medicine, science, and philosophy can be secularized a bit so that it is simply the products of a conscious and socially cooperative, reflective creature that are infused with dignity due to their purposefulness and utility. Dignity may be something that is conferred rather than a property that inheres in things themselves.

The essays collected here should contribute to an understanding of what dignity means, why it can or cannot be invoked, and what or who has dignity. The stakes involved in the debate are enormous, ranging from at what state in development human life demands respect to core notions of human rights that undergird the prosecution of crimes against humanity and genocide. Perhaps the philosopher George Santayana offered the best advice in thinking about the arguments in this book: "Our dignity is not in what we do but what we understand."

Arthur L. Caplan

REFERENCES

Burns, L. (2007) 'Gunther von Hagens' Body Worlds: Selling Beautiful Education,' *American Journal of Bioethics*, 4, no. 7: 12–15.
Macklin, R. (2003) 'Dignity Is a Useless Concept,' *British Medical Journal*, 327: 1419–20.

Acknowledgments

Our debts to colleagues, friends, and family are great. First, and most obviously, we are grateful to each contributor for his or her fine work. Our thanks are also due to the editors of the *Annals of Bioethics* series for their expertise and helpful peer review process, both of which notably improved the volume. In collaboration with Richard McClelland, we wish to thank Ray Russ and the Institute for Mind and Behavior for permission to publish a condensed version of the following: Richard McClelland (2011) 'A Naturalistic View of Human Dignity,' *Journal of Mind and Behavior*, 32, 5–48, © Institute of Mind and Behavior.

Specific acknowledgements from Stephen Dilley are as follows:
I would like to thank Sharon Nell, dean of the School of Humanities at St. Edward's University. I also owe much to Fr. Lou Brusatti, former dean, for his support in the initial phases of this book. Likewise, I am in debt to Adam Pyles and Kate Rosati, administrative assistants in the dean's office, for generous help on so many things. I am also grateful to Mark Cherry, a philosopher at St. Edward's who first encouraged me to propose a volume of new essays for the *Annals* series. More generally, I have enjoyed many substantive conversations with Mark as well as Peter Wake, Jack Green Musselman, and Bill Zanardi, my colleagues in the St. Edward's Philosophy Department.

I would be remiss if I did not express my gratitude for Nate Palpant, my co-editor. Nate has been an outstanding colleague at every stage of the volume, from initial brainstorming to the final touches. I am grateful for his partnership.

My greatest debts are to Andrea and Madeline, my wife and daughter. Andrea has been supportive of the project from its inception, deftly finding ways to coordinate our schedules and manage our household in order to allow me time to work on the volume. More deeply, she understands the importance of ideas, their consequences, and what is required for a meaningful academic life. For seeing this vision clearly—and for balancing it so well in the midst of her own important writing—I am more grateful than I can say.

I dedicate this volume to Madeline, my daughter, who is three years old. She is passionate, expressive, creative, and opinionated—in short, full of life and zeal. Although physically small and helpless in so many ways, she is powerful in so many other ways, imparting strength to me and others by her love, laughter, and joyful discovery of the world. In her life below, I see the life above.

And from Nathan J. Palpant:

I would like to thank Suzanne Holland, professor of religion at University of Puget Sound, who provided stimulating conversation and helped coauthor our piece on the human embryonic stem cell controversy. Denise Dudzinski in the Bioethics Department at the University of Washington has provided many valuable and formative conversations about this topic. I would also like to thank Mark Cherry for providing guidance and helpful critiques that have been useful in the development of this volume.

Steve Dilley has been an excellent colleague with whom to work from the original discussions to the final developments as we have put this volume together. I am particularly thankful for his willingness to guide me through his expertise in ethics and philosophy. His insights have provided much of the basis in my understanding of ethics as I have attempted to amalgamate my scientific knowledge with ethical issues in contemporary biomedicine.

I dedicate this volume to my wife, Darien. Her patience and constant support have been essential and appreciated. I cannot provide sufficient words to express my gratitude for the gift and joy she is to me.

Setting the Stage

1 Human Dignity in the Throes?
An Introduction to the Volume

Stephen Dilley and Nathan J. Palpant

A close inspection . . . shows that appeals to dignity are either vague restatements of other, more precise, notions or mere slogans that add nothing to an understanding of the topic.

—Ruth Macklin

"I don't know what you mean by 'glory'," Alice said.

Humpty Dumpty smiled contemptuously. "Of course you don't—till I tell you. I meant 'there's a nice knock-down argument for you!'"

"But 'glory' doesn't mean 'a nice knock-down argument'," Alice objected.

"When *I* use a word," Humpty Dumpty said, in a rather scornful tone, "it means just what I choose it to mean—neither more nor less."

"The question is," said Alice, "whether you *can* make words mean so many different things."

"The question is," said Humpty Dumpty, "which is to be master— that's all."

—Lewis Carroll

There is no such thing as human dignity, and this is a book about it.[1] Dignity talk is presently all the rage in bioethics and beyond—from fictional texts like Huxley's *Brave New World* to historical case studies like the Nuremburg trials to a host of political fora, including United Nations panels and the President's Council on Bioethics. A literature search on "human dignity" using Harvard's culturomics database (Michel et al. 2010) shows that references to the term have increased significantly in recent decades, a surrogate marker of the intensified interest on the topic. But like many frequently cited concepts, consensus remains elusive—not only in the *application* of human dignity to specific procedures and cases but also in the *ground* of the concept and even the *content* of the concept itself. Indeed, what has marked this increased interest is the complexity, urgency, and seemingly irreconcilable diversity of opinions about human dignity.

These discussions have not been limited to the Anglophone world. The German word for human dignity, *Menschenwürde*, has been under close scrutiny as well. As Dieter Birnbacher has written, some thinkers fear that human dignity will lose its normative weight and become "a piece of empty rhetoric full of connotation but devoid of denotation" (1996: 109). Birnbacher wrote this warning in 1996, and his sentiments have been echoed in recent years, indicating that the concept of human dignity still stands in need of clarification.

Closer to home, the exchanges over pronouncements by the President's Council on Bioethics and UNESCO's 2005 Universal Declaration of Bioethics and Human Rights are characteristic of the general melee about human dignity. When these entities published reports using human dignity as a sine qua non for controversial matters in bioethics and human rights, they were met with criticisms from an array of academic, political, and multinational organizations. Many dissenters focused on the fundamental problem of defining human dignity, questioning whether it was a legitimate concept for establishing substantive positions on controversial matters. Ruth Macklin, to cite just one example, famously argued that dignity is a "useless" concept meaning nothing more than "respect for persons" or "respect for autonomy," which are well-established and more precise terms (Macklin 2003: 1419–20).

To appreciate the poignancy of the debate, consider Margaret Somerville's analysis of the Universal Declaration of Human Rights, the UNESCO Declaration on Bioethics, and the Universal Declaration on the Human Genome and Human Rights. She observes that these bodies attribute a range of specific properties, implications, and relations to dignity. For example, recognition of human dignity implies the following metaphysical or ethical principles, among others:

1. Recognition of equal rights
2. Affirmation of the worth of the person
3. The equality of all people
4. Respect for cultural rights
5. Human beings are more than their genetic characteristics—that is, genetic reductionism is inconsistent with respect for human dignity

According to Somerville, these documents also discuss practical moral implications arising from the notion that human beings have dignity. For example:

1. We must improve health, but only within the confines of respecting human dignity
2. The human dignity of genetic research subjects must be respected
3. We ought to avoid genetic discrimination, which offends human dignity
4. Respect for dignity prohibits practices such as reproductive cloning
5. States must defend human dignity

6. Breaches of human dignity need to be identified
7. We need to promote respect for dignity, especially in light of scientific advances that can threaten dignity
8. Respect for culture must not be used as an excuse to infringe upon human dignity

Yet Somerville sharply notes, "Despite the extensive use of the concept of human dignity, it is *nowhere defined* in these instruments" (2010: 2–4, emphasis added). On the surface this seems troubling indeed. In these documents, great matters stand or fall upon human dignity—from respect for human rights to the equality of all people to guidelines for scientific research. Yet in these documents, the concept itself remains opaque. One might worry that so much seems to rest on so little.

As Leon Kass has pointed out, we are caught between caring "a great deal about human dignity," especially as we battle "slavery, sweatshops, and segregation," *and* regarding the notion with suspicion. "Among us," Kass explains, "the very idea of 'dignity' smacks too much of aristocracy for egalitarians and too much of religion for secularists and libertarians. Moreover, it seems to be too private and vague a matter to be the basis for legislation or public policy" (2008: 298). We instinctively put great stock in dignity and, on second thought, wonder if we ought to do so.

So does human dignity amount to anything? If it is a substantive notion, what is its source or ground? And what is its role in bioethics, if any at all? In an effort to make these questions tractable, distinctions have multiplied like party guests. Deryck Beyleveld and Roger Brownsword, for example, separate "human dignity as empowerment," which centers on respect for autonomy, from "human dignity as constraint," which prohibits certain (degrading) individual choices, from "dignified conduct," which admits of higher and lower degrees (Meilaender 2008: 273–74). These and other distinctions have raised questions about what types of dignity, if any, ought to be regarded as intrinsic and which types, by contrast, ought to be regarded as extrinsic (or instrumental). The former concept denotes an inviolable and essential property of a thing, while the latter denotes a contingent feature of a thing, subject to increase, decrease, or absence altogether. A further dispute concerns the relationship between intrinsic and extrinsic notions of dignity as well as their relative importance.

At a deeper level, the core concept itself remains contested. Some scholars, like Robert Andorno, think the concept *is* sufficiently clear and robust, *pace* the criticisms of Macklin and company. He argues that dignity cannot be replaced by "respect for autonomy," as Macklin supposes, because

> respect for persons' autonomy is just a *consequence* of human dignity, not dignity itself. If people would not have any inherent value, then there would be no reason to respect their autonomous decisions. In other words, the (usually implicit) ultimate ground for respecting

people's autonomy (as well as any other basic human good or interest) is the intrinsic worthiness that we attach to every human being. (Chapter 7, this volume, original emphasis)

Other scholars advocate *additional* concepts in order to help clarify or bolster dignity. Somerville, for example, commends appropriating the concept of the "sacred." She explains:

if we accept the sacred, whether the religious sacred or the secular sacred, we will accept that certain aspects of life are sacrosanct; not everything that could be done to life, in particular, human life, may ethically be done to it and certain moral and ethical principles that should be respected and should govern our conduct flow from that. (2010:13, emphasis removed)

Or perhaps, as Samuel Kerstein recommends, we ought to return to a Kantian position, since "treating persons as ends in themselves entails . . . respect[ing] the special value inherent in persons, that is, their dignity" (2009: 151). Of course, a host of other scholars disagree, variously arguing for different additional concepts to support dignity or outright replacement of the notion entirely.

Disputes have also broken out over which worldview best grounds and protects human dignity. Atheist philosopher Daniel Dennett, for example, argues that human dignity can still be a "sacred value" even within a materialistic universe (2008: 39–59). Rather strikingly, Dennett is the same thinker who triumphantly proclaims that Darwin's theory is a "universal acid" that "eats through just about every traditional concept, and leaves in its wake a revolutionized world-view, with most of the old landmarks still recognizable, but transformed in fundamental ways" (1995: 63). Robert Kraynak, on the other hand, contends that sacred values have no place in an accidental universe and that, in contrast to Dennett's materialism, "the Bible and Christian theology may make the strongest case for human dignity" (2008: 62). These thinkers and others grapple over an age-old question: If God is dead, to paraphrase Sartre, are all things permitted? In that case, must humans always be treated with dignity? These questions are as momentous as they are difficult.

In the end, it is safe to say that despite dignity's long pedigree—from the ancient Hebrews, Cicero, and the Church fathers to Kant and a host of contemporary thinkers—widespread disagreement still persists. So diffuse is the consensus that some pundits even contend that, in the words of one, "Across all disciplines, human dignity is an underexplored topic" (Castiglione 2008: 676). While all sides in the debate are talking about human dignity, few agree on its source, content, or what implications it has—if any—for public policy, laboratory research, and professional medical conduct.

To make some headway, a closer examination of the concept of human dignity itself is in order—not just its alleged meaning in discrete contexts, but its *fundamental meaning* and *source*, if any, as well as its *rhetorical* (or functional) meaning in specific ethical controversies. Accordingly, this volume brings together a collection of essays that rigorously examine the concept and rhetoric of human dignity from its metaphysical foundations to its polemical deployment in the public square. Part I of the volume analyzes human dignity through a worldview lens, exploring the source and meaning of human dignity from naturalist, postmodernist, Protestant, and Catholic vantages, respectively, letting each side explain and defend its own conception.

Part II moves from metaphysical moorings to key areas of macro-level influence: international politics, U.S. law, and biological science. These chapters examine the legitimacy of the concept of dignity in documents by international political bodies, the role of dignity as a second-order legal concept in U.S. jurisprudence, and the implications—and challenges—for dignity posed by Darwinism.

Part III moves from macro-level topics to concrete applications by examining the rhetoric of human dignity in specific controversies: embryonic stem cell research, abortion, human-animal chimeras, euthanasia and palliative care, psychotropic drugs, and assisted reproductive technologies. In this section, each chapter analyzes the rhetorical use of human dignity by opposing sides in a particular area and assesses the utility, if any, of this rhetoric in the debate. In light of this assessment, each chapter provides an argument about how the concept of human dignity can clarify or further the debate in fruitful ways or, alternatively, about how a novel (or different) concept, set of concepts, or general approach can be a *more* productive means of framing or guiding the debate. Thus, the chapters in Part III do not simply diagnose the rhetoric of human dignity in contemporary bioethical debates but also commend a helpful way to illuminate the discussion. Collectively, Parts I, II, and III explore human dignity from its fundamental moorings to its polemical use in bioethical disputes.

In the end, this volume faces hard questions. Is dignity a vacuous notion? Or is it an inviolable feature of human nature? How does the nature, scope, and meaning(s) of human dignity change when contextualized in postmodern, naturalistic, or theistic worldviews, respectively? Is dignity a crucial ethical concept that ought to govern bioethical policy? Or is it simply a distraction that is best cast aside? Controversy about human dignity, like death and taxes, will likely be with us for some time. But progress can only happen, if at all, with a free exchange of ideas. Having outlined the general contours of the volume, a closer inspection of this exchange is now in order.

SETTING THE STAGE

No contemporary exploration of human dignity would be complete without a historical examination of the concept's long and textured past. David Calhoun helpfully prepares us for the current discussion by leading a guided tour of the history of human dignity. He argues that the concept of human dignity constitutes a *tradition*—in this case, a 2,000-year-old ongoing effort to exploring the nature and value of human existence within an enduring framework. This framework has two primary features: human exceptionalism and the *imago Dei*. The former denotes that "humans are different from and superior to all other living things, primarily due to the power of reasoning," while the latter denotes that, of all creatures, human beings reflect God's own nature in a unique way. Calhoun contends that these two elements emerged independently in classical paganism and Judaism, were synthesized in early Christianity, and then were developed in a continuing way through the Middle Ages and Renaissance.

Turning his attention to the modern era, Calhoun explores the modifications made to the traditional concept of human dignity by Descartes and Kant, among others, as well as the outright attacks by Schopenhauer, Nietzsche, Darwin, and the like. In the contemporary scene, Calhoun examines the attempted restoration of the concept by more recent thinkers, including Jacques Maritain, Leon Kass, John Paul II, and company. Calhoun concludes by drawing on his historical analysis to offer reflections about how the history of the concept points the way forward to specific applications in present-day matters.

PART I: THE SOURCE AND MEANING OF HUMAN DIGNITY IN WORLDVIEW CONTEXT

This section of the volume treats the foundational source and meaning of human dignity at the worldview level. Christopher Tollefsen leads off with a well-written Catholic perspective. The concept of human dignity, he notes, is repeatedly emphasized in recent Catholic papal and magisterial teaching and, as a result, has become a central concept in the scholarship of a number of Catholic thinkers. Yet Tollefsen acknowledges that human dignity has come under fire. As such, after expositing a version of the Catholic view of dignity, he takes aim at three prominent objections. The first holds that the idea of dignity is a species-ist notion, a way of illicitly giving preference to human beings over the rights of nonhuman animals. The second objection, a familiar one, contends that the notion of dignity is too vague and that its vagueness serves as a rhetorical cover for views that are intellectually anemic. The final objection comes from within Catholicism itself, arguing that the Catholic idea of dignity is simply mistaken, perhaps by overemphasizing individual autonomy.

By way of reply, Tollefsen allows that the Catholic view of human dignity does indeed give preference to species membership but not unreasonably so. He also contends that, while dignity sometimes serves as a mere rhetorical gloss, it does not do so in Catholic thought, where it is conceptualized in a way that supports robust moral norms. Tollefsen also maintains that a Catholic conception of dignity preserves individual freedom *within*—rather than against—a framework of justice, the common good, the individual's obligation to serve both God and humanity. Finally, Tollefsen closes by arguing that a Catholic view of human dignity has a substantive role to play in the public square, one that can crucially support the foundations of (public) moral, political, and religious thought.

Unlike Tollefsen, who defended his view against objections, Paul Copan follows in Chapter 2 by going on the offensive, arguing that the metaphysic articulated in the Judeo-Christian tradition—with its emphasis on the *imago Dei* as well as the incarnation, death, and resurrection of Christ—offers the *best* metaphysical ground for human dignity. Copan critiques a range of nontheistic worldviews to make his case. He argues, for example, that traditional Eastern worldviews typically either conceive of humans as mere bundles of ever-changing properties rather than enduring selves or regard the distinction between humans as illusory, making individual rights impossible. Moreover, Copan contends that a naturalistic worldview also supplies an inadequate metaphysical mooring for inviolable human dignity. As Bertrand Russell acknowledged, in such a worldview, human beings are ultimately "the outcome of accidental collocations of atoms." But if humans are the product of nonmoral entities and forces, then why suppose they possess *intrinsic* dignity and worth? Copan also argues that postmodern perspectivalism falters because if there are no meta-narratives, then there can be no "objective, universal notion of human rights and intrinsic human dignity." But this leaves humans with only "local" notions—an endorsement of mere tribalism, which, historically, has been a source of violence.

Copan also articulates a distinctly Protestant understanding of human dignity (which he regards as compatible with a traditional Catholic view), emphasizing the work of Dietrich Bonhoeffer (Lutheran) and Max Stackhouse (Reformed) in particular. Copan then applies this conception of dignity to the treatment of the mentally handicapped and the elderly (especially in relation to active euthanasia). He argues, contrary to utilitarian or functionalist approaches, that "the least of these" in the human family have full dignity. Finally, Copan contends that a Protestant notion of human dignity can function well in a pluralistic society. Those who are not "people of the book" can still affirm elements common to our humanity—since all humans are God's image-bearers—and thus function properly when they live according to the way they have been designed. Thus, while not everyone may *believe* in a Protestant notion of dignity, they can nonetheless *abide* by it.

Mark Dietrich Tschaepe shifts away from Christian theism in Chapter 3, taking a distinctive postmodern turn. In his clear and well-written way, Tschaepe attacks the common assumption that a substantive understanding of human dignity cannot be sustained once we enter what Jean-François Lyotard called the "postmodern condition." The concept of human dignity, Tschaepe maintains, can function meaningfully in multiple ways within a postmodern context, especially in the fields of medicine and bioethics. Tschaepe offers three postmodern perspectives in support of his contention, inviting the reader to recognize their individual and collective value.

He first examines Lyotard's view, whose work *The Postmodern Condition* (1979) is considered a definitive postmodern text. For Lyotard, human dignity is not an intrinsic property of persons but, as Tschaepe explains, "the authority to speak and be heard on the level of local discourse"—an authority one person confers upon another when freely acknowledging another's view. Tschaepe then analyzes the views of Richard Rorty, an American postmodernist. Rorty develops a view of human dignity that includes strong connections between pragmatism, postmodernism, and emotivism, contending that dignity consists of an emotional bond a person feels toward another. Third, Tschaepe explores the thought of Larry Hickman, who advocates "post-postmodernism." Hickman's view includes postmodernism but also "biological, environmental, and communicative" elements which, when combined with instrumentalism, accommodate concepts like dignity insofar as they effectively solve ethical problems. Dignity, in short, is valuable for its problem-solving power.

Tschaepe then applies all three perspectives to two specific areas of bioethics: human-animal chimeras and voluntary euthanasia. He maintains that, in their varied ways, all three thinkers offer illuminating and helpful approaches to human dignity. The postmodern condition, Tschaepe concludes—far from denying or attenuating the concept of human dignity—extends and multiplies the conversations about human dignity in meaningful and attractive ways.

Finally, Richard McClelland rounds out Part I by providing a detailed and careful exploration of human dignity from a naturalistic (or agnostic) perspective. He notes that standard conceptions of dignity (typically involving some notion of equality or the prohibition of certain kinds of actions toward others) are conceptually underdeveloped or poorly grounded. This trend contributes to the mistaken idea, according to McClelland, that dignity does *not* have any determinate content or that it *requires* classical theism or Kantian rationalism as its ground. Instead, McClelland argues that a "naturalized" agnostic worldview—which draws heavily from a nontheistic understanding of evolutionary history—provides both rich content and firm ground for human dignity.

McClelland's unique account fundamentally contends that human dignity has biological meaning and that this meaning is bound up with the evolution of cooperative behavior among humans. Cooperative behavior, in turn, draws on at least three kinds of activities in humans' evolutionary history: reciprocity, punishment, and communication. These three provide content (or meaning) to the concept of human dignity in complementary ways. The first, reciprocity, undergirds human dignity by serving as a basis in human evolutionary history for what have become modern notions of equity and/ or equality, which are closely associated with human dignity (especially in contemporary international documents on human rights). The second, punishment, undergirds human dignity by providing an evolutionary context in which human beings learned to relate to their conspecifics, eventually forming conceptions of prohibitions of certain actions (e.g., torture) as well as a duty to punish wrongdoers. The final feature, dignity as a communicative signal, plausibly arose as a way for early humans to assess threats by rapidly processing facial patterns or whole body postures of other persons.

Having given a detailed account of the biological and evolutionary meaning of human dignity, McClelland proposes a fully "scientific" study of dignity, which includes rigorous analysis of cooperation using methods and techniques already well established in biology, physiology, and neuroscience. Such a research program can further illuminate the nature of dignity, a task that is increasingly important in an interconnected world in which human beings must learn to cooperate on a global scale.

PART II: THE POLITICS, LAW, AND SCIENCE OF HUMAN DIGNITY

This section of the volume moves from worldview considerations to areas of broad application and influence, focusing on the concept of human dignity in international politics, U.S. law, and Darwinian science, respectively. Roberto Andorno leads off by examining human dignity in international politics, with a special emphasis on bioethics. He notes that, since the end of the Second World War, the notion of human dignity has become a central legal and political concept that provides the ultimate rationale for the unconditional respect that every human deserves, both in the context of international law and in bioethics. By way of example, Andorno argues that human dignity and human rights are historically and conceptually coupled in the Universal Declaration of Human Rights (1948) and that they were, from the very beginning, closely related to medicine. He also contends that since the end of the 1990s, human dignity has occupied a central place in the intergovernmental documents addressing bioethical issues. Moreover, the meaning of human dignity in these documents is sufficiently clear: it not only denotes the inherent worthiness of individuals but also the need to preserve the integrity and identity of humankind as a whole.

Of course, some critics oppose any intergovernmental coalition that agrees upon a *single* conception of human dignity and also regards this conception as binding upon *all* other nations. Such a claim, say critics, is an affront to individual cultural sovereignty. But Andorno responds that a universal understanding of human dignity is perfectly compatible with respect for cultural diversity, as long as the latter is not conflated with cultural relativism. Additionally, he claims, such universality does not necessarily entail a form of Western cultural imperialism but rather attempts to guarantee a minimum standard of justice for all people in the world. Thus, human dignity not only has a definite meaning but a critical role in international politics and bioethics as well.

O. Carter Snead next turns our attention to the law and human dignity. He explores whether, and to what extent, human dignity operates as a juridical principle in U.S. public bioethics. That is, he examines just how the concept functions in a U.S. legal context in which the regulation of science, medicine, and biotechnology are aimed at the preservation or promotion of public (and private) goods. To make his analysis concrete, Snead investigates the legal and political contexts of abortion, embryonic stem cell research, and euthanasia. He argues that, in these cases, as well as more broadly, human dignity is not a concept that fits comfortably into the U.S. politico-legal framework as a first-order juridical principle. Nevertheless, it is often invoked at a second-order level, typically to add moral force to claims drawing on common legal concepts such as equality, liberty, and justice. Snead also notes that different thinkers use human dignity to bolster these other concepts in a variety of different ways—sometimes for diametrically opposed perspectives.

Given that dignity has been invoked for opposing positions and that it is a not a first-order principle in U.S. jurisprudence, one might wonder if it is replaceable by the more familiar terms of equality, liberty, and privacy (or their combination). Snead argues to the contrary. He contends that dignity—intrinsic dignity, that is—stands as "an inviolable threshold rooted in who people are as members of the human family. It is distinctive from liberty, privacy, and even equality, in that it makes claims on us all to treat others in a manner befitting a human person." Thus, while dignity occupies a less visible role in U.S. law than equality, liberty, justice, and privacy, it offers something vital—something that captures an essential part of being human—that these other concepts alone (or in tandem) do not.

Having examined the international political and legal issues surrounding human dignity, David Calhoun rounds out Part II by addressing what Darwinian science implies for human dignity. He begins by stating that the traditional concept of human dignity, which arose from pagan antiquity and the Judeo-Christian religion, centered on human exceptionalism and the *imago Dei*—that humans were different in kind from other animals and that they were created in God's own image. Darwin's theory of evolution, however, challenged both of these fundamental grounds for human dignity, opposing human uniqueness with common descent and rejecting the *imago Dei* by replacing special creation with the mechanism of natural selection. Some Darwinists further complicated matters by associating evolutionary theory with a particular philosophical outlook involving materialism, atheism, or scientism. In short, Darwinism appeared to support a form of nihilism that destroyed the basis of human dignity.

Accordingly, Calhoun considers whether human dignity can regain its moorings in the wake of evolutionary theory. He examines three basic approaches. The first is an evasion strategy that seeks to *avoid* the implications of Darwinism for human dignity. Examples include creationism or Gould's famous NOMA proposal. Second, reconception strategies, like

Darwinian existentialism or Darwinian progressivism, attempt to reground values and reestablish dignity *within* a Darwinian framework. Calhoun contends that both of these strategies fall prey to empirical, conceptual, or practical difficulties.

Calhoun recommends a third strategy, restoration, which he believes offers a rapprochement between evolutionary science and human dignity. He argues that an emergentist account of human cognition, which is compatible with common descent, shows that human beings are qualitatively superior to animals, thus grounding human exceptionalism. Further, Calhoun argues that this approach is fully compatible with the view that God is a "continuously creating" ontological cause of the physical universe. Since God sustains all that is, all things bear the mark of the divine. And since human beings reflect God's rationality and agency in a special way, they can be said to bear the *imago Dei*. Thus, Calhoun concludes, while Darwinism may appear to be devastating for human dignity, when appropriately conceptualized and augmented, it need not be so.

PART III: THE RHETORIC OF HUMAN DIGNITY IN BIOETHICS

This final section of the volume moves into concrete controversies in bioethics: embryonic stem cell research, abortion, human-animal chimeras, euthanasia, psychotropic drugs, and assisted reproductive technologies. These essays focus primarily on examining the *rhetoric* of human dignity—how the concept is used by rival factions politically, socially, polemically, and so on—and whether, in light of this rhetorical melee, the concept of human dignity is fruitful or replaceable.

The first three essays examine the concept of human dignity in issues related to early human life and procreation. Audrey Chapman leads with a critical analysis of how the concept of human dignity has been used in the debates about reproductive technologies. Chapman's assessment suggests that human-dignity rhetoric is typically either used normatively as a means of empowerment or as "a broad social or moral position that a particular type of activity is contrary to public morality or the collective good." Within these broad approaches, however, Chapman argues that, in most cases in which a particular technology is considered, the notion of human dignity is often carried out in a "subjectless vacuum" and lacks specific criteria for assessing the ethical implications of the technology under consideration. She also notes that much of this ambiguity in defining human dignity has to do with contentious foundations regarding moral evaluations of early human life, issues of naturalness in procreation, and the right to make private reproductive choices. Chapman concludes that human-dignity rhetoric, as it presently stands, may not be useful in bioethical dialogue over human reproductive technologies.

What can be done? To move the discussion forward, Chapman proposes four requirements for a meaningful treatment of human dignity in light of specific reproductive technologies: (1) a well-defined conception of human dignity with specific criteria that can be used for evaluative purposes, (2) a clear specification as to whose dignity may potentially be affected, (3) a nuanced discussion of the specific reproductive technology being assessed with a clear distinction between the technology itself and the way it is likely to be used, and (4) a distinction between first-person dignity and third-person dignity—that is, a distinction between the effect of the technology on the subject and the manner in which the subject is viewed by others, along with their assessment of the implications of the particular technology for the human species as a whole.

Next, Scott Rae turns our attention to the rhetorical use of human dignity in the abortion debate. For those who oppose abortion and support restrictive abortion laws, Rae points out that dignity almost always assumes or attempts to establish that embryos and fetuses are, in an anthropological sense, persons from conception forward. The encyclicals of the Catholic Church (particularly *Dignitas Personae*) and various Protestant commentators, Rae points out, have been chief among the pro-life groups who place great weight on the concept of dignity.

On the other side, Rae observes that pro-choice advocates are preeminently concerned with the rights and decisional autonomy of the woman, claiming that, in the context of the abortion debate, these desiderata are central to a woman's dignity. Rae notes that both the pro-life invocation of fetal dignity and the pro-choice subversion of it only make sense if the prior issue of who belongs in the human community has been settled.

Rae follows with an analysis of contemporary juridical interpretations of human dignity that have been a basis for either allowing or restricting abortion. The 1992 *Casey* decision attempted to balance competing versions of human dignity, asserting "symbolic" dignity to the fetus while ultimately deciding in favor of maternal dignity and decisional autonomy. In the *Gonzales v. Carhart* decision, the court used dignity in a way that it had not used it before—appealing to the dignity of the pregnant woman *to restrict abortion*.

Although Rae believes human-dignity rhetoric has been useful in the debate, he raises concerns. For example, one major block to effective discourse hinges on legitimate competing moral claims about maternal versus fetal dignity, which are rooted in the problem of the metaphysical and moral status of the fetus. Amid these issues, Rae points out that symbolic ascriptions of dignity (as seen in the *Casey* decision) and efforts to use dignity as a rhetorical trump card are not useful. Rae concludes that effective dialogue around human dignity requires being clear about one's metaphysical and anthropological commitments that form the basis for human dignity.

Rounding out this tripartite assessment of human-dignity rhetoric in issues related to early human life, Nathan Palpant and Suzanne Holland examine the human embryonic stem (hES) cell controversy. They open with a historical observation about the origins of research on hES cells in the 1950s at the inception of in vitro fertilization. As research on early embryos has become a more public issue since that time, Palpant and Holland note that much of the contemporary debate on hES cells has been influenced by the abortion controversy, with the emphasis resting predominantly on evaluating the metaphysical and moral status of the embryo.

Palpant and Holland focus their evaluation of human-dignity rhetoric around Lennart Nordenfelt's three varieties of dignity. First, the "dignity of moral stature" is the result of the moral deeds of the subject. Palpant and Holland point out that the core issue with this notion of dignity is the integrity of our personal and communal sense of self respect. In this context, a moral agent's decision to support or oppose research with early human embryos has implications for the agent's sense of self-respect, making this conception of dignity a critical facet of the hES controversy. Second, the "dignity of identity" refers to the level of the integrity of the subject's body and mind. This notion forms the core argument for therapeutic use of hES cells because it seeks to protect an individual's dignity in order to preserve the value of the individual's ongoing biological well-being. Last, the "dignity of *Menschenwürde*" pertains to all human beings to the same extent and cannot be lost as long as the subject exists. In this conception, human dignity is inviolable. Palpant and Holland observe that much of the literature on the hES cell controversy draws upon this definition of dignity. Unfortunately, the multifaceted interpretations of the dignity of *Menschenwürde* make it difficult to use the term as a common ground for discussion and policy. Moreover, scholars disagree about the *basis* of intrinsic dignity—invoking various concepts such as personhood (e.g., Kant), potentiality, species identity and uniqueness, and human rights.

In light of this survey, Palpant and Holland conclude that, because appeals to human dignity are numerous and discordant, the rhetoric itself hinders rather than helps the formation of a workable public policy on human embryonic stem cell research. Although human-dignity rhetoric may be important and even essential in other contexts, they advocate two alternative rhetorical strategies for clearer deliberation in the hES cell controversy. First, the concept of healing spans time and ideology, providing a substrate for discussion on embryo uses in in vitro fertilization and embryonic stem cell therapy. Second, Palpant and Holland advocate evaluations of moral complicity both in the use of technologies involved with human embryos and in the development of each current (and future) source of pluripotent stem cells.

Moving from discussions on human dignity at the beginning of human life, Thomas McCormick follows with an examination of the rhetoric of human dignity at the end of life. He points out that, in medical care, normative ethical duties to individuals at the end of life are often couched in terms of human dignity. For example, McCormick observes that the Death with Dignity Act, currently practiced in Oregon and Washington, was established largely under the rhetorical guise of respecting individual dignity and autonomy in end-of-life decision making.

McCormick next turns his attention to three landmark cases, *Quinlan*, *Cruzan*, and *Glucksberg*, that led to significant changes in clinical care of dying patients. In all of these cases, McCormick argues that appeals for change in clinical practices were defended and opposed by appeals for respecting human dignity. He contends, however, that in these (and other) end-of-life cases, appeals to the notion of dignity are often strongly reinforced by an appeal to individual autonomy. Individual autonomy itself is grounded in the larger concept of "respect for persons," which goes beyond autonomy by including those whose limitations may inhibit or prevent autonomous choices and actions. As such, while McCormick does not advocate dispensing with the notion of human dignity altogether, he favors focusing on respect for persons instead.

This latter concept, he believes, will serve as an effective working principle in caring for persons at the end of life. Respect for persons, for example, allows for ethically acceptable alternatives in end-of-life care within a pluralistic society where valuing communities may disagree. Moreover, the concept itself is reasonably clear: McCormick cites the Belmont Report, which holds that respect for persons encompasses "two basic ethical convictions: first, that individuals should be treated as autonomous agents, and second, that persons with diminished autonomy are entitled to protection." An additional advantage is that these two considerations seem to supply (at least) minimum requirements for protecting the dignity of persons. And even though the concept of dignity ought to be secondary, McCormick maintains that it is still important in fostering compassionate treatment and an ideal relationship between doctor and patient.

The latter portion of Part III diverges from evaluations of dignity at various stages of human life and toward analysis of dignity in technical and pharmacological manipulations of human life. John Loike leads with an assessment of human dignity in the context of human-animal chimera research. Stem cell and genetic technologies allow scientists to introduce human cells and genes into animals to engineer human-animal hybrids. Loike begins by arguing that, historically, chimera research has led to the creation of animal models that resulted in marked advances in therapy for diseases such as AIDS, hepatitis, and cancer. Current and future applications of chimera research, Loike notes, include brain research, reproductive medicine, and organ development for autologous transplantation.

Loike focuses his philosophical evaluation of dignity in chimera research on two related issues: the assessment of *Homo sapiens* as a unique species and the apparent need to respect human dignity. Recent studies have elucidated the capacity to transfer behavioral characteristics between species. In light of this technology, Loike asks whether there are appropriate ethical limits to engineering human-animal chimeras—especially the obligation to avoid creating chimeras that exhibit either human behavioral traits or contain human reproductive organs or cells. Turning to the concept of species, Loike points out that there is no authoritative definition of species, making it difficult to assert moral evaluations on the basis of species boundaries. Loike then outlines the variant interpretations of species and human dignity in the chimera literature (Catholic, Jewish, secular, etc.), elaborating on each perspective's justifications for support of (or opposition to) various technological developments pertinent to chimera research. Loike also considers the public policy side, examining legislative actions in various countries for regulating chimera research and the varying degrees of concern over this technology implied by different legislative actions. Given the wide range of interpretations about species boundaries and the concept of human dignity, Loike argues that discussions on chimera research are convoluted and lack moral clarity.

Even so, Loike supports the notion that respecting human dignity, whether defined within a secular or religious perspective, is an important ethical consideration in determining the types of chimera research that should be pursued. In support of the National Academies of Science guidelines for university and corporate institutional review boards, Loike endorses the perspective that human–animal chimeras with enhanced cognitive function or containing human reproductive organs should be discouraged.

The volume closes with an analysis by William Cheshire on the relationship between human dignity, psychotropic drugs, and the brain. Even though discussions about dignity are rare in the literature on psychotropic drugs, Cheshire points out that the capacities associated with dignity (such as rationality, memory, and autonomy) have neuropsychological correlates in the brain that are subject to influence by a wide assortment of psychotropic drugs. Discussions about dignity, then, are highly relevant to the ethical use of psychotropic drugs. Accordingly, Cheshire explores how consideration of human dignity informs—or ought to inform—drug-induced mood, memory, and intelligence enhancement, among other topics.

On a more philosophical level, Cheshire points out that a person's view of human nature informs how the concept of dignity impinges upon the proper use of drugs capable of modifying brain chemistry. Materialistic appraisals find in the language of human dignity no additional meaning beyond what can be described in terms of molecular neurochemistry and neuronal circuits of efficient causation. Practical appraisals that rely on utilitarian philosophy equate dignity with operational terminology such as "autonomy" and tend to emphasize human performance and instrumental usefulness. Perspectives that embrace a robust conception of human dignity, on the other hand, tend to recognize in humankind a further worth beyond words—an intrinsic nature with special status not reducible to particles or utility.

Affirming the importance of human-dignity rhetoric, Cheshire argues that where the skeptic finds the language of human dignity impenetrably vague, that lack of specificity permits a conceptual open-endedness appropriate to the reality of human nature, a reality that deserves profound respect. Thus, although challenges abound, Cheshire concludes that "[p]ursuing an understanding of human dignity ultimately can enrich society, not because it is easy, but because it is difficult."

Having surveyed the chapters in this volume, it is now time to turn the book over to the contributors themselves.

NOTE

1. For the idea for this opening line, I am in debt to Steven Shapin's *The Scientific Revolution* (1996: 1).

2 Human Exceptionalism and the *Imago Dei*
The Tradition of Human Dignity

David H. Calhoun

The concept of human dignity plays an increasingly central role in contemporary ethics, particularly in the subfields of politics, international law, and biomedical ethics (Kretzmer and Klein 2002; President's Council on Bioethics 2008; Marshall et al. 1967; Kass 1974; Kass 1985: 31–32; Davis 2008). Appeals to dignity in contemporary ethics are normally traced to documents articulating principles of international law and policy in the aftermath of World War II. The landmark 1948 Universal Declaration of Human Rights, for example, asserts that "recognition of the inherent dignity and of the equal and inalienable rights of all members of the human family is the foundation of freedom, justice and peace in the world" (United Nations General Assembly 1948; see McCrudden 2008).

Despite the growing importance of the concept of human dignity, the term is, by complaint and admission of many scholars, ill defined. Government and intergovernmental documents, including the Universal Declaration of Human Rights, invoke the concept without offering any clear account of its meaning.[1] One consequence of this state of affairs has been a flood of scholarly treatments of the concept of human dignity (notable examples include Kass 1974; Kretzmer and Klein 2002; President's Council on Bioethics 2008; Meilaender 2009).

All is not well, however. Many critics have found contemporary discussions of dignity to be plagued by difficulties, ambiguities, incoherencies, and rhetorical vacuousness. Ruth Macklin famously called dignity a "useless concept," arguing that the idea is just a placeholder for respect for autonomy (Macklin 2003). Others, such as Harvard evolutionary psychologist Steven Pinker, detect the malign agenda of the religious right behind the rhetoric of dignity (Pinker 2008). A number of scholars acknowledge problems with the concept of dignity but argue for one or another ways of improving the clarity and precision of the concept (Johnson 1998; Sulmasy 2006b; McCrudden 2008).

One pathway for a better understanding of the concept of dignity is to explore its historical development. While many writers on the subject of dignity acknowledge the importance of the classical and Christian conceptual roots, most treatments briefly address historical matters before moving on to questions of application, as one might expect for a concept in practical ethics (see, e.g., the helpful discussions of Kass 1974; Schulman 2008).[2] Some writers on the subject have addressed the question of historical development, but of those most have focused on particular thinkers, such as Cicero or Kant, or specific historical movements, such as Renaissance humanism or the recent politics of bioethics (Cancik 2002; Shell 2008; Trinkaus 1973; Ekert 2002; Davis 2008). A few authors have offered more systematic and detailed treatments of the concept of dignity (very different examples include Baker 1947 and McCrudden 2008). The historical sources of dignity do not determine our apprehension and application of the concept; at the same time, they can inform our current understanding and practice and can shed light on the ambiguities and apparent contradictions in our conceptual framework.

I argue that the concept of human dignity constitutes a tradition, a continuous history running over 2,000 years. By calling the Western treatment of dignity a tradition, I mean that it is an ongoing project of exploring the nature and value of human existence within an abiding framework.[3] That framework is a claim of human *distinctiveness*, advanced in two threads: (1) that humans are different from and superior to all other living things, primarily due to the power of reasoning, what we might call the "human exceptionalism" criterion, and (2) the *imago Dei*, the view that humans reflect the divine image in a way unique in the created order. I will show how elements of the traditional concept emerged independently in classical paganism and Judaism, were synthesized in early Christianity, and then were developed in a continuing way through the Middle Ages and Renaissance. I will sketch out the modern modifications of the traditional concept and examine the late modern attack on the tradition and its restoration in the 20th century. I conclude with a few observations about how the history of the tradition points the way forward to contemporary application of the concept of dignity.

CLASSICAL SOURCES FOR HUMAN DIGNITY: FROM SOCIAL HONOR TO UNIVERSAL HUMAN VALUE

In the most basic sense, *dignitas*[4] and related terms[5] in Latin and Greek[6] concern social honor, position, or rank; the characteristics of excellence and achievement that qualify one for rank; and the recognition one receives as a consequence of both status and achievement (see Trinkaus 1973: 136; Kass 1974: 71–72; Berger 1983: 174; Taylor 1992: 46–47; Lorberbaum 2002: 55–56). From the beginning, the focal meaning of dignity is distinctiveness, which asserts the status and value of persons against a comparative background of others who are of lesser status and worth. The rank-value link is replicated in definitions of the derivative English term *dignity*: "the quality of being worthy or honourable; worthiness, worth, nobleness, excellence" (*OED* s.v. "dignity").

Of course, any reader familiar with modern discourse of dignity will recognize that social rank hierarchy is not the same as universal human dignity. The dignity captured by the basic use of *dignitas* and other rank terms in Latin and Greek, and the basic sense of the English term *dignity*, is the rank and honor accorded to *exceptional* human beings. By contrast, the concept of *human* dignity concerns the rank or status accorded to all human beings despite their differing levels of achievement or excellence (Kass 1974; Meilaender 2009).

One promising clue regarding the emergence of the classical ideal of dignity is Aristotle's portrayal in his catalogue of virtues of the "great-souled man" or μεγαλόψυχος, the person who manifests the virtue of character sometimes translated "magnanimity" (*Nicomachean Ethics* 4.3). The scope of the virtue of magnanimity concerns the right attitude toward honors and dishonors (*Nicomachean Ethics* 4.3.1123a35–1123b25), which is the locus of the concept of dignity. Aristotle describes the magnanimous man in some detail as a person of moral gravity and seriousness, excellence of action and character, and sufficient self-awareness to know what is due him and what is beneath him. Aristotle describes him as having "slow movements, a deep voice, and calm speech. For since he takes few things seriously, he is in no hurry, and since he counts nothing great, he is not strident" (*Nicomachean Ethics* 4.3.1124b27–1125a16, Irwin trans.). Clearly, Aristotle is not simply describing one virtue among many in the catalogue here; rather, this is a description of a truly good person whose character serves as a kind of "adornment" to all the other virtues (*Nicomachean Ethics* 4.3.1123b25–29, 1124a1–4). The magnanimous man exemplifies in his bearing and behavior something very close to what we mean by dignity.[7] He does this as a moral exemplar, unlike the ordinary run of human beings.

The excellence of magnanimity is the proper exercise of self-governance and self-knowledge regarding social rank. This virtue is not possessed simply by achieving rank or recognition but by being properly ordered as a human being with respect to rank or recognition. In short, the magnanimous man is a sort of paradigm of virtue: he exhibits the proper function of human beings by rightly ordering his choices, actions, and habits by reason (*Nicomachean Ethics* 1.7).

The special distinctiveness of the magnanimous person is made possible by the general functional distinctiveness of human beings. To be human is to be rational, which is itself the capacity, through formation of right habits, to be properly governed by reason, and thereby integrated psychically and ethically. While all humans fail to actualize this capacity fully, it is nevertheless in principle possible for every human being to achieve some measure of excellent self-governance.[8] Virtue is not natural or automatic, but the development of virtuous habits and character is a power rooted in human nature (see *Nicomachean Ethics* 2.1.1103a19–1103b1). Despite the different degrees to which individual human beings cultivate or employ their rational powers, they naturally and universally possess a capacity for rational activity.[9] In an important sense, the magnanimous man excels and thereby outranks the ordinary mass of human beings and simultaneously affirms the fundamental promise—the *dignity*—of human nature, and therefore of all human beings.

Identification of human nature with rationality is an abiding theme of ancient Greek philosophical anthropology. Human reason is central to the pre-Socratic quest for the principles underlying the cosmos, as, for example, in the Pythagorean idea that reason allows humans to ferret out the numerical principles that govern the harmonies of music and of the celestial bodies and the Heraclitean view that the rationally accessible λόγος is the dynamic principle of cosmic order (McKirahan 1994: 91–113, 116). The capacity for reasoning provides the foundation for the systematic accounts of human nature offered by Plato and Aristotle. It is reason that allows humans to grasp the principles of organization or causes of reality (Plato, *Phaedo* 96a–100b; Aristotle, *Metaphysics* 1.1), and to critically examine self and others (Plato, *Charmides* 166c–d; *Apology* 29e, 36b–c, 37e–38a; Aristotle, *Nicomachean Ethics* 10.7.1177a27–1177b2).

The pinnacle of human rationality in the ancient view is the ability to inquire into the nature of being itself, even divine being. This theme, explored by Heraclitus and Parmenides (McKirahan 1994: 116, 119, 151–53), finds majestic expression in some of the most famous passages of Plato's works (*Republic* 6.507a–7.519c; *Symposium* 210e–211c). Aristotle develops the connection between reason and the divine in a two-pronged way: reason gives human beings the ability to understand divine things, but it also allows us to participate in divine life, for the ultimate principle of the universe is pure rational activity (*Metaphysics* 12.7–9, esp. 12.7.1072a14–31; *Nicomachean Ethics* 10.7–8).

The capacity of human beings for divinization—for participation in divine reason—became the basis for the Stoic conception of universal human brotherhood or cosmopolitanism. Given the rational nature of human beings, and the fact that reason or *logos* is the divine ordering principle of the cosmos, human "ontological equality" follows (Arieli 2002: 14; Nussbaum 2001b: xxi; Baker 1947: 72–74). Marcus Aurelius, the Stoic emperor, put it as a syllogistic chain: "If mind is common to us all, then so is the reason which makes us rational beings; and if that be so, then so is the reason which prescribes what we should do or not do. If that be so, there is a common law also; if that be so, we are fellow-citizens" (*Meditations* 4.4, quoted in Long 2008: 56).[10]

Drawing from and synthesizing these ideas, Cicero founds the explicit tradition of human dignity by articulating a doctrine of *hominis excellentia et dignitas*:

> the nature of man has precedence over cattle and other beasts; those feel but pleasure. . . . if we consider what excellence and dignity is in the nature of man, we'll recognize how shameful it is to be dissolved in

luxury and to live in a spoilt and weak way, and how virtuous in a moderate, continent, severe, and sober way. (Cicero, *De Officiis* 1.105–7, trans. by Cancik 2002: 20–21)

Cicero adapts the social hierarchy term *dignitas* and makes explicit the implicit universality of human distinctiveness in the Greco-Roman philosophical tradition.[11] In Cicero's use, *hominis dignitas*, human dignity, draws together the set of elements in classical human exceptionalism: it concerns an inherent quality bestowed by nature and shared equally by all humans (Cancik 2002: 21, 24–27), realizable to different degrees, manifested in rationality, and marked by human superiority over other living things.[12]

Cicero makes no theological appeal in the formulation of human dignity in *De Officiis* (Cancik 2002: 27). However, his view that "nothing is better than intelligence and reason," and that mind is exemplified in perfect divine rationality (*The Nature of the Gods* 2.38–39), shows that human dignity is inseparable from divine majesty. Indeed, the gods "possess the same rational faculty as is possessed by the human race" (*The Nature of the Gods* 2.79). Further, the antecedents of Cicero's view in Plato, Aristotle, and the Stoics embed the notion of human dignity in a cosmic context that includes human relationship to divine reason. Thus we see the notion of human dignity in the philosophical anthropology of classical paganism articulated in terms of human exceptionalism and a link to divine reality.

A critical aspect of classical human distinctiveness concerns the human-animal relationship, or, to put it another way, the relationship between human intellect and human animality. We might think that affirming human distinctiveness would require asserting a radical dichotomy between humans and animals. Such a view was taken by Stoics such as Epictetus, who contrasted irrational animals to rational humans capable of contemplating the works of God (*Discourses* 1.6.19–21).[13] However, distinctiveness does not require *radical* difference. After all, the magnanimous man is distinctive without being nonhuman.[14] Similarly, human beings can be distinctive without being nonanimal.[15] Plato's modified Pythagoreanism allowed soul transmigration between human and animal bodies, affirming human-animal continuity (*Phaedo* 81d–82b, 83d–e, 113a; see McKirihan 1994: 81, 84–88; Baker 1947: 7–9; Nussbaum 2001a: 1514–15).[16] Nevertheless, the Pythagoreans and Plato maintained that the sort of rational soul subject to transmigration into human and possibly animal bodies is qualitatively different from all other living things.[17] If this is an odd form of *human* exceptionalism, it is exceptionalism nonetheless.[18]

Aristotle employed the traditional language of "soul" for the form or principle of organization of living things, but he conceived of soul as a principle of organization rather than a thing in its own right that is separable from body (Aristotle, *On the Soul* 2.1.412a17–19). The soul, the principle of organization, makes possible the various life functions of living things by dynamically ordering them in particular ways. In Aristotle's famous formulation, human beings are "political animals"; like animals in their functions of sensation, locomotion, and desire but unlike them in their capacity for understanding, intelligible speech, and just and unjust action (*Politics* 1.2.1253a1–18; Steiner 2005: 57–62). The three-level psychic hierarchy of plants, animals, and humans acknowledges the fundamental continuity of living things while maintaining human functional exceptionalism based on human intellect (*On the Soul* 2.2–3).[19]

Despite their differences, the three streams of classical philosophy that most influenced subsequent thinking about human nature agreed in principle about human exceptionalism. Human beings are distinct from the rest of nature, even if they share important ties with other forms of animal life. At the heart of the distinctiveness is intellect, which is not merely natural but in important respects divine. While some individual human beings can achieve exceptional status by virtuously cultivating reason and rational self-governance, all human beings are distinctive by virtue of being rational. Thus, a quality that initially designated exceptional status in the human community by the recognition and acknowledgement of others came to mark an intrinsic and essential feature of humanity: its dignity.

SHARING IN THE GLORY OF GOD: THE JEWISH AND CHRISTIAN SCRIPTURAL TRADITION

Strictly speaking, the phrase "human dignity" never occurs in the Bible in Hebrew or Greek (Cancik 2002: 21, fn 7; Sulmasy 2006a); once again to trace the origins of the doctrine of human dignity we must inquire more widely.[20] As is the case in classical paganism, in the Hebrew language, social-rank terms typically designate a position of honor or status, such as the status of honor held by a king (see, e.g., Genesis 31:1, 45:13, Esther 5:11; Job 19:9; Proverbs 3:35). However, several critical evaluative terms, most notably *kavod*, or "glory," extend beyond social rank to refer to the majestic and radiant presence of God himself (Exodus 24:16–17; see also 1 Kings 8:11; 2 Chronicles 7:1–3; Psalm 19:1; Psalm 24:8–9; Isaiah 6:3; Ezekiel 3:23). Insofar as *kavod* conveys the idea of the value intrinsic to a being—particularly God—it can be translated "dignity" as well as "glory" (Arieli 2002: 10).

From its opening pages, the Jewish scriptures advance the striking claim that God, through a gracious act, extends his glory to human beings. Of course, this does not mean that human beings manifest the full luminescence[21] and majesty of the divine presence. For the glory of man, his dignity, is to reflect God by imaging or reflecting him: "God created man in his own image, in the image of God he created him; male and female he created them" (Genesis 1:27 NIV; see also Lorberbaum 2002). This reflection of God's glory is found not only in the majesty of the sovereign but is universally present in human beings, especially as they exercise designated dominion over the created order (Genesis 1:26).[22] The divine image is inherent in the created nature of human beings.[23]

The Psalmist offers a celebration of God's power and majesty that links the *imago Dei* to human exceptionalism:

> When I consider your heavens, the work of your fingers, the moon and the stars, which you have set in place, what is man that you are mindful of him, the son of man that you care for him? You made him a little lower than the heavenly beings [*elohim*, lit. "gods"] and crowned him with glory [*kavod*] and honor [*hadar*]. You made him ruler over the works of your hands. (Psalm 8:3–8 NIV)[24]

On this view, the cosmos is hierarchical, with God as ultimate sovereign and human beings, "a little lower than gods," delegated sovereigns over the created order, including all other living things (Kass 2007: 45). That delegated sovereignty depends on human spiritual kinship with God. One critical element of this is human reason, which allows human beings, within limits, to understand the principles of creation, as creation speaks intelligibly about God just as God intelligibly speaks nature itself into existence (Genesis 1; 1 Chronicles 16:23–33; Psalm 19:1–6; Proverbs 3:13–20; Proverbs 8:1–2, 22–31; but compare Job 38; Isaiah 40:18–26; Isaiah 55:8–9).

The notion of human distinctiveness, and the notion of human dignity that it grounds, is amplified in early Christianity. The Christian scriptures proclaim that the honor given by God to human beings is so great that he takes on human form to serve his beloved ones (John 1, John 3:16), in service that is self-giving and sacrificial (Philippians 2), which has as its objective to fundamentally transform his children into friends rather than servants (John 15:12–15) and to draw those who belong to him into a companionship that is a sharing of the divine nature (2 Peter 1:4). The valuation by God extends to all as they are reconciled to him through Christ, such that "there is neither Jew nor Greek, slave nor free, male nor female, for you are all one in Christ Jesus" (Galatians 3:28 NIV). Lest we think that this equal dignity only extends to those within the bounds of the Christian community, the teaching of Jesus clearly asserts that the extension of human value is universal (Matthew 5:44–45; Luke 10:25–37).

The notion that reason is a particular locus of human distinctiveness is carried over from the Jewish scriptures. This is especially true regarding the resonance between human reason and the divine principle of creation and revelation. Echoing the Genesis creation account, John 1 identifies Christ with the articulate principle of intelligence responsible for creation and affirms that God sheds light and glory (δόξαν) on even sinful and rebellious humans (John 1:5–14; see Colossians 1:15–17).[25] Saint Paul asserts that human beings are culpable for their failure to respond to God's self-revelation, since nature plainly reveals God's "invisible qualities" (Romans 1:19–20 NIV). While human philosophical systems can obscure God's self-revelation and yield ignorance rather than understanding, it is God's wish to fully disclose himself by making accessible "the mind of Christ" (1 Corinthians 2:16; see 1 Corinthians 1:20–25).

We find a point of convergence between the Jewish-Christian account of dignity and that of the classical pagan tradition in Saint Paul's Mars Hill message to the Athenian philosophers. Paul employs Stoic imagery on the precise question of the divine spark in human nature, noting that God

> made from one man every nation of mankind to live on all the face of the earth, having determined their appointed times and the boundaries of their habitation, that they would seek God, if perhaps they might grope for Him and find Him, though He is not far from each one of us; for in Him we live and move and exist, as even some of your own poets have said, "For we also are His children." (Acts 17:25–28 NASB)

Significantly, Saint Paul emphasizes Stoic cosmopolitanism to assert the universal extent of human distinctiveness. Since all are descended from Adam, and since Adam reflected divine glory, human beings are not divided by tribes or races; all human beings share in the dignity of the *imago Dei* and the possibility of community with God.[26] The glory of human beings is a kind of fame or honor, thought quite unlike the honor of social recognition (Lewis 2001: 36). Human praise for God yields in turn to divine praise for his beloved children, the glory that is itself dignity.

HUMAN DIGNITY IN THE CHRISTIAN TRADITION FROM ANTIQUITY TO THE RENAISSANCE

Christian intellectuals of the patristic period were formed in the system of classical pagan education and made a concerted effort to blend classical ideas with Christian ideas. While there were vigorous debates about the appropriateness of using pagan materials, leading early Christians were deeply influenced by classical anthropology and ethics (Lindberg 1986: 24). We know that Saint Ambrose, Saint Jerome, and Saint Augustine had access to Cicero's *De Officiis*;[27] it, and the idea of human nature that it transmitted from the classical world, shaped and informed the early Christian development of the idea of human dignity.

Saint Augustine framed human distinctiveness in terms of the *imago Dei*. He noted that anthropomorphic theology led to the mistaken, even idolatrous, view that humans physically image God (*Confessions* 3.6.12, esp. fn 30, 6.3.4–4.5, 7.1.1, 7.9.15).[28] Instead, Augustine argued that creation in the image of God foreshadows the renewal of the mind in Christ (13.22.32; see 1 Corinthians 2:16) and so links essentially to the human capacity to know that is the basis for human superiority over the beasts and dominion over creation (13.23.33–34). Summing up, Augustine observed, "we see *man,* made to Your *image and likeness,* dominating all the irrational animals by reason of that same image and likeness, that is by the power of reason and understanding" (*Confessions* 13.32.47).

Another important expression of human dignity is found in the treatment of Saint Thomas Aquinas, who grounds human dignity in the "infinite dignity" of God (*Summa Theologica* I, Q. 25, Art. 6, Repl. Obj. 4). One critical development of Thomas's account is a link between personhood and dignity: the capacities of reason and free action are found in persons, beings capable of free and reasoned agency. God is the supreme person, and human beings reflect the divine image by their subordinate personhood: "It belongs to man's mode and dignity that he be uplifted to divine things, from the very fact that he is made to God's image" (*ST* IIB, Q. 175, Art. 1, Repl. Obj. 2; see also I, Q. 29, Art. 3, Repl. Obj. 2; I, Q. 59, Art. 3; III, Q. 2, Art. 2, Repl. Obj. 2). Consequently, Thomas noted that the "dignity of a person is circumstance that aggravates a sin" (*ST* IIA, Q. 89, Art. 3). Sin offends against the dignity of human beings by vitiating the distinctiveness of rationality and undermining the natural division of humans and animals (*ST* IIB, Q. 64, Art. 2, Repl. Obj. 2; see also IIB, Q. 147, Art. 1, Repl. Obj. 2; III, Q. 1, Art. 1).[29] Dignity is therefore rooted in reason, in free action, in reflection of God—in short, in personhood (Finnis 1998: 179).

With renewed access to classical works in the Renaissance, humanist scholars were able to reconnect Jewish-Christian notions of human dignity to their classical antecedents. Perhaps the most eloquent was Pico della Mirandola, whose argument for the life of humanistic learning has come to be known the *Oration on the Dignity of Man* (1486). Pico's argument for human dignity is in two parts. First, he acknowledges the significance of standard justifications for "the pre-eminence of human nature," which cluster around the place human beings hold between animals and angels in natural hierarchy (Pico 1956: 3–4). While Pico allows that the standard reasons are weighty ones, he believes that the real ground for human dignity must be located in free human self-definition. God made Adam a "creature of indeterminate image" and provided that "whatever place, whatever form, whatever gifts you may, with premeditation, select, these same you make, have and possess through your own judgment and decision. . . . you . . . may, by your own free will, to whose custody We have assigned you, trace for yourself the lineaments of your own nature" (Pico 1956: 6–7). In a unique twist, therefore, the uniqueness of human nature is its malleability. Human rationality facilitates choice beyond this or that set of concrete options to the glorious dignity of self-definition and self-transformation.

The account thus far has brought into focus a clear tradition. Human dignity concerns a twofold account that grounds human distinctiveness, and a consequent dignified place in cosmic hierarchy, on human exceptionalism and the *imago Dei.* Exceptionalism is primarily rooted in human reason, moral choice, and personal agency. The *imago Dei* is the reflection of God inherent in human nature, demonstrating divine love and affirmation and legitimizing human status near the top of the cosmic hierarchy. In both of these elements we find an objective basis for human worth.

MODERNITY TO THE ENLIGHTENMENT: DIGNITY CHALLENGED BY MECHANISTIC SCIENCE AND SECULARIZING TRENDS IN PHILOSOPHY AND LAW

Despite the upheavals of the Protestant Reformation, Counter-Reformation, and other changes of the late Renaissance and early modern periods, the traditional notion of human dignity survived essentially unaltered. Violently contentious theological and political disputes did not fundamentally challenge either human rational exceptionalism or the *imago Dei*. The persistence of the key elements of the tradition can be seen in the creation narrative of Milton's *Paradise Lost* (1674), for example, which maintains a view of human distinctiveness based on "Sanctitie of Reason" and the divine image (*Paradise Lost* 7.505–23).

At roughly the same time, however, trends in European intellectual culture, particularly in the areas of natural philosophy and theology, spurred major shifts in the grounds of the traditional concept of human dignity. Four important 17th- and 18th-century developments shaped the reception of the traditional idea of human dignity in the modern period, providing context for deeper critiques of dignity in the 19th century.

First, the "new science" of the early modern period conceived of nature in mechanistic terms.[30] Where ancients and medievals largely regarded nature as teleological, governed by ends intrinsic to natural objects (Aristotle, *Physics* 2.1; Carroll 2000: 327–30), early modern scientific thinkers and practitioners such as Bacon, Galileo, and Descartes generally understood natural change in terms of physical particles moving and interacting and primarily regarded proper study of nature in terms of empirical investigation of efficient and material causes (Bacon 1964; Descartes 1993: 24–34; Ashworth 2003).[31] Borrowing the basic cosmology of ancient atomism, they reenvisioned the physical world as a "mechanistic universe of lifeless, indivisible atoms moving randomly in an infinite void" (Lindberg 2007: 365; see also Kass 2007: 40; Hahn 1986).[32]

The machine model of the cosmos itself led in two possible directions. If natural mechanisms were insufficient to account for natural objects and phenomena, then immediate divine agency was required. The path from the origins of the new science to teleological theistic arguments of the sort advanced by William Paley (Paley 1809) is therefore a straight one. Of course, the alternative view, that natural mechanisms were sufficient to account for physical phenomena, led in the other direction, to the claim that there was no need for God in a systematic scientific account of nature, as when LaPlace famously declared his abandonment of the God hypothesis (Hahn 1986: 256).

Second, the mechanistic conception of nature not only complicated ideas about divine agency, it made problematic the place of human beings, especially in comparison to animals. Some thinkers, such as Hobbes and La Mettrie, were willing to embrace the view that humans are concatenations of physical particles like all other physical objects.[33] Most modern thinkers, however, even those deeply committed to physical mechanism, were not. Descartes is an instructive example: he fully endorsed the mechanist account for almost all of reality as part of his embrace of modern science but asserted that human beings must be in some way different. Such a dichotomy was necessary to explain the clear difference between the mental powers of human beings and the mechanical nature of biological life (Descartes 1993: 26–33; Descartes 1970: 36–37, 53–54, 63–64, 206–8, 244–45). In consequence, Descartes proposed two independent, equally real domains: the physical, which is mechanical, extended, and law-governed, and the mental, which is nonmechanical, nonextended, and free (Descartes 1993: 26–34, 97–104). Modern dualisms can therefore be understood as a rearguard action to save human exceptionalism.[34] It would be misleading to assert that all early modern thinkers other than the handful of materialists are metaphysical dualists, but broadly dualistic thinking certainly became a prominent outlook after Descartes.

Invocation of some sort of nonmechanical mind was necessary to account for human powers, but no such claim was necessary for animals, because animals, the thinking went, could be understood completely in mechanical terms. The machine was not only a model for understanding reality, it was a way of accounting for life itself. Following the research of English physician William Harvey on blood circulation, Descartes asserted that living bodies are machines; they are clockworks (Descartes 1993: 26–32; Descartes 1970: 36–37, 53–54, 63–64, 206–8). If, Descartes argued, the heart is a kind of furnace, burning with a "flame without light," it will give off "animal spirits" that rise to the brain and fuel life functions (Descartes 1993: 29–30, 31; Descartes 1970: 37, 146). Life, then, is not a spiritual phenomenon, it is a chemical reaction, a physical apparatus, that can be explained in mechanical material terms.[35]

For Descartes, then, the only salvation for human distinctiveness in contrast to animals was that not all human functions could be explained in mechanical terms. Descartes identified two "machine tests," two criteria for distinguishing mere machines—soulless mechanisms—from machines operated by minds: (1) communication of *ideas* and (2) creative agency (Descartes 1993: 32–33; Descartes 1970: 53–54, 206–7, 244–45). Descartes regarded it as obvious that animals would fail these tests and humans would pass them. While animals communicate, they do not communicate *ideas*; they merely reveal by mechanical means their internal mechanical states. Further, animals behave in complex ways, but they are governed, as Descartes says, "through the disposition of their organs" (Descartes 1993: 32). For Descartes and other mechanists, the obvious consequence of this *bête-machine* or mechanical animal doctrine is that animals do not think or reason.[36]

The effect of broadly dualist conceptions of reality and anthropology was to radicalize and harden the difference between humans on the one side and animals or "brutes" on the other. Human beings maintained their place of honor and dignity at the height of creation, but that place was, so to speak, no longer so high, because the mechanist view flattened the remainder of physical nature. While many scientists, philosophers, theologians, and ordinary people of the 17th and 18th centuries rejected Descartes's "animal mechanism" view, especially the apparent implication that animals were not subject to pain, the general human-brute dichotomy was well established in public culture and discourse up through the mid-19th century. For example, the great moral crusade of the 1700s to 1800s in Europe and the Americas, slavery, concerned in large part the question of whether "savage races" should be counted as genuinely human.[37]

Third, on a very different front, the decline of institutional authority and ensuing skepticism of an increasingly independent age chipped away in subtle ways at religion. While genuine and publicly vocal atheists were relatively rare, a number of challenges to religion and religious authority emerged (Wilson 1998: 32–34). A broad pattern of secularization shifted a great deal of social power and influence away from religious institutions toward guilds, commercial concerns, and, ultimately, nascent states (Taylor 2007). Religious ideas were destabilized, with the inevitable result that the philosophical anthropology associated with those ideas was destabilized as well. Critically, Enlightenment humanists "believed that the central concern of human existence was not the discovery of God's will, but the shaping of human life and society according to reason. For these thinkers, human dignity was not a function of man's allegedly divine origin, but of the ordering and rational possibilities of earthly existence" (Luik 1998).

While the norm in Europe at this time was still some form of supernaturalism, particularly a supernaturalism that underwrote human distinctiveness (perhaps most notably in the deism of many of the French revolutionaries and the ambiguous theism of references to "Laws of Nature and of Nature's God" in the United States Declaration of Independence), challenges to orthodox theism were implicit challenges to the *imago Dei* strand of human distinctiveness. While rational theology remained a going enterprise into the 19th century and even beyond (Gray 1861; Hull 1973: 5–6; Browne 1995: 130; Brooke 2009a; Livingstone 2003; Secord 2000; R. A. Richards 2009: 176–77; Roberts 2009), critics such as David Hume (Hume 1988) explored a range of alternatives regarding God and nature. Charles Taylor dubs this overall trend "providential deism," an "anthropocentric shift" that heightens human responsibility at the expense of divine prerogative, leaving to God only the roles of creator and final judge (Taylor 2007: chap. 6). One might argue that this trend enhanced human dignity by shifting authority and autonomy from God to human beings. Paradoxically, however, it enhanced human autonomy only at the cost of the traditional ground for human dignity, the *imago Dei*.

Kant's ethics of autonomy illustrates Taylor's anthropocentric shift. Kant appealed to many of the traditional grounds for human dignity as foundational elements of his ethics: free agency, rationality, personality. But he did so in a way that clearly detached these values from their traditional theological moorings. He made Pico's idea of self-definition fundamental:

> The power to set an end—any end whatsoever—is the characteristic of humanity (as distinguished from animality). Hence there is also bound up with the end of humanity in our own person the rational will, and so the duty, to make ourselves worthy of humanity by culture in general, by procuring or promoting the power to realize all possible ends, so far as this power is to be found in man himself. (Kant 1964: 51, 2.1.1.Intro, 8.1a; see also 99, 2.1.1.1.2.1 §11)

The human capacity to project possible courses of action is therefore a new humanized form of divine creative power. On the basis of this power, humans are and should be autonomous—self-ruling.[38] As a result of projective choice, rationality, and personhood,

> Humanity itself is a dignity; for man cannot be used merely as a means by any man (either by others or even by himself) but must always be treated at the same time as an end. And it is just this that comprises his dignity (personality), by virtue of which he assumes superiority over all the other beings. (Kant 1964: 132, *Metaphysics of Morals* 2.1.2.1.2 §38; Long 2008: 51)[39]

Fourth, in a development closely related to the focus on rational human autonomy, the foundations for emerging political ideals, including the modern notion of rights, shifted from religious to increasingly secular rational grounds. Many important modern political theorists, such as Samuel Pufendorf, adapted and developed classical and medieval ideas into the Enlightenment framework, often in a way that recognizably reflected their origins in traditional ideas of human rationality and freedom if at the same time radicalizing those original sources (Pufendorf 1964; Pufendorf 1994; Cancik 2002: 30–33).[40] Human beings retained the position of honor and dignity at the summit of the natural order, but whether that order was created and in what sense became an open question, at the very least. Perhaps most importantly, the rationality central to the traditional conception of human distinctiveness became an independent basis for grounding the social order (Taylor 2007: esp. chap. 4; Arieli 2002: 7, 30–33, Cancik 2002: 37; Saastamoinen 2010). With the calls for rights came new, increasingly democratic forms of social organization that were marked by significant shifts in thinking about authority, hereditary class, and monarchical power.[41] Social reform movements, such as those concerning rights of religious exercise and women's rights, called assumed orders and hierarchies into question, often appealing to conceptions of human dignity to buttress their arguments.[42]

THE TRADITION SHATTERED: THE LATE MODERN FATE OF HUMAN DIGNITY

In late modernity, the narrative of human dignity becomes tortuously complicated. Put bluntly, shortly after Kant there was no longer a coherent tradition of thinking about human dignity, but instead there were multiple branches extending in different directions. Of course, dissenters from the traditional view of dignity existed prior to the late 19th century. As early as 1651, Hobbes, for example, set aside the traditional meaning of dignity and, in *Leviathan*, reverted to the aboriginal sense of social rank (*Leviathan* 1.10). By the late 1800s, however, something of a different character emerged: a series of substantial modifications and attacks shattered the broad unity of the tradition.

Briefly, we can identify five streams diverging from the tradition over the last century and a half. Of course, there are oversimplifications and overlaps here, but these categories capture the conceptual geography of thinking about human dignity from late modernity to the present.

Adapters. Beginning most clearly with John Stuart Mill and including figures such as William James, adapters were mainstream philosophers not committed to Kantian or traditional philosophy who employed the language of dignity, often in nontechnical ethical contexts, in their philosophical systems. Perhaps reflective of trends in philosophy from the mid-1800s to the early 1900s, there was a tendency to speak of dignity as a self-regarding sentiment, involved in some way with moral development or an ideal of moral perfection. Mill, for example, described personal dignity as "that feeling of personal exaltation and degradation which acts independently of other people's opinion, or even in defiance of it" (Mill 1985: 95–96), and affirmed a universally human "sense of dignity" (Mill 1985: 212; see Mill 1977: 243, 279).[43] William James used the term *dignity* in *Varieties of Religious Experience* to characterize the emphasis of "healthy-minded religion" in contrast with the focus on human depravity in traditionalist religion (James 1985: 81). He also speaks of "the feeling of the inward dignity of certain spiritual attitudes" (James 1956: 187). In the usage of both Mill and James, dignity is a psychological state rather than an inherent quality of human persons. Significantly, it is a state that admits of degrees of more or less, even one to which one can ascend of from which one can fall, suggesting it is more akin to the concept of varying degrees of character excellence than to universal and irrevocable human dignity.[44]

Debunkers. Starting with Arthur Schopenhauer, a series of modern thinkers discovered something incoherent, misguided, or otherwise amiss in the traditional concept of human dignity. Schopenhauer, who scores points for the memorable quality of his invective, targeted Kant's view of dignity as a unqualified good, complaining that "that expression, *dignity of man*, once uttered by Kant, afterward became the shibboleth of all the perplexed and empty-headed moralists who concealed behind that imposing expression their lack of any real basis of moral, or, at any rate, of one that had any meaning." In short, he regarded Kantian dignity as conceptually incoherent (Schopenhauer 1965: 100–101).[45] Nietzsche picked up the cudgel and attacked the concept of dignity on a number of fronts: it is an empty phrase used by the mediocre to praise themselves; it is a vacuous term used to pacify slave classes, the effectiveness of which is near exhaustion; it is a kind of anesthesia to deaden us to the wretchedness of our lives, which is a "slavery that hides from itself"; it is the epithet by which we once convinced ourselves of our favor with God (Nietzsche 2006: 70, 78, 88–89, 432, 482). He could not resist putting into the mouth of Zarathustra's shadow a mocking hymn of praise to "European dignity" (Nietzsche 1961: 319).[46] More recent debunkers include Freud, who celebrated the dethronements of human dignity prompted by the Copernican revolution and his own psychoanalytic revolution (Freud 1920: 246–47; see also Freud 1974: 139–40), and Skinner, who longed to dispense with the unscientific and unhelpful notions of dignity and freedom and replace them with mechanical schedules of reinforcement to shape and control human behavior (Skinner 1971). Many debunkers have dismissed dignity by simply disregarding it as an outdated concept, as did the positivists of the early 20th century.

Saboteurs. Closely related to the debunkers are a significant group of thinkers who attack traditional anthropology, and thereby indirectly undermine the philosophical framework for human dignity. Paramount in this group is Darwin. While Darwin personally avoided conflict with Christianity, he was acutely aware of the fact that evolution by natural selection destabilizes the traditional concept of human dignity.[47] With respect to traditional human dignity, the thesis of common descent of all living things flattened nature and implicitly challenged human exceptionalism (Darwin 1882: 65–66). Explaining biological organisms, species, and processes by natural mechanisms appears to supplant divine agency and consequently undermined Paley-style natural theology (see, e.g., Paley 1809; Darwin 1859: 243–44, 275–76, 469–70, 472, 480–82, 488–89; Huxley 1887; Dawkins 1986: 3–6; Haught 2010: esp. 11–27). The overall effect on dignity of the Darwinian revolution was something like sabotage. On the one hand, Darwin affirmed the basic consistency of evolution and human dignity, even in its austere Kantian form, while on the other, sometimes in different rhetorical or communicative contexts, he expressed pessimism about traditional anthropology and ethics (Darwin 1882: 46, 109–10). Twenty-one years prior to the publication of *On the Origin of Species*, while Darwin was exploring the ideas that would become the theory of natural selection, he noted in a research notebook that on the assumption of common biological descent, the "whole fabric" of human distinctiveness "totters & falls" (Darwin 1987: 263 [Notebook C 76–77]). Shortly after publication of the *Origin*, in a letter to his friend and mentor Charles Lyell, he starkly concluded, "I am sorry to say that I have no 'consolatory view' on the dignity of man" (F. Darwin 1887: vol. 2, 262; Desmond and Moore 1991: 505). It is striking how many committed Darwinians have concluded that any form of human distinctiveness is untenable after Darwin. Some conclude this reluctantly, some eagerly, but they see it as an implication of Darwinism (a representative sample: Arnhart 1988; Rachels 1990; Rue 1994; Dennett 1995; Provine 1988).[48] Right or wrong, many have concluded that dignity cannot work in contemporary intellectual culture, especially in the light of the anthropology of modern natural science.[49]

Cautionary prophets. Some saw that a problem with traditional anthropology and ethics and found this worrisome. While the more aggressive existentialists such as Nietzsche viewed the collapse of traditional theology and anthropology as the promise of a new dawn, others, most notably including Heidegger and Sartre, regarded the modern threat of nihilism—including the collapse of a traditional conception of dignity—as something of a catastrophe. In *Being and Time* (1927), Heidegger emphasized the two strands of traditional philosophical anthropology: human rational exceptionalism and the *imago Dei* (Heidegger 1962: 74–75 [I.1.H48–49]). However, Heidegger argued that this anthropology, along with other philosophical and scientific attempts to understand the human condition, had failed to disclose the particular sort of being characteristic of human beings. Although Heidegger abandoned the traditional language of dignity, it is clear that in seeking the meaning of being he was exploring the possibility of finding meaning and value for human beings in the face of what he took to be the collapse of a tradition. In "The Question Concerning Technology," he suggested that for man the "highest dignity of his essence" is found in the act of safeguarding or watching over the ways that the world is revealed, especially with respect to how technology pre-frames our ways of relating to the world (Heidegger 1977: 32, 103). Sartre noted that the modern shift in focus from questions of being (what a person is, character) to doing (the series of an agent's acts) problematizes human dignity (Sartre 1966: 527 [Part 4]). In his famous lecture, "Existentialism Is a Humanism" (Sartre 1985), Sartre pled for existentialism as a framework for appreciating the dignity of human freedom and agency.

Caretakers. Despite the existence of something of a historical lacuna in the tradition of dignity from the mid-19th century to the early 20th century, the tradition did not completely disappear. It was maintained and preserved by various groups in intellectual enclaves, such as neo-Kantians, defenders of traditional philosophical anthropology, and religious proponents of the *imago Dei*. For example, in *The Sickness unto Death*, one of his most explicitly anthropological works, Kierkegaard repeatedly emphasized the contrast between humans and animals (Kierkegaard 1980: 15, 121). Noteworthy caretakers of the last century might include Dietrich Bonhoeffer (Bonhoeffer 2005), Jacques Maritain (Maritain 1951; McCrudden 2008: 662), Gabriel Marcel (Marcel 1963), Leon Kass (Kass 1974; Kass 1985), and the Roman Catholic popes who addressed fundamental questions of human dignity in encyclicals and other official documents, such as Leo XIII, Pius XI, John XXIII, and John Paul II (Leo XIII 1891: §§20, 40; Pius XI 1937: § 51; John XXIII 1963: §§ 10, 34–35, 38, 44, 47–48, 145; John Paul II 1991: §§ 3, 5, 11, 13; McCrudden 2008: 662).[50]

Perhaps I should mention another stream, one that has emerged in the last century:

Restorers. Early in the 20th century, many diplomats and officials active in international organizations began to see the regime of rights, autonomy, and minimal principles of respect characteristic of the League of Nations as insufficient for managing international relations. Appeal to a principle of inherent value in human persons seemed necessary, even if the meaning of that value was unclear and even if the principle did not come with a ready-made justificatory strategy. As a result, increasing attention was focused on the idea of human dignity as a principle for governing relations (Morsink 1993; Dicke 2002: 112–14; Arieli 2002; Chaskalson 2002; McCrudden 2008). Something similar happened in the 1960s and early 1970s, as debates about how to treat terminally ill patients exhausted the ethical resources of respecting patient autonomy. Restorers, many of them working in practical contexts of law, international relations, patient care, and bioethics, are the catalyst for philosophical focus on dignity in the last three decades. Their working assumption seems to be that a robust concept like human dignity is imperative for contemporary practical ethics. Hence we find ourselves seeking to understand, revive, and renew the tradition of human dignity.

HUMAN DIGNITY: OBSERVATIONS FROM THE HISTORICAL ACCOUNT

This sketch of a historical narrative of human dignity suggests a number of tentative observations and conclusions.

The tradition of dignity. I have in this chapter made the case that discussion and development of the concept of dignity constitutes a tradition, an ongoing framework for understanding human value in the light of human nature and capacities. While it is a complex tradition, embracing ancient Jews, Aristotle, Renaissance humanists, Kantians, Kierkegaard, Catholic theologians, and contemporary legal and ethical theorists, it nevertheless exhibits a dynamic unity.

Dignity and human distinctiveness. Although there have been many ways of articulating and defining human dignity for over 2,000 years, the core meaning is distinctiveness, understood in terms of human rational exceptionalism and a rich human-divine relationship that merits claim of the *imago Dei.*

Human dignity, ethics, and philosophical anthropology. Those with the greatest practical interest in human dignity seek to employ it as an ethical principle, as a criterion for making judgments about value. However, for almost all of its history human dignity has been a concept of philosophical anthropology rather than systematic ethical theory. In other words, the concept of human dignity is more directly concerned with what human beings *are* than with what human beings should or must *do.* Of course, there is a dynamic relationship between being and doing: Aristotle (and existentialists) would point out that what we are conditions what we do, and what we do inevitably shapes what we are. Nevertheless, we should keep in mind that dignity is less a guide for action than it is a mode of understanding, revealing, and relating to human persons.[51]

The vulnerability of dignity. The period of diminished cultural influence of the traditional concept of human dignity from the late 1800s to the early to mid-1900s shows that the traditional concept of human dignity is vulnerable. The increasing secularism characteristic of modern Western democracies threatens the basis for the *imago Dei.* A number of factors, including mechanistic science, materialist metaphysics, and aggressive forms of biological reductionism, endanger human exceptionalism. Even if dignity is an essential part of our conceptual apparatus for ethical practice in law, international relations, and biomedical ethics, it remains a vulnerable concept (Dennett 2008).

Religion and dignity. Some have argued that dignity is an essentially religious concept, perhaps linked inextricably to classical theism (Gelernter 2008 regards this as a good thing, while Pinker 2008 finds it deeply problematic).[52] While it is true that the early development of human dignity occurred in a context of decisive influence by the Jewish-Christian theological tradition, it is also true that it emerged independently in classical paganism. Even in classical paganism, however, the notion of dignity was linked to the divine dimension of human nature. While this does not preclude articulating a secularized variant of the traditional conception of human dignity, it should at least inform our attempts to explore such a possibility, particularly in the contemporary context of religious pluralism and secularism (Somerville 2009: 53–66; Somerville 2010; Rue 1994; Goodenough 1998; Raymo 1999).

The power of the concept of dignity. The dramatic revival of the concept of dignity in law, international ethics, and biomedical ethics in the 20th century suggests that it is a powerful resource for reflecting on human character and action. The default modern criteria for ethical judgment are rights and autonomy. However, dignity features an intuitive and emotive resonance that goes beyond abstract principles of right. Further, the resources of late modern ethics are widely seen as insufficient to provide a check on what humans are willing to do to one another, particularly in the spheres of politics and international relations and medical ethics. This suggests that even a deeply problematic concept of dignity is superior to emotivist, utilitarian, contractarian, or existentialist conceptions of ethical goodness. One possible reason for this is that dignity roots questions of treatment in recognition of the being of persons, not in judgments of rational agents, calculative rules of action, and the like.[53] Further, there is an implicit objectivity in the concept of dignity that many modern ethical principles lack.

The dual relevance of agent and patient in ethics. Rights language tends to focus attention on protecting a person who stands as the patient of an agent's potential action. Rights are therefore limitations on agency. However, the history of dignity shows that genuine ethical action requires focus on both the constraints on one's other-regarding actions, and the self-regarding aspect of pursuit of excellence of character. The tradition of human dignity includes attention to both agent and patient and therefore serves as a rich resource for reflection on human action.[54]

Conceptual puzzles about dignity. One of the theoretical problems with using the concept of dignity as a criterion for ethical practice is that it seems to blend conflicting meanings (see, e.g., Kass 1974; Gelernter 2008: 393–94; Sulmasy 2006b; Sulmasy 2008; Meilaender 2009). For example, dignity is something a person has inviolably, and yet we also think of it as something that we can impair by actions, as when a person diminishes his or her dignity by cruelty or greed, for example. We often regard dignity (like honor) as a status earned by one's actions and acknowledged by others; and yet we think of dignity as something that exists independently of the attributing acts of others, and indeed cannot be taken away by others. In the face of such a range of meanings, it is easy to see why a critic might conclude that the concept is simply incoherent, or at the least employed in equivocal ways (see Lee and George 2008: 409–10).

Tracing the history of the concept of dignity, however, helps us to understand some of its inherent conceptual tensions. Such a history helps us to see that the concept features at least three related axes. The first, which we might call the *qualitative axis*, focuses on whether dignity is cultivated by degrees or possessed whole. The second, the *ontological axis*, concerns whether dignity is something one earns by what one does or is a result of being a certain sort of thing. The third concerns whether one has dignity as an inherent quality or as a result of an act of bestowal by another; we might call this the *attributive axis*. As these axes illustrate, dignity can be conceived in a variety of ways. However, attention to the historical development of the concept helps us understand the variations. The early form of the classical pagan thread emphasizes the acts in which one engages and the character one has, and therefore regards dignity as a quality one cultivates and for which seeks recognition by others. Following the ancient sources, I suggest that we call this conception the dignity of exceptional excellence.[55]

By contrast, the fully developed Jewish-Christian view (which incorporates and enhances the pagan view) emphasizes the special bestowing act of God upon human beings as the source of dignity, and hence regards it as inherent and inviolable. Further, it identifies the powers necessary for exceptional excellence as basic to the human kind. This is the dignity of universal human value.[56] The convergent concept of dignity maintains both strands in a complex mix. However, the conditions for pursuing excellence and cultivating dignity lie in the inherent human capacities that make deliberation and choice possible. Therefore, they—and the inherent inviolable dignity that describes them—are prior to and necessary for any successful or unsuccessful activation of them.[57] The dignity of universal human value— the essential and inviolable worth of human beings—is prior to the dignity of exceptional excellence.

NOTES

I would like to acknowledge the helpful input of my colleague Douglas Kries for the material in this chapter.

1. Jacques Maritain, who contributed to the deliberations that produced the Universal Declaration, argued that the lack of an explicit conceptual foundation for the notion of dignity was no mistake, but a conscious recognition that shared values could be based on a number of competing justificatory grounds (Maritain 1951: 76–79; McCrudden 2008: 677–78). Similarly, Morsink (1984) argues that a coherent philosophical framework can be articulated for the Declaration but admits that such a philosophical framework neither is articulated in the Declaration nor was the likely basis for the views of the adopting delegates.
2. Some suggest that dignity is essentially a modern concept, hinted at in ancient and medieval sources but only developed as a meaningful point of reference in modernity, and especially by Kant (see Sulmasy 2006a). I agree that Kant is a critical figure in the history, but, as we will see, he is working within a recognizable tradition.
3. Human dignity is not itself a "tradition of enquiry" in the sense used by MacIntyre (see MacIntyre 2007: xii–xiii), but it certainly is a concept that belongs to such a tradition.
4. The etymology of *dignitas* is disputed, with some sources tracing it to the Sanskrit *dac-as*, "fame," and others to the Sanskrit *dic* and related Greek root δεἰκ-, "to bring to light," "to show," "to point out" (see Trinkaus 1973: 136; Kass 1974: 71–72).
5. We cannot conduct a historical examination of the concept of dignity with narrow attention to the word *dignitas*, because the tradition that gives rise to the notion of human dignity precedes Roman literature; a study of the concept of dignity is not a study of a Latin word (Baker 1947). Study of the concept of dignity not equivalent to study of the Latin word *dignitas*; neither is it

study of the terms narrowly defined as dignity in different languages. Daniel Sulmasy is certainly correct in asserting that the *language* of *dignitas* has a clearly circumscribed history in Stoicism, Cicero, Renaissance humanism, Kant, and modern practical ethics (Sulmasy 2006a; Sulmasy 2008: 470–71). But the tradition of the concept of human dignity is not confined to this narrow literature.

6. In addition to *dignitas*, important social rank terms in Latin would include *honoris* (Lewis and Short 1879); key antecedents in Greek are τιμή, "honor," ἄξιος, "equal in value" or "worthy," and πρεσβεία, "seniority" (Liddel and Scott 1940), the latter of which can easily be translated "rank" or "dignity" in many contexts, as for example when it is used to describe the status of the supreme ideal reality, the Good itself, in the famous central passage of Plato's *Republic* (6.509c; see Liddel and Scott 1940 s.v. πρεσβεία). Sulmasy takes σεμνότης as the standard Greek equivalent for "dignity" and observes that it appears not at all in Aristotle's *Nicomachean Ethics* and only a handful of times in the *Eudemian Ethics*, where it describes an obscure virtue that is a mean between servility and unaccommodatingness (Sulmasy 2006a). To the contrary, however, a variant, σεμνός, is used to describe the lofty status— the *dignity*—of the divine intellect in *Metaphysics* 12.9.1274b18. All of this evidence shows that we should cast a wide net to understand the nature of dignity in classical literature.

7. Nussbaum explicitly links Aristotle's ideas of greatness of soul and dignity in her translation of *Nicomachean Ethics* 1.10.1100b30–33: While great misfortunes can damage human flourishing by pain and by limitations on activity, "nonetheless, even in such circumstances the noble shines through (*dialampes to kalon*), if a person bears many great misfortunes with dignity, not because he doesn't feel them, but because he is noble and great of soul" (Nussbaum 2001b: xxiv, 333). The Greek phrase Nussbaum translates "bears with dignity" is φέρῃ εὐκόλως, "bears contentedly," "bears calmly"; Irwin renders this "bears with good temper." Adam Smith also associates virtue, magnanimity, and dignity. In a passage that seems to echo Aristotle's *Nicomachean Ethics*, Smith identifies dignity with the human capacity to bear misfortune and pain with equanimity (Smith 1984: 244–45, 271).

8. Strictly speaking, Aristotle would argue that excellence also requires goods of fortune, including health, a stable community, and sufficient wealth. The role of fortune in cultivation of human goodness is the subject of Nussbaum's *Fragility of Goodness* (Nussbaum 2001b). The vulnerability of humans to fortune, and the consequent interplay of dignity and neediness, is addressed in Nussbaum 1998.

9. Aristotle's claim that all men by nature desire to know (*Metaphysics* 1.1.980a21) is not an empirical assertion based on generalizing from individual human cases, but instead identifies a potentiality present in all human beings as a result of human nature. Aristotle further recognizes that this capacity can be impaired by circumstances, disease, or accident. It is characteristic of the human natural kind, however, and therefore should be understood as universally human.

10. The ancient compiler and biographer Diogenes Laertius summarizes Stoic cosmopolitanism in largely the same terms as Marcus Aurelius; see Diogenes Laertius, *Lives of the Eminent Philosophers* 7.1.87–88.

11. The seriousness and gravity, the stateliness of personal carriage in Cicero's account of *dignitas* closely parallels Aristotle's description of the μεγαλόψυχος or "great-souled man," as discussed above (*Nicomachean Ethics* 4.3). This serves as independent justification for taking Aristotle's account of the μεγαλόψυχος as the original classical reference point for human dignity.

12. Nussbaum argues that contrasting Aristotle and the Stoics on the question of universal human dignity reveals defects in Aristotle's ethical thought: "The first and most striking defect is the absence, in Aristotle, of any sense of universal human dignity, a fortiori of the idea that the worth and dignity of human beings is equal. Perhaps there is actually an internal tension in Aristotle's thought: for at times he stresses (as I shall emphasize) that every natural being is worthy of awe. But it must be admitted that in his ethical and political writings distinct rankings of human beings are recognized: women subordinate to men, slaves to masters. For the Stoics, by contrast, the bare possession of the capacity for moral choice gives us all a boundless and an equal dignity. Male and female, slave and free, Greek and foreigner, rich and poor, high class and low—all are of equal worth, and this worth imposes stringent duties of respect on all of us" (Nussbaum 2001b: xx). The topic of Aristotle's purported ethical and political elitism is a very complex one. Nonetheless, I agree that Nussbaum is right to detect a "tension" in Aristotle's account, as analysis of the magnanimous person illustrates. While Aristotle tends to emphasize the exceptional excellence attained by virtuous individuals, however, his account presupposes the potential for human excellence present in every human being. It is precisely this potential that Cicero identifies as the basis for human dignity, and it is telling that Cicero emphasizes the "dual role" of universal functional promise and individual excellence of achievement (*De Officiis* 1.107).

13. Similarly, Cicero, who identified himself as an Academic philosopher but reported Stoic views quite sympathetically (Long 1986: 231), asserted that the only living things that are truly rational are gods and men, for animals lack language and understand nothing (*The Nature of the Gods* 2.133). Concerning the Stoic human-animal dichotomy, Nussbaum comments: "The Stoic recognition of the dignity and worth of humanity is based upon factors in humanity that distinguish humans from 'the beasts.' This is constantly evident in the Stoic recognition of human dignity. Even as they build up the human, making it something of precious and boundless value, so they denigrate the animal, making it something brutish and inert, something lacking in dignity and wonder" (Nussbaum 2001b: xxiii).

14. Aristotle repeatedly notes that virtue is godlike, but that moral perfection (along with associated characteristics, such as complete self-sufficiency), is properly and fully divine. A human being is neither a beast nor a god, but something in between. While human beings should strive for divinization, we should at the same time acknowledge our humanness (Aristotle, *Nicomachean Ethics* 7.1.1145a23–30, 10.7; *Politics* 1.2.1253a27–34; Meilaender 2009: 4).

15. For a much more extensive set of arguments for the compatibility of human distinctiveness and human-animal continuity, see my "Prospects for Human Dignity after Darwin" in this volume.

16. The implicit humor of Plato's account of human-animal soul transmigration makes it difficult to determine if Plato means for human-animal continuity to be taken seriously. Even if it is, however, it is clear that rationality is central to the Platonic conception of soul.

17. For discussion of Plato's lack of attention to the human-animal question, see Steiner 2005: 55–57.

18. In theory, if soul is the independent principle of life as the Pythagoreans and Plato's Socrates assert (*Phaedo* 105c–d), then there is nothing to prevent transmigration of souls not only between human and animal bodies, but to plant bodies as well. Needless to say, this view is problematic on a number of levels, most obviously from the fact that plants and most animals lack any manifestation of the rational powers that Plato associates with soul.

19. Strictly speaking, human beings are not the highest level of the hierarchy. Insofar as the heavenly bodies and the Unmoved Mover have life, the life of perfectly actualized being, they clearly surpass human life (*Metaphysics* 12.7–8, esp. 1072b25–29; Nussbaum 2001a: 1517 fn 43).

20. It is telling that Saint Jerome, translating the scriptures into Latin in the Vulgate, employed the term *dignitas* only in a handful of passages, mostly dealing with the conventional sense of social rank hierarchy (e.g., Ezekiel 24:25, Proverbs 14:28), despite his familiarity with the elevated use of the term in Cicero.

21. We have already noted that "light" is a basic feature of glory in Hebrew; this visibility is characteristic of *dignitas* as well, insofar as it "can be seen, it has splendor, it shines; it is an ornament" (Cancik 2002: 23). Recall that one possible etymological source for *dignitas* is the Greek root δείκ-, "to bring to light," "to show," "to point out" (Kass 1974: 71–72), and that Aristotle speaks of the excellence of the magnanimous man as an "adornment" to all of the other virtues (*Nicomachean Ethics* 4.3.1123b25–29).

22. "The theological message underlying the first chapters of Genesis is that it is not the king alone who reflects the divine image, but that humanity—every human being—is created or born in his image" (Lorberbaum 2002: 55).

23. Some might object that the account of human reflected glory in Judaism—and by extension in Christianity—does not advance an *inherent* quality of goodness; after all, it is given by God in much the same way that subjective social valuations of individuals are attributed extrinsically. This, however, misunderstands the related concepts of *creation* and *nature*. While it is true that the glory of man is in Judaism and Christianity the result of God's valuing activity, this valuation is not a mere subjective attribution of value by a third-party agent. Rather, God's creation of human being in the *imago Dei* is the structuring of a being with inherent qualities, including the quality of reflected divine glory or dignity.

24. The Psalmist beautifully celebrates the attention that God lavishes on individual human beings in Psalm 139:13–17. We find an echo of this in the teaching of Jesus that God cares about and numbers the hairs on human heads (Matthew 10:28–31).

25. Heidegger notes that glory (δόξα) is, in both pagan and Christian thought, much more than celebrity added as a supplement to a thing. Instead, it is the light that reveals a thing, it is grandeur, it is the highest mode of being (Heidegger 1959: 102–5). It is being that reveals by appearance. This suggests that divine glory is not merely some quality superficially added to human beings but is instead that by which humans are in fully revealed. The self-revelation of God is at the same time a disclosure of human being.

26. It is quite plausible that the universal claim of human value played a key role in the explosively rapid expansion of Christianity across the Mediterranean basin (P. Johnson 1976: 3–63).

27. Ambrose not only read Cicero's *De Officiis*, he used it as a model for a work of his own on ecclesiastical offices, *De Officiis Ministrorum* (see Lenox-Conyngham 2005). Cicero formed part of the "bloodstream" of Western culture, appearing in multitudes of manuscripts, commentaries, and editions from late antiquity to the early modern period (Arieli 2002: 27–28; Long 2008: 56).

28. The notion of God as physical, and the consequent interpretation of the *imago Dei* as literal, has antecedents in the rabbinic interpretive tradition (Lorberbaum 2002: 56). Augustine's stance on the issue in the *Confessions* shows that this view was not treated sympathetically in early Christian theology.

29. Meilaender notes that Thomas's argument here seems to undermine the idea of universal and irrevocable dignity (Meilaender 2009: 7–8). However, Thomas's attention at this point is on the dignity humans are able to cultivate or impair

by their choices and actions, the sort of dignity exemplified by the magnani-mous man. As we have seen, dignity in this sense presupposes beings of a cer-tain kind, with capacities of rationality and deliberative choice. Human choices cannot obliterate that natural dignity, even when we act in ways that deeply obscure it. Meilaender makes a similar argument, distinguishing between "human dignity," which concerns actualization of our species-specific func-tional capacities, and "personal dignity," the presumption of respect held by any individual person. While human dignity in his sense can be effaced by vicious actions, personal dignity is ineffaceable. I return to discussion of the relation between these different accounts of dignity at the end of this chapter.

30. It is true that there are antimechanist conceptions of nature in the modern era, such as natural magic and Romanticism, but these have the effect of increasing the relative autonomy of nature just as mechanism does (see R. J. Richards 2009: esp. 104–5; Brooke 2009a: 202–3; Ashworth 2003). In any case, while movements such as natural magic and Romanticism affected spe-cific modern thinkers such as Newton and Darwin in significant ways, the overall thrust of modern science has been mechanistic.

31. Specific early modern natural philosophers, such as Boyle and Bacon, allowed some place for final causes in the study of nature, but the overall effect was toward understanding nature as an autonomous mechanism with purposes present only when imposed by God from outside of the natural order (Ashworth 2003: esp. 79–80; Lindberg 2007: 365). Indeed, for many such thinkers, the presence of purposiveness in nature is direct evidence for divine agency, in versions of teleological theistic argumentation of the sort that culminates with Paley (Paley 1809). The general movement away from natural teleology was dramatically pronounced by the early 20th century, as biologists in the wake of the "Darwinian synthesis" abandoned any appeal to purposiveness in natural processes (Provine 1988).

32. In a striking irony, one of the primary classical sources for early modern mechanism was Cicero (Lindberg 2007: 365).

33. Hobbes's affirmation of human-animal difference is, given his materialism, perhaps unexpected. Noting the Aristotelian doctrine that some animals are social like humans, Hobbes affirms the unique character of human sociality due to a series of distinctions: humans seek dignity and honor where animals do not; individual human interests depart from the social good while in social animals they converge; humans, unlike animals, can act according to reason and thereby judge their own actions and the actions of others; while animals communicate, they cannot communicate ideas; humans are disturbed by ease while animals are not; and, finally, social agreement among animals is natu-ral, while in humans it is the product of artificial covenant (*Leviathan* 2.17).

34. John Cooper has argued that a broadly construed dualism—including views such as metaphysical dualism, Platonic idealism, and even the ancient Jew-ish view of a soul-body unity with the hope of personal survival in resur-rection—is a common ancient anthropology (Cooper 2000). As Cooper's analysis illustrates, the main motivating concern of this broad dualism or duality is explanation of the possibility of survival of the person after death. By contrast, the arguments of Descartes, Locke, and other modern figures show that modern dualisms are motivated much more clearly by an attempt to understand human functional powers, especially reason, as exceptions to the mechanical cosmology of modern science. Contemporary monists such as Joel Green take issue with both dualism and modern reductive materialism, arguing that orthodox Christianity and modern science are compatible with one another so long as survival of death is construed in terms of resurrection rather than survival of an indestructible soul (Green 2008, esp. chap. 5).

35. Speaking of the structures of animal bodies, many of which he has studied by dissection, Descartes claims, "I have found nothing whose formation seems inexplicable by natural causes. I can explain it all in detail, just as in my *Meteors* I explained the origin of a grain of salt or a crystal of snow" (Descartes 1970: 64).

36. Critics often infer from Descartes's claim that animals are mere automata the view that animals cannot suffer or experience pain. Cottingham (1978) defends Descartes against this "monstrous thesis" but admits that the claim that purely mechanical beings can experience pain or any other "mental" content is strictly speaking inconsistent with Descartes's metaphysical dualism. Harrison (1992) suggests plausibly that Descartes was agnostic on animal capacity for pain. At the very least, he is inconsistent on the matter. For example, he claims that animals have "organic sensation" and that brain folds are the basis for memory in both humans and animals, even if humans have a different and "altogether spiritual" form of memory; in other places he asserts more unequivocally that the evidence shows that animals do not think or have subjective experience, including the experience of pain (Descartes 1984: 287–88; Descartes 1970: 76, 244–45).

37. The famous emblem of the British abolitionists, a medallion depicting a naked black man in chains, originally produced by Josiah Wedgwood (Charles Darwin's grandfather), asked the question "Am I not a man and a brother?" (Desmond and Moore 2009: 6–7; see also Wedgwood Museum 2010).

38. The concept of autonomy is not incompatible with the notion that human beings are dependent, in the sense meant by Aristotle's dictum that a human being is by nature a "political animal," πολιτικὸν ζῷον (*Politics* 1.2.1253a2; *Nicomachean Ethics* 1.7.1097b11) and developed by MacIntyre 1999 and Arnhart 1998. It is at the very least, however, a difference in emphasis. The description of autonomy by Kant minimizes the interdependence that is basic to human life, not only at the beginning and end but constantly.

39. As Arnhart points out, Kant must invoke a dualistic split between the human will and mechanical nature in order to preserve the freedom of will necessary for autonomy (Arnhart 1998: 83).

40. Particularly significant for the topic of dignity is Pufendorf's emphasis of human reason and judgment: "the dignity of man far outshines that of beasts by virtue of the fact that he has been endowed with a most exalted soul, which, by its highly developed understanding, can examine into things and judge between them, and, by its remarkable deftness, can embrace or reject them. And for this reason the actions of man are put in a class far above the motions of beasts, which are but a reflex of their senses, without any previous reflection" (Pufendorf 1964: vol. 1, 38).

41. One very concrete effect of the shift toward democracy is that social value systems based on honor will decline, as Berger illustrates with the case of *Don Quixote* (Berger 1983; see also Taylor 1992: 46–47). Berger argues that dignity *replaces* honor when the culture of honor collapses at the end of feudal culture; I would argue for a more evolutionary shift of emphasis, since dignity is present as a value concept from pagan and Jewish antiquity forward. But it is certainly the case that a democratic conception of social and political life, rooted in Enlightenment assumptions, will highlight and emphasize the universality of human dignity that was often more promise than reality in earlier periods of history.

42. Mary Wollstonecraft's *Vindication of the Rights of Woman*, published in 1792, provides one such example. Wollstonecraft leavened the text with references to dignity, claiming that false conceptions of "female manners" had robbed women of dignity, and declared that "It is time to effect a revolution

in female manners—time to restore to them their lost dignity—and make them, as a part of the human species, labour by reforming themselves to reform the world" (Wollstonecraft 1891: 84, 95).

43. Mill rightly credits the role of pagan antiquity in grounding of contemporary morality, including ideas of honor and personal dignity, but denies—as I see it, mistakenly—antecedents in the "religious part of our education" (Mill 1977: 256).

44. Patrick Lee and Robert George point out that one's *sense* of one's dignity is distinct from one's condition of dignity. A person experiencing the symptoms of a painful terminal illness might not *feel* particularly dignified, but her psychological state is separable from the objective features of her distinctiveness as a human being (see Lee and George 2008: 410–11).

45. Schopenhauer's rather obscure abusive epithet, "shibboleth," is suggestive of his point. The term derives from a biblical episode in which the Gileadite tribe needed to distinguish friend from foe in conditions of war (Judges 12:1–6). They hit upon the device of a test word, one pronounced differently by the enemy Ephraimites. The test word, *shibboleth*, has come to stand for a demarcation term, a test word used by a party or group to identify ins and outs. Schopenhauer's point was that "dignity of man" came to be a signal of an in-group, not really conveying any meaning, but reassuring and encouraging the proponents of human exceptionalism. In so doing, Schopenhauer anticipated modern critics who find the term vacuous. Schopenhauer used the term shibboleth in another context to claim that Pantheists are similarly guilty of saying nothing of significance, because an impersonal God is no god at all. Better to admit to atheism than to coddle an in-group with talk of an impersonal divine force (see Schopenhauer 1901: 215–17). Dennett likes the term shibboleth too, and would doubtless suggest it is an excellent example of a meme (Dennett 1995: 467, 341–52).

46. Complex thinker that he is, Nietzsche is also capable of making positive use of the idea of dignity, as he does when remarking that we have "our greatest dignity in our meaning as works of art," and that true dignity is to serve as a means for the emergence of genius (Nietzsche 2006: 58, 94). While Nietzsche seeks to debunk the conventional notion of dignity, he is happy to suggest that austere and demanding expectations for oneself constitute a kind of dignity.

47. Darwin repeatedly asserted that evolution by natural selection was compatible with traditional religion, including religious anthropology (F. Darwin 1887: vol. 1, 288–89, 304, 317 fn; F. Darwin 1887: vol. 2, 353–54; Darwin 1876: 421–22; Huxley 1887: 202–3; Brooke 2009a; Brooke 2009b). At the same time, he was clearly aware of the potential challenges posed by his scientific views to all manner of traditional ideas (Darwin 1958: 85–87).

48. The reverberating cultural effects of Darwinian evolution on ethics, politics, and social policy is the subject of the controversial book *From Darwin to Hitler* (Weikart 2004). Regardless of the views held by Darwin himself, there is little doubt that ideas inspired by and developed from a broadly Darwinian outlook played a critical role in the late modern collapse of the traditional concept of human dignity.

49. For a more detailed treatment on the effects of Darwin and Darwinism on the traditional concept of human dignity, see my "Prospects for Human Dignity after Darwin" in this volume.

50. Some have assigned a particularly significant role to 19th-century Kantian theologian Antonio Rosmini in interpreting and translating Kantian ideas about human dignity in the Roman Catholic theological context (Sulmasy 2006a; Cleary 2008). Sulmasy cites the influence of Rosmini on Pope Leo XIII as evidence for "very late" emergence of the contemporary sense of

"human dignity" (Sulmasy 2006a). Whatever Rosmini's importance, however, the historical record shows that a broad notion of human value and significance based on distinctiveness was neither new in the 19th century nor uniquely formulated by Rosmini.

51. Baker (1947) is instructive on this point. Published at the time of the post–World War II revival of the concept of dignity, Baker's historical survey of the concept of dignity focuses almost entirely on philosophical anthropology. Baker clearly believes that anthropology has implications for ethics and value theory, but understands that dignity is essentially a view of human beings.

52. It is without doubt true that Jewish and Christian thinkers, and especially Roman Catholics, have been instrumental in promoting the idea of human dignity. Official documents of the Catholic Church, such as the *Catechism*, are richly larded with the language of dignity (see United States Catholic Conference 1997: 1.2.1.308, 1.2.1.1.6.356, 1.2.1.1.6.357). Similarly, a search for the term "dignity" on the English-language site of the Vatican returns thousands of hits.

53. Meilaender stresses the notion of personal dignity, that inhering in persons, as the central and inviolable notion of modern practical ethics. He contrasts the sort of dignity grounded in persons, which he takes to be noneffaceable and not subject to comparison from individual to individual, with the dignity of differing degrees of excellence, which he calls "human dignity" (Meilaender 2009: 7–8). While he sees the two conceptions of dignity as contrasting concepts, I believe that the historical record shows that the concept of personal dignity arises out of the features of human distinctiveness.

54. Many have noted that resurgent interest in virtue ethics in the past half-century amounts to something of a revival of this tradition in Western thinking (key texts would include Foot 1978; MacIntyre 2007, 2001b). One possible reason for this resurgence is widespread recognition of the limitations of rights language, especially in practical ethics and international law. While some would argue for replacing modernist rights language with the traditional language of dignity, most seem to acknowledge the value of integrating dignity and rights (see, e.g., United Nations General Assembly 1948; Nussbaum 2001a).

55. What I call here the dignity of exceptional excellence is close to what Meilaender calls "human dignity" and to what Sulmasy calls "inflourescent dignity" (Meilaender 2009: 4–6; Sulmasy 2008: 473–74).

56. The dignity of universal human value is similar to Meilaender's "personal dignity" and Sulmasy's "intrinsic dignity" (Meilaender 2009: 6–7; Sulmasy 2008: 473).

57. Compare Meilaender's argument for the priority of personal dignity over human dignity (Meilaender 2009: 7–8).

Part I

The Source and Meaning of Human Dignity in Worldview Context

3 A Catholic Perspective on Human Dignity

Christopher Tollefsen

Reference to "human dignity" and "the dignity of the human person" occurs repeatedly in recent Catholic papal and magisterial teaching, beginning roughly with the encyclical *Rerum novarum* of Leo XIII, continuing in the documents of the Second Vatican Council, and eventually coming to dominate the writings of Pope John Paul II. As a result of that increased emphasis, concern for the idea of human dignity has also come to characterize the scholarly work by Catholics of recent decades that has been devoted to articulating and arguing for that teaching (see, e.g., Gormally 2004; Hittinger 2006; Sulmasy 2007, 2008; Neuhaus 2008; Lee and George 2008). Human dignity, and the overshadowing of that dignity in modernity, are considered the key concepts for an adequate philosophical anthropology, for working out acceptable solutions to contemporary bioethical dilemmas, and, more broadly, for understanding the condition, and even crisis, of man in the contemporary world. It would appear, in fact, that contemporary Catholic social and moral teaching, and even its ecclesiology, is grounded, ultimately, in this very idea.

It is perhaps inevitable that a concept such as human dignity, given, as it is, so much prominence in recent Catholic thought, should be subject to a backlash, both by secular thinkers, dubious of Catholic positions on a host of controversial issues, and even by some Catholics, suspicious of the uses or abuses of the concept within their own tradition. This chapter will investigate the core ideas of the concept of human dignity in Catholic thought and defend that idea against three strands of criticism. The first two are made by secular thinkers, including, most prominently, bioethicists. The first of these secular complaints is that the idea of dignity is, essentially, a species-ist notion, a way of illicitly giving preference to beings like us over nonhuman animals and of attributing excellence that is manifestly not present to beings of diminished or undeveloped cognitive capacity.

The second secular criticism is that the notion of dignity is too vague, and that its vagueness serves as cover for the intellectual deficiencies of the positions it is meant to support. Dignity, on this view, serves merely as a rhetorical device by which conservative Christians, especially Catholics, can help themselves to conclusions to which they are already committed.

A third strand of criticism comes from within Catholicism itself and holds that the modern Catholic idea of dignity is simply mistaken. Robert Kraynak in particular has recently argued that the Catholic "personalism" of John Paul II, Jacques Maritain, John Finnis, and others has been unacceptably Kantian in its outlook, with disastrous consequences, in particular, an increasingly individualistic and subjectivist assertion of rights.

In my dialectical exchange with these three types of criticism, I will articulate a version of the Catholic view on human dignity that does indeed give preference to species membership, but not, I will argue, unreasonably. In consequence, all human beings, even those of diminished or undeveloped cognitive ability, are privileged morally over every subhuman animal, yet are fundamentally equal with one another. I will argue further that, while the idea of dignity can, and sometimes does, serve only as rhetorical cover where arguments are missing, it need not, and that Catholic thought on dignity, especially the thought of Pope John Paul II, is attuned to precisely that which must be articulated in order for the idea of dignity to give birth to substantive moral norms. In this essay, I concentrate especially on norms governing the field of bioethics and, in particular, norms concerning the so-called human life issues such as abortion and euthanasia, norms concerning reproductive technologies, and norms grounding the right to health care. Finally, I will argue that the evaluative foundations and conclusions in which dignity is implicated are indeed best described as "personalists" describe them; and that while they generate robust rights claims, those claims are neither subjectivist nor individualist; the resulting account, I argue, both corrects and complements contemporary liberalism.

THE DIGNITY OF THE HUMAN PERSON

I shall begin with the thought of Pope John Paul II, in his Encyclical *Evangelium vitae* (Pope John Paul II 1995), for there the pope appears to be engaged in a summation of Catholic thought on the subject of human dignity precisely in order to combat the overshadowing of that dignity in modernity, and especially in the medical context. Abortion, euthanasia, and even capital punishment are described as threats to human dignity, and the foundational nature of that dignity for respectful treatment of the human person is reiterated often.

John Paul identifies, in *Evangelium vitae*, three aspects to the dignity of the human person (Pope John Paul II 1995: nos. 34–38). First, the human person has dignity because of his source: he is made by God, and entirely gratuitously. Second, because of his nature: he is made in God's image. And finally, man has dignity because he is made for God, a destination the nature of which is transformed in Christ, by whom we are called to share in God's life, as his adopted sons and daughters. In what follows, I examine each of these key ideas, the understanding of each of which penetrates the understanding of the others. These ideas summarize the pope's thought, and magisterial teaching more generally, on the core constitutive dignity of the human person, a dignity that is inalienable, and not lost even by actions which are not consonant with that dignity: "Not even a murderer loses his personal dignity, and God himself pledges to guarantee this" (Pope John Paul II 1995: no. 9).

Made by God as a Gift

God is the author of *all* creation, not only of man. How, then, can man's creation be special? *Gaudium et Spes*, from the Second Vatican Council, gives us the answer, in describing man as "the only creature on earth that God has willed for itself" (Second Vatican Council 1965b: no. 20). This raises a natural question: what is it about man that makes it *possible* for his existence to be willed for itself? The answer to this can only emerge gradually through discussion of dignity in its three aspects, but it is clear from Genesis, and has been held true throughout the Catholic tradition, that God willed the existence of the rest of earthly creation for the sake of man. *Gaudium et Spes* thus articulates the following view: all earthly creation exists for the sake of man, but man exists for his own sake; and in this is partly to be found man's dignity.

Moreover, in that willing of man's existence for his own sake, God can rightly be said to have willed man's existence, and, of all earthly creation, *only* that existence, as a gift. God did not need to create man, so man serves no instrumental purpose, unlike the rest of creation, which is created only for the good of man himself. Man's creation thus is entirely gratuitous. This, too, reflects an aspect of human dignity. But the glory of both this willing-as-gift and the willing of man for his own sake (two interpenetrated notions) are magnified by the fact that the source of the creation is God himself. God who is all good and all powerful is nevertheless capable of creation for the sake of the good of what is created, and the goodness of the creation takes its value, its dignity, from its source as well as from the way in which the source has willed.

Made in God's Image

Of all the reasons commonly proffered as explaining man's dignity, the most commonly mentioned is the second—that man is made in the image of God. This claim, I will argue, has both descriptive and normative components; it can only be fully understood insofar as it is recognized to point ahead of man's given nature to the nature he should take on in action, a nature that is other regarding and eventually to be completed in Christ.

The descriptive component is often, and accurately, understood in the following way: man is made in the image of God precisely because his existence is that of a person. Man, that is, is possessed of reason and will and is capable of rational thought and free choice (for an articulation and defense of this view, see Lee and George 2008). We see this dual emphasis in the documents of the Second Vatican Council: in *Dignitatis Humanae*, the council writes that our "dignity as persons" is our existence as "beings endowed with reason and free will and therefore privileged to bear personal responsibility" (Second Vatican Council 1965a: no. 2). John Paul similarly summarizes: "The biblical author [in the book of *Sirach*] sees as part of this image . . . those spiritual faculties which are distinctively human, such as reason, discernment between good and evil, and free will" (Pope John Paul II 1995: no. 34).

Already, in these descriptions, we see anticipation of the way in which the image of God is normative, for reason and free will, in man, are necessary precisely so that we may bear that privilege of "personal responsibility." This responsibility was identified centuries before by Saint Thomas Aquinas as our participation in God's eternal law; unlike the rest of earthly creation, which follows God's law blindly, because made to do so by God, human beings are given a share in divine reason, and the faculty of free choice, precisely so that they may guide and constitute themselves as the persons they are *to be* in accordance with God's plan (Aquinas 1920: I–II, Prologue). It is in this active shaping of their own lives in accordance with reason that human beings are *like*—in the image of—the divine. But the orientation of this power must be identified; that is, what is the substantive content of the deliverances of reason? This question concerns the content of the natural law, a question to which I will turn later in this chapter.

The descriptive content further shades into the evaluative in magisterial and papal reflections on human dignity as those are carried out in light of the first Genesis account. In that account, man is made in God's image as male and female, and this was a subject of intense interest for John Paul II in his reflections on marriage. In their complementary maleness and femaleness, spouses are uniquely enabled to extend what the Second Vatican Council teaches is perhaps the most important way that human beings image God—namely that they are capacitated, by their personhood, and enjoined, by reason and the divine will, to love. Marital love is a profound imaging of the divine precisely because by it spouses realize a unity similar to that of the unity of the divine persons; they are one flesh, as the divine persons are one God. And they mirror the fruitfulness of the divine love, which goes out from the father to the son, and then from the two in the procession of the Holy Spirit, in the embodying of their love in another, viz., a child. Martial union and fruitfulness are so much like the triune divine life that it is not out of place to speak of the "dignity of marriage," as we speak also of the dignity of the person.[1]

We reach here as well a starting point for the fully normative task of persons insofar as they are made in the image of God, which *Gaudium et Spes* puts as follows:

> God did not create man as a solitary, for from the beginning "male and female he created them" (Gen 1:27). Their companionship produces the primary form of interpersonal communion. For by his innermost

nature man is a social being, and unless he relates himself to others he can neither live nor develop his potential." (Second Vatican Council 1965b: no. 12)

We find this thought more strikingly articulated as the conclusion of a point already noted: "man, who is the only creature on earth which God willed for itself, cannot fully find himself except through a sincere gift of himself" (Second Vatican Council 1965b: no. 24).

We see here two things. First, the earlier point about one aspect of man's dignity being found in his source—that is, God's gratuitous creation—here is joined to the second point about man's dignity as being made in the image of God. For God, in creating a being for its own sake, creates that being as *able to love*—that, we here find articulated, is what it means to *be* a being capable of existing for its own sake. And it is thus only in loving that man realizes, or fulfills, the nature by which he has dignity. Thus, man's dignity presents him with a task (Beabout 2004; I will return to this theme later), and we could say that his constitutive dignity—the dignity he has in virtue of what he is—thus sets on man a requirement to strive for existential dignity (Gormally 2004), the dignity of excellence *as* a being capable of love and self-gift.

A final point should be made about the dignity of man as image of God, to which I will return later as well. It is tempting to interpret the language of "image of God" as it bears on man's personhood in a purely spiritual way. That is, one might think that it is only as thinking and willing that man is a person in such a way that the body is other than that person. This form of dualism is resisted by the Catholic tradition, as the following passage from *Gaudium et Spes* makes clear:

> Though made of body and soul, man is one. Through his bodily composition he gathers to himself the elements of the material world; thus they reach their crown through him, and through him raise their voice in free praise of the Creator. For this reason man is not allowed to despise his bodily life; rather he is obliged to regard his body as good and honorable . . . the very dignity of man postulates that man glorify God in his body. (Second Vatican Council 1965b: no. 14)

Made for God

Man, as created in the image of God, is also created "capable of knowing and loving his Creator" (Second Vatican Council 1965b: no. 12). We can discern three levels of this knowing and loving, each of which manifests the dignity of man.

First, even apart from revelation, there are adequate reasons to conclude that a personal and creative God exists who is responsible for not only man but man's nature, including the reason which makes possible man's orientation toward his own fulfillment. Thus, a relationship of cooperation is, it may be inferred, offered by God to man: to do as reason requires is to cooperate with God in a project, the end of which is that fulfillment available to man on earth. Thus, a form of friendship with God is offered to man and may be said to be his destiny and, thereby, his dignity. That friendship, moreover, is deeply personal, for God calls each human person to a particular life; this personal call, man's *vocation*, has been especially emphasized by recent magisterial and papal documents and is likewise central to man's dignity.

Second, that offer of friendship is extended, in revelation, to a promise of the Kingdom: men are offered the possibility of everlasting life in a communion with other saints, with angels, and with the persons of the Trinity. This eternal life is, moreover, a bodily life, and the Second Vatican Council alluded to this, in the passage quoted above, in explaining the dignity of man's bodily life.

Third, the offer is extended once more, in baptism, through which we are not simply cleansed of sin and made newly worthy of eternal life but by which also we enter into the divine life of the son and thus become adopted sons and daughters of the father. We enter, that is, into the divine life of the Trinity itself, and this divination finds its completion in the beatific vision, an immediate knowing of God in his divinity for which no created intellect is adequate.

This threefold destination—friendship with God, eternal life in the Kingdom, and divine life in the Trinity—can all be summarized by speaking of man as "made for" God, although each in different ways. And this being made for God is the third of the aspects of man's dignity identified by John Paul II.

Before turning, in the next section, to the first of the three criticisms of the idea of human dignity, it is worth noting that these three aspects of human dignity—being made by God, in His image, and for God—though presented by John Paul II, other popes, and the Magisterium within the rich framework of Catholic teaching, nevertheless are *all* witnessed to by natural reason as well as specifically Catholic teaching. That the existence of all contingently existing beings can find an adequate explanation only in the existence of a non-contingently existing being that nevertheless freely creates is a truth available to philosophy and is recognized unreflectively, if inadequately, by almost every culture in history. That man's quest for fulfillment requires the active cooperation of that creative being is similarly available to natural reason (Grisez 2008). And, finally, the special nature of man, particularly insofar as he is a reasoning and choosing being, has been recognized throughout intellectual and social history even by those, such as Plato, with no access to revelation. This is not to say that revelation and faith add nothing; nor is it to say that the truth is these matters is easily or always obtained. But the idea of human dignity that supervenes on the convergence of these three truths about man is far from sectarian or inaccessible to human reason.

THE CHALLENGE OF MORAL INDIVIDUALISM

Among the most striking aspects of the Catholic view of human dignity is its universalism and its egalitarianism. That is, first, all human beings are held to possess human dignity, regardless of their age, stage of development, cognitive capacity, or any other accidental property. What is sufficient for possessing this dignity is simply membership in the species *Homo sapiens*. Thus, if, as seems scientifically demonstrable, human embryos are individuals in that species, then human embryos possess human dignity. Similarly, if patients in a persistent or permanent vegetative state remain human beings, then such patients also possess human dignity. Finally, even murderers, as we have seen, because they remain individuals of the species, do not lose their human dignity (recognition of this truth has developed through time in the Church; Saint Thomas Aquinas, for example, did not hold this view). If there is an overriding theme of John Paul II's encyclical *Evangelium vitae*, it is, in fact, the "dignity of every human being" (Pope John Paul II 1995: no. 25).

The equality of dignity in all persons is likewise well attested. *Gaudium et Spes*, for example, notes that

> Since all men possess a rational soul and are created in God's likeness, since they have the same nature and origin, have been redeemed by Christ and enjoy the same divine calling and destiny, the basic equality of all must receive increasingly greater recognition. (Second Vatican Council 1965b: no. 29)

This universal and equal nature of dignity has concrete practical implications, among which is the principle forbidding the instrumentalization or subordination of the life of one human being to that of another. The weak, the young, the old, the poor, the vulnerable—since all are of fundamentally equal worth, it is a violation of dignity to sacrifice one for another; yet, because members of the rest of animal creation do not possess this fundamental form of dignity, and are, in fact, made for man and under his dominion, it is licit to act toward animals in ways that would never be permissible with respect to human beings: raising them for meat, using them in medical experiments to their detriment, deliberately sacrificing their lives for the sake of human well-being.

Further, for reasons that have been touched upon, and that will be addressed at greater length in the next section, the dignity of the human person requires respect for the bodily life of all persons, a respect held to be incompatible with direct killing. Thus, even killing that is done "for the good of the patient," as euthanasia is often claimed to be, is impermissible on the Catholic view. John Paul's words on this are, in fact, especially striking:

> Therefore, by the authority which Christ conferred upon Peter and his Successors, and in communion with the Bishops of the Catholic Church,

I confirm that the direct and voluntary killing of an innocent human being is always gravely immoral. (Pope John Paul II 1995: no. 57)

These claims put the Catholic Church at odds with much of contemporary bioethics, most practitioners of whom are wedded to a view that James Rachels has called "moral individualism"—the view that "The basic idea is that how an individual may be treated is determined, not by considering his group membership, but by considering his own particular moral characteristics" (Rachels 1990: 173; quoted in McMahan 2002: 354).

Moral individualism has recently been defended by the philosopher Jeff McMahan, who, not coincidentally, takes as a spokesman for the view he is criticizing the Catholic legal philosopher John Finnis (Finnis 1995). McMahan makes two criticisms—one moral, one metaphysical—which I can only briefly address here; a full response must await another occasion.

McMahan's moral criticism is based on an analogy. Suppose that God creates someone without original sin; she is innocent in a moral sense.

Ought God to send her to Hell because she belongs "to a kind of being characterized by" original sin. . . ? . . . I think that most people would think that a just God would determine her treatment by reference to whether she is actually sinful rather than by reference to the nature of typical members of her kind. And if God would act this way, presumably we should as well. (McMahan 2002: 358–59)

Thus, individuals ought to be treated based on their own particular moral characteristics.

McMahan's analogy founders, however, as can be seen by the following two-step argument. The first step has to do with McMahan's understanding of nature and its relation to sin and punishment. On most accounts of sin, original or otherwise, sin and its effects are not essential properties of a being. That is, although we talk of fallen nature as a result of human sin, we do not think that human beings underwent a substantial change as a result of original sin—we do not think they underwent, that is, a change of nature. Yet we do think that punishment is deserved as a result of sin. So the treatment of sinful man as deserving of punishment is not based on a judgment of human nature but a judgment concerning nonessential properties, specifically guilt and culpability; if those properties change, then the judgment regarding desert may change. Yet such a change would not impugn the possibility of some (nonpunitive) form of treatment that is justified by nature, and which does *not* change on the basis of change in nonessential properties.

This, then, is the second step of the argument. Are there forms of treatment that we think are justified not on the basis of particular and "actual" properties but rather on the basis of nature? This way of asking the question is itself somewhat flawed, for a nature *is* a particular and actual property of a being that has it. But the sense of the question is clear: are there forms of treatment that we believe should be justified because of the kind of being something *is*, rather than anything the being has done, or achieved? The answer appears to be affirmative. Unlike punishment, which generally responds to actual guilt, our affirmation of a being's possessing fundamental, inalienable, and absolute human rights seems predicated more properly on the nature of the being than on any nonessential properties, contingencies, or achievements. So the Catholic teaching that dignity, which is recognized as the ground for absolute and inalienable rights, pertains to a being's species membership seems justified.

McMahan's second criticism goes to this very claim, however, for he does not see how all members of the species, even infants and the radically cognitively disabled, *could* have the sort of nature that possesses dignity unless that nature possessed a soul; and this notion McMahan, in turn, professes to find unsatisfactory. He thus finds no grounds on which to say that all human beings possess dignity.

As will be apparent from a previous quotation, from *Gaudium et Spes*, the Catholic Church does teach that humans are complex beings constituted by both body and soul, though these two components are unified so as to form one being. Soul is considered by most Catholic thinkers to be the animating and organizing principle of a living body and, in the case of a rational animal, the principle that makes possible the spiritual acts of which human beings are capable. This conception is distinctly influenced by Aristotle's hylomorphism, and while Catholics, following Aquinas, assert that the immaterial soul is *subsistent*—that is, capable of an independent existence—they deny that it is a *substance*, a complete being with a nature; only the body-soul composite it that. But it is important to note that Catholic belief in the soul is a *consequence* of their recognition of the presence of the same properties that are thought to justify human dignity. That is, the properties that ground dignity are judged to be impossible in a being that was not itself possessed of a nonmaterial principle, which, by making a material being to be more and other than *merely* a material being, could be considered the actualizing principle of that being (cf. Brugger 2008, 2010; Lee and George 2009). So the Catholic view is that possession of a certain nature—a rational nature—is a sufficient condition both for an immaterial soul and for possession of dignity. These claims might thus be put in the form of two conditionals:

1. If all human beings possess a rational nature, then all human beings have souls.
2. If all human beings possess a rational nature, then all human beings have dignity.

And Catholics also assert the antecedent to both conditionals:

3. All human beings possess a rational nature.

In consequence, one might deny the first conditional without that denial amounting to an argument either against the second conditional or against the assertion of the antecedent to both conditionals.

McMahan acknowledges that Catholics assert the first conditional. He does not find it convincing, however, for he believes that there is "no reason to suppose that self-consciousness and rationality require a non-physical explanation if the simple consciousness of animals can be accounted for in wholly physical terms" (McMahan 2002: 11). This is not much of an argument against the first conditional, since materialist programs are, at best, far from capable of making good on their promises of complete natural-istic explanations, *and* because the view described here denies that human capacities are the same as the capacities of other nonhuman animals. And it is no argument at all against the second conditional.

In fact, McMahan spends much of his time arguing much more directly against the antecedent of the two conditionals, for he does not think that cognitively undeveloped, damaged, or retarded human beings *do* possess the properties that are sufficient for the possession of dignity, viz., a ratio-nal nature. And McMahan further thinks that that those who do assert the antecedent must be committed to a Cartesian concept of the soul, as an inde-pendent rational agent, rather than an Aristotelian hylomorphic conception of the soul as the organizing principle of an organic body, for, according to McMahan, the hylomorphic soul "determines how the body *is* organized, not how it might later be organized" (McMahan 2002: 13).

I think that here McMahan is mistaken; the Aristotelian concept of a soul, utilized by many Catholic thinkers in understanding the relationship between soul and body, is that the soul is a principle not just of present orga-nization, in the sense of structure, but of a thing's nature. And a nature really is understood not by focusing on properties that a being happens to have at a particular time slice but at the properties that things of that sort develop, through their own initiative, through time. This is why the debate about soul really comes back, as the earlier discussion of hell and original sin did, to a question about human nature. The Catholic view, to reiterate, is that it is in the nature of a human being—any human being—to develop itself, unless hindered by disease, environment, external threat, or other deficiency, to the point of being able to exercise its rational capacities, and that this nature is possessed even by human beings who have not yet, will not in fact, or are no longer capable of exercising those rational capacities.

The key thought is this: cognitively undeveloped, damaged, or retarded human beings are themselves still beings *of the same kind* as those who most fully manifest the activities characteristic of beings with a rational nature. They, too, are thus to be considered beings with a rational nature. And, in accordance with the first and second conditionals above, they, too, are held to be animated by an immaterial soul and to possess human dignity.

To conclude this part of the discussion, then: while it is true that the Catholic account of human dignity is more than friendly to—and, indeed, in some sense requires—an account of a soul, its deeper claim is that all human beings possess a common nature—a rational nature. The necessity of "soul" is motivated by a further, defensible, rejection of the sort of naturalism that would deny the irreducibility of the spiritual capacities of human beings to material causation. The deeper claim about a common rational nature is contested, but the claim is not intrinsically unreasonable; nor, in the eyes of Catholic thinkers, has it been shown false. The Catholic account thus stands as a genuine and reasonable opponent of the moral individualist account which denies that all humans have a nature of the same kind and, thus, denies any role for the notion of human dignity.

THE "STUPIDITY" OF DIGNITY

A second recent challenge to the importance of the notion of human dignity, made with particular attention to its Catholic supporters, has been put forth by Ruth Macklin, John Harris, and, perhaps most provocatively, Steven Pinker. Pinker puts the objection like this: "The problem is that 'dignity' is a squishy, subjective notion, hardly up to the heavyweight moral demands assigned to it" (Pinker 2008). And Pinker cites Macklin's claim that bioethics can get by without the notion, by relying exclusively on the concept of autonomy and the accompanying principle of informed consent (Macklin 2003).

Harris, too, criticizes the appeal to human dignity as "comprehensively vague," and does his best both to show that its most adequate gloss, the Kantian principle that human beings should never be treated as a means only, but also always as an end, has counterintuitive—and, indeed, ridiculous—conclusions, such as that blood transfusions should be outlawed (Harris 2005).

Pinker identifies three features of dignity that reveal why it is doomed to fail as a core moral concept: dignity is relative, fungible, and occasionally harmful. That is, different acts, classes of persons, and types of appearance are considered dignified or not in different social settings, cultures, or time periods; hence, the relativity of dignity. Further, we make trade-offs of our personal dignity, in deciding, for example, to strip for a medical exam or remain at attention during an airport screening; hence, its fungibility. And finally, when we (or, more likely, our rulers) "stand on" our dignity, to the detriment of those who are punished for their offenses against our dignity, we reveal dignity's positively harmful character. As Pinker says, "Totalitarianism is often the imposition of a leader's conception of dignity on a population" (Pinker 2008).

No doubt, such points are well taken; yet they have little, if anything, to do with the concept of dignity articulated in this essay. Still, this failure to engage with the "Catholic" conception, allegedly one of the very conceptions Pinker was concerned with in his essay, does illustrate one of two problems with dignity talk that the "stupidity and vacuity" charges really do illustrate. This is that dignity, like many other concepts, has a variety of different kinds of applications and usages, many of which are connected by way of the relationship of paradigm to analogous cases, others of which are connected by a relationship similar to Wittgenstein's "family resemblances." And when we move from one to another of these usages, various claims true with one usage cease to be so with another. Confusion and specious argumentation are the result if these shifts are not taken into consideration.

Thus, in this essay, I have concentrated, as the Catholic Church does, on the core of the nature of dignity as an intrinsic property possessed by human beings as possessed of the capacities for reason and free choice. Different authors, in attempts to indicate that *this* is the type of dignity to which they refer, modify the word in different ways: connatural dignity (Gormally 2004); intrinsic dignity (Sulmasy 2007); personal dignity (Lee and George 2008); or, as here, constitutive dignity. On the account shared by all these (Catholic) authors, it is clear that dignity cannot be lost except by virtue of a human being's ceasing to exist, as at death. Yet the notion is easily extended, for example, to what Gormally calls "existential dignity," the dignity or excellence to be had in living ethically up to the demands that are made on one in virtue of being a person with constitutive dignity. Since this is a normative sense of dignity, one can fail to live up to that dignity and thus "possess" dignity to a different degree than some other human being, even as one remains equal in constitutive dignity to all other human beings.

Similarly, it is clearly possible to treat a human being in ways not in accordance with his dignity, another type of normative failing. It is perfectly sensible to say this, however, *and* say that the person so treated does not lose any of his or her (constitutive, connatural, or intrinsic) dignity, or decline in dignity at all.

Finally, because dignity signifies a form of excellence that traverses onto-logical and normative considerations, the understanding and realization of which can be variable and socially dependent, it is clear that different societies (or persons) will have different appreciations for the notion, realize that notion in their social practices and ways of life in different ways, in, indeed, to different degrees. Some such societal (or personal) understandings and realizations will be deficient, perhaps radically so. The concept of dignity might, at such a point, be only very tenuously connected to the concept here articulated and might, as Pinker argues, be a source of positive harm, again without ever impugning the core idea of dignity identified by the Catholic tradition.

So while Pinker's objections serve to point out ambiguities and possible confusions, they do not serve to vitiate the possibility of dignity playing a key role at the foundations of ethics and bioethics. But there is still the related challenge, most clearly stated by Harris, that dignity is "comprehensively vague." Is it possible for the concept to be given more bite normatively, such that it plays a role in the generation of substantive, contentful moral norms? To conclude this section, I wish to show a way forward—one that will give an important but not foundational role to the concept of dignity in ethical deliberation.

The way forward on this matter is indicated, I believe, in some key remarks made by the Second Vatican Council's *Gaudium et Spes*, by Pope John Paul II in his encyclicals *Veritatis splendor* and *Evangelium vitae*, and, above all, by the insights of the Thomistic tradition of thought about the natural law.

As I have noted, the respect in which human persons have constitutive dignity is one that traverses both ontological and normative dimensions. Human persons are created in God's image and free and rational; but to partake in such a nature is to be presented with a normative task, characterized by *Gaudium et Spes* as that of living as a gift. Yet even living as a gift for another can only be possible for us as a fulfillment of our nature; it could not be good were it divorced from our capacities, the fulfillment of which is our actualization and perfection as persons.

Philosophically, then, the question is this: how is the normative dimension of our dignity—the task with which, as persons, we are presented—to be understood as itself a fulfillment of our nature, such that it is also our good? Note the two-fold direction here, towards the good of both others and ourselves, is itself indicated in the Council's words: "man, who is the only creature on earth which God willed for itself, cannot fully *find himself* except through a sincere gift of himself." (Second Vatican Council 1965b: no. 24, emphasis added)

This question can only be answered by reflecting upon our knowledge of our good, and the nature of what we know to be our good. Following other thinkers in the Thomistic natural law tradition, and some hints by John Paul II as well, I believe that our good is known through practical reason as that which is desirable for us, because it promises some perfection, in action, for us.

We can reflect upon the goods which are so known by noting that not all that we do is done for the sake of something further, as medicine is taken to achieve health. Rather, some ends are pursued for their own sake and give entirely intelligible answers to the question, "Why did you do that?" Such ends, which we can then understand as the promised goods of persons that are the grounds of our practical thinking, include human life, skilled performance, aesthetic experience, friendship, marriage, personal integrity, and a harmonious relationship with whatever greater than human source of meaning there may or may not be.

These ends, or goods, are aspects of the well-being of all persons; our normative task, as John Paul makes clear in his discussion of the Ten Commandments, is to protect and serve these goods in *all* persons; *our* fulfillment is to be found only in the communion that is instantiated through action for the good in the person of *others*; a life lived only for the good in one's person would be deficient precisely from the standpoint of human flourishing, failing to achieve some goods altogether, and achieving no good as successfully as it could be achieved in communion with others.

How can we return this discussion to the question of dignity and its moral content, or lack thereof? It is only with the horizon of human goods in practical focus that the dignity of human beings, and the demands of that dignity, can be fully known. For it is, as I noted in quoting Saint Thomas earlier, by our reasoned and free relation to these goods that we are beings who have a share in God's providence for ourselves—that is, have dignity at all; and it is by our relation to these goods in an *unrestricted* way—not giving arbitrary preference to ourselves or those near to us—that we respect the dignity of *all* human beings.

What should emerge from this discussion, then, is that it is true that, taken on its own, the notion of human dignity will not provide us with a rich, contentful, and clear account of the moral norms that should govern our action. Thus, absent a larger conceptual framework, which includes, at its fullest, normative, metaphysical, and religious elements, "dignity" will simply not do the work in, for example, bioethics that is demanded of it by its opponents. But fitted within a natural law understanding of the human good, the task that dignity sets for each human being relative to all others can begin to be articulated, minimally, as a demand never intentionally to act against the goods of human persons, in oneself or another; more robustly, as a demand to foster and pursue human goods in creative, fair, and sustained ways, in our own person and that of others.

This approach has immediate consequences in the domain of bioethics, some of which I shall address here, before turning in the final section to the remaining criticism of dignity.

Life Issues

I earlier noted that the dignity of the human person was not divorced, in its nature or consequences, from the bodily existence of such persons. While no corporeal being without a soul could possess the spiritual faculties necessary for dignity, we are not identical to our soul but are rather spiritual animals, bodily beings whose existence nevertheless transcends material limitations.

This bodily existence is reflected in the basic goods of persons, among which are the goods of life and health; skilled performance (which, in many cases, is bodily); aesthetic experience (in which aesthetic form is mediated by sensory experience); and marriage and friendship, both of which, but especially the former, have bodily dimensions. Of most importance here is the good of life and health; if human beings are bodily persons for whom life is a basic good, never to be intentionally acted against, then it becomes intelligible why, in John Paul II's work, abortion, suicide, and euthanasia were considered such great violations of human dignity. And, because our life is normatively to be lived in a relation of self-giving to others, a self-giving that is both our own fulfillment but also only to be understood in relation to the basic goods, our own violations of these goods in respect to others, as when we choose an abortion of an unborn human child or assist in the euthanasia or suicide of another, are likewise to be seen as violations of our *own* dignity, as *Gaudium et Spes* made clear: ultimately, "they do more harm to those who practice them than those who suffer from the injury" (Second Vatican Council 1965b: no. 27).

Reproductive Technologies

It is well beyond the scope of this essay to enter into the details of Catholic teaching on reproductive matters. But central to that teaching where assisted reproductive technologies, such as in vitro fertilization, artificial insemination, and human cloning are concerned, is the notion of human dignity, particularly as it is implicated in the idea that man's life—the life of each human being—is a gift from God, the result of divine love, and that this gift-of-love status must be replicated in our human attitudes toward, and practices of, procreation. Thus, when children are conceived as the fruit of marital embrace, in which husband and wife become one flesh as a re-instantiation of their mutual commitment to a complete sharing of lives, the Church sees the dignity of ensuing children as respected and protected: they are not chosen, per se, by the couple, but accepted *as* a gift, conceived in love with the cooperation of the divine.

By contrast, when would-be parents exercise a technological control over the process of reproduction and turn toward a making, rather than a begetting, of children, then they are seen to adopt an attitude of control and domination over the origins of human life that is, at least at that origin point, inconsistent with the radical equality of human dignity. Technological assistance is thus only acceptable, in the Catholic tradition, insofar as it aids married couples in procreation as an outcome of marital sexual intercourse: drugs and surgical treatments to ensure ovulation, for example, are permissible within this framework.

The Right to Health Care

Within an understanding of human dignity as a task, whose horizons are set by basic goods, including the good of life and health, human beings are presented with strong and exigent good Samaritan obligations. When a human being is in grave need and another can provide the necessary assistance without threatening her own welfare, there are *obligations* of charity. But health care needs are among the most significant that human agents possess; and, in some cases, those needs will be better addressed not by one-off attempts by individuals to come to the assistance of those in need but by the assistance of groups whose members are brought together by a shared commitment to meeting the needs of others. Yet even these voluntary or mediating associations require coordination, and occasionally positive supplementation in their efforts, lest some unfairly go wanting while the needs of others are met. There thus can be a positive role, demanded in justice, for the state to provide assistance along the axis of the good of life and health (Boyle 1977, 1996, 2001). And this obligation, for reasons that will be articulated further in the next section, can be seen from a different perspective as a right, on the part of those in need, to the necessary health care. Catholic teaching has thus embraced a right to health care in developed nations and pointed, as it does in its condemnation of abortion, to human dignity as a grounding concept here.

DIGNITY AND LIBERALISM

The final challenge to the account of dignity presented here—an account that has taken its cue, to be sure, from Catholic teachings and scholarship of the second half of the 20th century—comes from within the Catholic tradition. Robert Kraynak has acknowledged this recent emphasis on human dignity as a function of human freedom and reason, but argued that this represents a Kantian turn in Christianity, with potentially pernicious consequences. In particular, Kantian-inspired dignity claims are used to generate further rights claims (as, in the previous section, we could speak of the right to life, the rights of children to come into existence as the fruit of loving marital union, and the right to health care); and these rights claims constitute, in Kraynak's words, "essentially ungrateful claims against authority" (Kraynak 2001: 172).

Moreover, in the absence of robust authority claims, particularly religious authority claims, and with no more than a Kantian grounding in the self-legislating autonomy of the will to buttress it, the concept of human dignity can quickly become denuded of any more content than the demand that human beings be able to live their lives however they see fit. Kraynak thus sees contemporary liberalism as pervaded by skepticism about the good as an input of the dignity framework as well as inclined toward a soulless consumerism as an output of that same framework. Mediating this input and output is a commitment to liberal democracy, demanded by skepticism, resulting in mass society. Yet the Christian churches, and especially the Catholic Church, have embraced not just dignity but also dignity as a route to a defense of democracy; Kraynak argues that this embrace is simply too dangerously close to liberal dignity, skepticism, and democracy for Christians to accept the new emphasis with equanimity.

Kraynak's argument can perhaps be intellectually situated in the following way. There can be no doubt that the Catholic Church and its main leaders, teachers, and intellectuals of the past half-century have expressed in numerous ways dismay at the direction taken by contemporary liberal democracy: a widespread relativism and a rejection of religion have led to what John Paul called "the eclipse of God and man" and the "culture of death." Yet, at the same time, most of the key thinkers of this same period have argued for a political understanding, a moral understanding, and an ecclesiological understanding that have taken some of the very same concepts of contemporary liberal skepticism as key terms: freedom, equality, self-governance, and, as encapsulating all of these, dignity.

An argument could be made, therefore, that such Catholic thought represents the "true" liberalism: committed to the freedom and equality of all persons, yet critical of the *understanding* of these notions that pervades a largely secular society. Yet Kraynak wishes to critique these commitments from without, not, as the Church increasingly has, from within, by returning to an earlier (allegedly Augustinian) understanding of the relationship between the Church and politics and of the nature and dignity of the human person: less emphasis on freedom, more emphasis on the immortality of the soul and the call to holiness.

To my mind, the various criticisms—for example, those in an issue of *The Catholic Social Science Review* devoted to Kraynak's book—that have been made of Kraynak's project are convincing and largely confirm the direction that this essay has taken. Contrary to Kraynak's hesitance at the embrace of freedom, W. Norris Clarke has shown that freedom is pervasive in the Christian tradition in understanding the *imago Dei*; but Clarke similarly shows that this freedom is a key aspect of our special relatedness to God, and to the good that God has set for us (Clarke 2004; see also Clarke 1992). There simply is no freedom without the good, certainly not in a rootless ability to legislate for oneself one's ends and goods.

Likewise, Gregory Beabout has shown that this orientation of our freedom to our good is, at the same time, as I have emphasized, an orientation toward a task, the task identified by *Gaudium et Spes* as the "sincere gift of self." Moreover, Beabout finds in the Second Vatican Council ample evidence that our freedom is, as it was for Saint Thomas, "self-determining, but not self-legislating" (Beabout 2004). Thus, we are a far cry from ungrateful claims against authority when the Council undertakes to articulate, in the first chapter of *Gaudium et Spes*, "The Dignity of the Human Person," and goes on to make various rights claims on behalf of all persons.

Further, in the work of John Finnis, we see a convincing account of the nature of such rights that sees them as complementary to claims of duty, yet available "for reporting and asserting the requirements of justice *from the point of view of the person(s) who benefit(s) from* that relationship" (Finnis 1980: 205; see also Lewis 2009). There can be no articulation of valid rights claims, within such framework, absent a full understanding of the demands of justice and the common good; an individualistic framework is simply not at work here. Finally, in the natural law undergirdings of the normative task of human dignity, we find little to match the skepticism of contemporary liberalism. The Catholic conception of human dignity should thus be seen as the paradigm understanding of human dignity, refracted in contemporary society through a cracked and smudged lens: what is seen in contemporary liberalism is real, and the vision has led to genuine goods, such as religious freedom. But the Catholic vision is clearer, and has in its sights a concept that redounds only to the good of the human person and the glory of God.

The Catholic conception is, of course, a thick, substantive, and even "comprehensive" conception. It is thus opposed by those versions of contemporary liberal theory that hold that only "neutral" reasons, or reasons that can be disengaged from conceptions of human good and well-being, or from substantive "worldviews" may be raised, considered, and addressed in public discourse. This restrictive view of liberalism thus stands as a final challenge to the Catholic concept of human dignity, if that account is to have a role in the public square.

These liberal demands for neutrality, as many people have noted, are themselves far from neutral (Neuhaus 1984). They rest upon a particular conception of the human person and of human freedom, according to which it is an affront to human dignity to be "coerced" in accordance with reasons that are not one's own. This conception is not shared by those who believe, for example, that freedom is valuable for persons only insofar as it is oriented toward the truth, and that the freedom of unreasoned self-assertion is, in fact, damaging to human character and welfare. But if the demand for neutrality depends upon a particular substantive view, then it cannot be carried out without falling afoul of its own requirements.

Nor, on their own merits, are such requirements reasonable. To refuse to listen to reasons—evidence put forth to defend a claim as *true*—is unreasonable, and to refuse to allow such reasons to be put forth in the public square is unjust—it unfairly restricts some citizens' participation in the public conversation on arbitrary grounds. So our public conversation on matters of public weight and importance should be unfettered, as regards the kinds of reasons that are permitted. There is thus reasonable space, within liberalism, corrected of its skeptical and restrictive errors, for the contributions which the Catholic account of dignity can make. This concept is indeed an apt, and essential, concept for use at the foundations of moral, political, and religious thought.

NOTE

1. The unity between husband and wife is further sacramentalized as a sign of Christ's relation to His Church.

4 A Protestant Perspective on Human Dignity

Paul Copan

Some societies affirm honor killings or capital punishment for individuals converting out of the dominant religion. Are these mere cultural offenses for Westerners—much like belching after meals or eating dog meat? Not the least. Honor killings or stifling religious liberty (freedom of conscience) is a culture-transcending wrong and never justifiable. A culture or an individual is in moral error when approving them. If such violations of human dignity and worth are culture-transcending, which metaphysical outlook supplies the appropriate context for individual dignity and worth, regardless of culture or history? Which worldview best accounts for our outrage at such moral horrors?

A right is "a claim that people are entitled to make on others or on society at large by virtue of their status. Human rights are claims that people are entitled to make simply by virtue of their status as human beings" (Wiredu 2001: 298). Still, the more fundamental consideration concerns which worldview or metaphysic justifies such an entitlement. A right is not self-standing; it requires a ground or justification.

With increasing globalization and pluralistic sensibilities, a single widely held foundation for human dignity appears more elusive than ever. Some consider human rights a Western creation that non-Westerners may discard as alien and even imperialistic. However, in this essay I shall argue that human dignity and worth—the basis of *universal* human rights—are rooted in a good, triune, personal God and Creator who made human beings "in the image of God" (Genesis 1:27), who has made salvation available to all without exception through the incarnate, atoning, and resurrected Christ. Theism, especially biblical theism, offers the most fruitful metaphysical grounds for affirming human dignity. Though the various world religions in the main affirm human dignity or the importance of human well-being in some form, biblical theism has the metaphysical capital to actually ground this affirmation. Human dignity and worth—and thus human rights and accompanying duties and virtues—can be metaphysically secured in the *imago Dei* and completed by Christ's achievement. By contrast, I will argue that other prominent worldviews—especially naturalism, theism's chief rival—lack the resources to ground a robust and defensible conception of human dignity and worth.

In this chapter, in addition to arguing for the superiority of biblical theism as the ground for human dignity and rights, I will also map some of the contours of human dignity from a Protestant perspective as well as apply this concept of dignity to Down syndrome and end-of-life issues. Moreover, I will show how this understanding of dignity can function successfully in a pluralistic society. In short, I will explain and defend a Protestant view of dignity and show its wide applicability and accessibility. Before doing so, however, I begin by criticizing several prominent secular bases for human dignity and worth.

THE INCOMPLETENESS OF NONTHEISTIC ETHICS

While an exhaustive analysis of all secular foundations for human dignity and worth cannot be undertaken here, several representative approaches will be examined. More detailed treatments can be found elsewhere, including in the work of John Hare, for example, who argues persuasively that ethical systems ignoring a robust theistic foundation, such as secular Kantian deontology, Millian utilitarianism, Aristotelian virtue ethics, and the like—not to mention official documents regarding human rights—will necessarily be incomplete (Hare 2007; see also Craig 2009; Copan 2008, 2009; Linville 2009).

More generally, secular ethicists often remind the rest of us that they can have confident moral knowledge without belief in God. Indeed, it is not surprising that atheists and other nontheists have moral knowledge: they, too, have been made in the divine image—and have available to them salvation through Christ's atoning death. Typically, these ethicists are long on epistemology (on the rationality of certain moral beliefs and how proper moral judgments are to be made) but short on ontology (namely, the more fundamental metaphysical basis for affirming human dignity and worth). For example, the social contractarian John Rawls famously advocated a position of the inviolable rights of individual human beings: "Each person possesses an inviolability founded on justice that even the welfare of society as a whole cannot override. For this reason justice denies that the loss of freedom for some is made right by a greater good shared by others" (Rawls 1999: 3–4). Thus, the rights of others cannot be bargained away by cultural, political, or even some religious hierarchy. Yet Rawls provides no convincing metaphysical basis for inviolable human dignity and worth.[1]

More pointedly, naturalism has great difficultly grounding inviolable human dignity and worth. Given that naturalism accepts nonmoral entities and forces (like quarks and gravity) as ontologically fundamental, why think that human beings have inviolable dignity and worth? That is, why think human beings have such dignity and worth in a worldview where, to paraphrase Bertrand Russell, our origin, growth, hopes and fears, loves, beliefs, and achievements are nothing but "the outcome of accidental collocations of atoms"? As Russell counseled, all that remains is building our lives on the "firm foundation of unyielding despair" (Russell 1963: 41).

Little wonder that, in *The Nature of Dignity* (2008), naturalistic philosopher Ronald Bontekoe frankly acknowledges that naturalism undercuts any presumed human dignity:

> In the light of [naturalistic] evolution, we are left with nothing but hypothetical imperatives. . . . Human beings cannot be deserving of a special measure of respect by virtue of their having been created "in God's image" when they have not been *created* at all (and there is no God). Thus the traditional conception of human dignity is also undermined in the wake of Darwin. (Bontekoe 2008: 15–16)

A universe of electrons and selfish genes has no metaphysical wherewithal to produce beings possessing intrinsic dignity and worth. The wrongness of rape or torture for fun presupposes the more basic fact of human dignity and worth. As I will argue in the next section, rather than "atoms and the void," so to speak, a more harmonious and less-surprising context for the emergence of intrinsic dignity and worth is that of a good personal Creator.

Moreover, secularized Platonic approaches to morality falter as well. This approach assumes a massive cosmic coincidence—one that readily dissolves given theism. For example, atheist philosopher W. Sinnott-Armstrong places objective morality on the same level as the laws of physics (e.g., $e = mc^2$) and mathematics (e.g., $2 + 2 = 4$). When T. rex ruled the land before free moral agents did, the moral fact still obtained that free agents raping other free agents is wrong. Of course, Sinnott-Armstrong avers, moral *beliefs* and *attitudes* evolve, but moral *facts* remain ever the same (Sinnott-Armstrong 2009: 92–93). Yet such a claim invokes a massive cosmic coincidence. The secular Platonic moral realist in this case holds that (a) certain necessary (even timelessly true) *moral facts* exist,[2] and (b) *self-reflective, morally responsible, and intrinsically valuable beings* eventually appear on the scene (through unguided, highly contingent evolutionary steps) who both can recognize these preexisting facts and are obligated to them. This remarkably ad hoc scenario begs for explanation. This Platonic-like moral realm, it appears, was *anticipating* our emergence—a remarkable cosmic accident! A far simpler, less ad hoc explanation is available: a good God—the very locus of objective moral values—created human beings with dignity and worth. As we will see, this affords a far more elegant, smooth, and natural explanation.

Postmodern approaches to human dignity remain unpersuasive as well. Postmodern perspectives—for all their celebration of plurality and concern for the "other"—can only ground human rights as arbitrary, contextual, parochial, and culturally constructed elements of competing meta-narratives. The postmodern era, with its "incredulity toward metanarratives" (Lyotard 1984: xxiv), gathered strength in the wake of Auschwitz's horrors. Even so, Lyotard wondered, "Where, after the metanarratives, can legitimacy reside?" (1984: xxv). If the postmodern critique is universally true (!), then any objective, universal notion of human rights and intrinsic human dignity cannot be sustained. After all, as Michel Foucault argued, by digging back into history ("archaeology"), one can discover that "Man" is a socially constructed invention of the 18th century that will eventually disappear and take on a new form (Foucault 1994: xxiii). As a spokesman for postmodernism, Lyotard denies any natural right (1993). He favors plurality and dialogue, even if no universal standards of humanness, morality, or rationality exist. Yet, as Max Stackhouse and Stephen Healey observe, the postmodern romanticizing of diversity is itself problematic since plurality without some overarching unity is mere tribalism; indeed, "the lack or denial of comprehensive narratives is historically the source of violence" (Stackhouse and Healey 1996: 500). Additionally, postmodernism's repudiation of universality—including the unique status of humans as creatures of dignity and worth—becomes its own universal creed. Such a sweeping denial is apparently applicable to all epochs, cultures, and civilizations. This "enlightened" stance is benighted by its own perspectival provincialism. Ultimately, the postmodern project lacks the depth to adequately respond to modernism's moral and intellectual failures.

Similarly, the leading impersonal versions of the "Ultimate Reality" (e.g., *Brahman*) in many Eastern religious and philosophical traditions lack the metaphysical wherewithal to ground personal human dignity. Their impersonal or abstract Ultimate Reality—whether monistic or pantheistic—leaves us wondering how these contexts could ontologically support personal virtues such as love, compassion, and kindness. For example, Buddhism's no-self doctrine (*anatman*) affirms that humans are mere bundles of ever-changing properties rather than enduring selves. Or take Hinduism's Advaita Vendanta's monism. The self (*atman*) is identical to the Ultimate Reality (*Brahman*), resulting in the obliteration of any (illusory) distinction between the rights of this or that human being. Looking eastward for metaphysical grounding for human dignity and worth does not look promising.

Finally, I should note that the Euthyphro dilemma fails to undermine a theistic basis for human dignity and worth. The dilemma holds that morality must either be arbitrary (based on whimsical divine commands) or autonomous (existing independently of God's commands). By way of response, God's character—not God's commands—is the sufficient (and necessary) locus of goodness, thus evading the horns of the dilemma. God's character is the moral terminus; any divine commands are rooted in something deeper and more stable: God's good character, of which human beings are a finite reflection ("the image of God").

In sum, not only does the Euthyphro dilemma falter, but secular perspectives like naturalism, postmodernism, secularized Platonism, Buddhism, and Advaita Vendanta monism fail to adequately ground human dignity and worth. By contrast, as I will argue, Protestant-informed theism provides a robust ground for dignity and worth.

PROTESTANT-INFORMED THEISM AND HUMAN DIGNITY

Thomas Aquinas wrote that certain "laws" exist. There is an *eternal* law, which God alone knows and by which God created and governs the universe. Since we are created in the *imago Dei* as moral, volitional, and reasoning beings, we are capable of recognizing a self-evident *natural* law, reflecting God's eternal law in the created order. All humans can know natural law apart from the *divine* law, which God has given to us through special revelation (the Bible and Jesus Christ). God has placed within us a disposition to have moral knowledge. Unless we suppress our conscience, we *naturally* know basic moral truths—these appear self-evident to anyone with a reasonably functioning conscience.

Unlike Roman Catholicism, Protestantism's appropriation of natural law has been somewhat ambivalent, wavering between qualified acceptance (within Lutheranism) and outright, vehement rejection (e.g., Karl Barth, Jacques Ellul, Stanley Hauerwas). This does not mean that natural law cannot be appropriated within Protestantism (Braaten 1992). So in this section, I want to illustrate how a Protestant-informed theism rests squarely on the previous discussion of the theistic metaphysical foundations for human dignity and worth, the recognition of which is simply an extension of natural law.

Two Concerns about Natural Law

Protestant opponents of natural law have voiced two chief concerns, neither of which is insurmountable. The first concern is that natural law appears to be free-floating and utterly independent of God. The second concern is that natural law implicitly denies the fallenness of humankind and the doctrine of original sin.

However, these concerns do not ultimately present theological barriers to natural law. First, natural law is not independent from God but rather is rooted in, and necessarily linked to, a good God who has created human beings in the divine image. Second, the image of God, though affected by human fallenness and sin, is not erased; indeed, human fallenness anticipates the redemptive work of Jesus Christ, the "second Adam" who commences a new creation. We can still apprehend and follow the natural law, even if we do so imperfectly. As such, humans are capable of recognizing the dignity and worth of others. While I will address later *how* a Protestant view of dignity and worth can function well in a pluralistic society, the crucial point for now is that, as beings created in God's image—who have His law "written on our hearts"—we not only have dignity and worth but can recognize this fact in ourselves and in others.

Two Protestant Viewpoints on Human Dignity

Given that the dignity and worth of human beings, according to Christianity, rests directly on the *imago Dei*, I shall give a brief overview of two Protestant thinkers—one Lutheran, the other Reformed—who take differing, yet complementary and mutually illuminating, viewpoints on what the image of God means.

Dietrich Bonhoeffer: An Eschatological Approach

Lutheran theologian Dietrich Bonhoeffer offers fruitful directions for ethical discussion and grounding human dignity. Bonhoeffer does use "essentialist language" in his *Ethics*, noting the human body has an "'innate' right" or "inherent right of inviolability" and thus rejecting any claim to "an absolute right of free disposal over the bodily members that have been given to me by God" (Bonhoeffer 1955: 155, 181). Our right to self-preservation notwithstanding, any conscious injury to another's body "constitutes the destruction of the first natural right of man" (Bonhoeffer 1955: 159). Our bodily life is "passively received" as a result of God's will (Bonhoeffer 1955: 155).

Furthermore, the "natural" is divinely preserved until—and properly understood in light of—the eschatological Christ event, the focal point of creation and history. For Bonhoeffer, understanding the image of God does not come from the text of Genesis 1. Rather, our grasp of human identity (and thus dignity) must be rooted in *Christology* and *eschatology*. In terms of Christology, Bonhoeffer begins with Jesus Christ as the focal point for understanding what true humanity is. His Incarnation affirms human worth and dignity.

This conviction, I hasten to add, is also affirmed by Catholic and Orthodox alike. For instance, the Catholic encyclical *Gaudium et Spes* (1965) affirms that the mystery of our true humanity is revealed in the incarnation of Christ, the new Adam; we are fully persons in the person of Christ, who restores the fallen image of God.

To talk about "being good" and "doing good," suggests Bonhoeffer, incorrectly assumes that the self and the world are the self-sufficient ultimate reality, since, on this view "there is no belief in God as the ultimate reality" (Bonhoeffer 1955: 188, 189). Grasping human identity can only be done by looking outside of ourselves to Christ. Human life is a gracious gift of God, and our humanity is fulfilled in Christ.

In terms of eschatology, our humanity will not be fully realized until it is redeemed in the eschaton, reaffirming our origin (cf. Reed 2007: 67–90). Christ is the completer of humanity, the archetypal human whose death and resurrection usher in a new creation (2 Corinthians 5:17). Thus, as remarkably fruitful as the "image of God" is for understanding human rights and dignity, this idea is not some free-floating ontological concept. Rather, the image of God in humans is properly understood and is redeemed by the ultimate human being, Jesus of Nazareth, *the* image of God.

Max L. Stackhouse: An Essentialist Approach

Max Stackhouse, though acknowledging the eschatological, more strongly affirms the essentialist perspective, but he ends up in much the same place as Bonhoeffer. Stackhouse affirms the intrinsic connection between theology and anthropology or God and human dignity (and, by extension, human rights). Indeed, "a moral plumb line hangs prophetically over every political community" (Stackhouse 1986: 11). Ironically, those who deny or doubt universal human rights "press toward a universal relativism" because all such perspectives are rooted in sociohistorical contexts cannot themselves defend their own view except as a mere cultural preference (2004: 23). The biblical tradition affirms a "profound individualism"—a certain "moral inviolability of each person" (2004: 27). This contrasts with the communitarian or socio-logical urge (e.g., "Asian values") that can threaten this inviolability.

According to Stackhouse, the image of God can be understood thus: by virtue of divine creation, "we have some residual capacity to reason, to will, and to love that is given to us as an endowment that we did not achieve by our own efforts" (Stackhouse 2004: 27). Despite human fallenness, the image of God implies that each person has dignity, "whatever his or her condition, status, or behavior." This image cannot be violated "without vio-lating the source of that which is right and true, just and loving" (Stackhouse 1986: 14). Furthermore, each person is called to use these God-given capaci-ties to engage in particular relational networks as part of our role to be just, loving agents on earth (Stackhouse 2004: 27).

Christ's incarnation is a model for the dual affirmation of the contextual and historically conditioned as well as the transcultural and universal. Jesus of Nazareth is the eternal God who has become flesh to live among us (John 1:14), dies for all humanity, and is now the resurrected and ascended Lord of all.

The doctrine of the Trinity—with Father, Son, and Holy Spirit as intrin-sically and mutually giving and loving persons—resists an imposed Con-stantinian civil religion of the "Christian" majority oppressing professing non-Christians within its borders. Rather, the Trinity grounds the human enterprise of community and social engagement for the public good. Such an engagement is an outworking of the fact that God is other-concerned and relational rather than self-enclosed. The doctrine of the Trinity "makes the world safe for pluralism" (Stackhouse 1996: 69).

Having explored, first, the metaphysical basis for human dignity and worth (a good, personal God) and, second, what the image of God involves, we can bring these together by looking at the implications of these two realities.

THE MEANING OF HUMAN DIGNITY

Human dignity is an entailment of the divine image. What is the meaning of human dignity, and what responsibilities flow from human dignity? Rather than go into detail on two or three categories, I shall roughly sketch out a range of topics that examine the meaning of human dignity—its key concepts and the duties or obligations that fall to individual humans and their communities.

Human Life as a God-Given Gift

Human life is no mere metaphysical surd or accident of nature. The Psalmist affirms that even in his mother's womb he was "fearfully and wonderfully made" (Psalm 139:14, NASB). Israel's prophetic literature also uses related language to express the value of the unborn (Isaiah 49:5; Jeremiah 1:5). Later in the New Testament, even John the Baptist was filled with God's Spirit and was said to have "leaped for joy" while in his mother's womb (Luke 1:15, 44; cf. 1:42; 2:21). Indeed, all children should be welcomed as a "gift of the LORD" (Psalm 127:3). The Psalmist elsewhere calls on God not to forsake them when they are old and gray (71:9, 18). In Proverbs the elderly or gray-haired are affirmed as worthy of honor (Proverbs 16:31; 20:29): "Listen to your father who begot you, and do not despise your mother when she is old" (23:22). Thus, from the womb to the end of life, human beings possess dignity and worth as God's image-bearers. Each human being is a creature specially made by God. Neither age nor function diminishes the dignified status a human being has. With God, there are no "potential human beings" (as opposed to "actual human beings"—no more so than a woman being "somewhat pregnant").

Qualifying "The Sanctity of Human Life"

Yahweh declares in Deuteronomy 32:39, "There is no god besides Me." It is this God who has authority to "put to death and give life." This relates to creation, but it also applies to redemption as well. Saint Paul asks: "do you not know that your body is a temple of the Holy Spirit who is in you, whom you have from God, and that you are not your own?" As all humans are not only God's image bearers but potential recipients of redemption, Paul reminds his audience: "you have been bought with a price: therefore glorify God in your body" (1 Corinthians 6:19–20). Whether we speak of biological or spiritual life, both are gifts from God, and the proper human response ought to be gratitude rather than clutching life as our possession or that which is owed us by God.

Humans are not divinely owed 70 or 80 years; moreover, the "sanctity of life" ought not to be taken to idolatrous ends, falsely assuming that life ought to be sustained at all costs or that we are the sovereign masters of our fate. As creatures uniquely made and endowed by God, we do not have absolute ownership of our own (or our child's) life, but rather stewardship of it.

Individual Rights Not Opposed to Community Well-Being

Because human life is a gift from God, certain rights will flow from this status. Individual human rights cannot be set in opposition to community or societal well-being. Just as the triune God creates human beings ("male and female") to co-rule with him, the human is not some atomized, detached individual, but he belongs to a broader community and culture that gives shape to and builds character into our lives. For example, if a deformed baby is born to parents, their first concern should not be their "rights" or "fulfillment," but rather being the right kind of parents for this gift entrusted to them by God. Stanley Hauerwas correctly remarks that the language of "rights" will "teach us little about the kinds of skills we need to train ourselves to take on that kind of task" (Hauerwas 1977: 140). Indeed, when freedom and its enhancement becomes an end in itself, "we lose any account of human life that gives content and direction to freedom" (Hauerwas 1977: 14).

Indeed, a common, but morally misguided, "reproductive freedom" or "control over my body" is often used to treat others unjustly—namely, the unborn, who have no say in the matter.

And when children are brought into the world as the parents' source of "self-fulfillment," this mind-set will leave us with little skill or knowledge about how to become nurturing and loving parents (Hauerwas 1977: 140).

Death Is Not the Ultimate Evil

In Christian theology, death is not a dignified event. Death is an indignity to our humanity, and one ought not to seek it. Yet death has lost its sting through the bodily resurrection of Jesus, whose new embodied life is a precursor to the believer's. Thus, one need not live in fear of death, as though it is the ultimate evil. The follower of Christ may have confidence in the face of death, recognizing that death need not be postponed at all costs.

God has created human beings for this purpose: to live in harmonious relationship with God and fellow creatures. What is right, then, is that which promotes the kingdom or rule of God. The exceptionless rule for each human being—indeed, for the benefit of each human being—is this: "One ought to promote the Kingdom of God" (Layman 1991: 52, 53). Life lived under God's rule means a life of service to God and others, and we can live life with courage and grace in the midst of its challenges and ambiguities. Yet the time comes when we no longer fight death, but accept its inevitability (Hauerwas 1977: 36). Moreover, though we consider self-sacrifice praiseworthy and virtuous, laying down one's brief life for another makes little rational sense without God (and thus immortality).

The Place of Suffering

The model of Christ is one of suffering for the sake of others and for doing the will of God. On this view, some occasions call us to bear suffering, quite contrary to naturalistic ethical perspectives which simply beckon us to maximize pleasure and minimize pain. Not all suffering is to be avoided, and preventing all suffering is not obviously a virtue.

Some medical practitioners forget that they must always care, even when they cannot cure. And when some patients are euthanized to "spare" them further suffering, the tragic irony is that their lives should be extinguished to achieve this. Furthermore, the fear of the living is that the chronically ill will live on, and it is the living who often want to be spared of hardship, not the ill.

Suffering itself can be very subjective, and more often than not, suffering is internal and psychological rather than external and physical. A child with Down syndrome may be quite happy-go-lucky and hardly suffering. While we should not seek suffering, life can have great value and depth in the face of great suffering. The valuable life is not necessarily free of suffering (Hauerwas 1977: 167–68).

All Humans as God's Image-Bearers

The image of God entails that humans have intrinsic dignity and that (non-human) animals do not. The co-founder of People for the Ethical Treatment of Animals Ingrid Newkirk is incorrect in her notorious assertion: "A rat is a pig is a dog is a boy." For one thing, why think *anything* has value given a backdrop of valueless, unguided, material processes that were produced by the Big Bang? And why think that the value of each is equal? Do these creatures have absolutely *no* difference in value? Moreover, we should reject Peter Singer's distinction between *persons* and *humans*. While nonhuman persons exist (e.g., the divine persons of the Trinity, angelic beings), there is no human who is not also a person. Yet Singer thinks that infants with Down syndrome, hemophilia, spina bifida, or other "defects"—as well as elderly patients with Alzheimer's disease (except Singer's own mother, Cora)—can justifiably be killed. In fact, Singer pronounces pigs or chimpanzees to be more worthy of life than these humans, who are not persons. Contrary to Princeton University's own anti-abuse/-harassment policy against threatening or injuring persons with any "handicap," Singer curiously appears to single out those who can't fight back—namely, the young and the old.

Human dignity—rooted in the *imago Dei* and reinforced by the Incarnation and resurrection of Jesus of Nazareth for the redemption of humanity—entails that humans are ends rather than means; they have, as Kant put it, "worth," not "price." Thus, (Christian) theism rejects the standard utilitarian ethic, in which human individuality and individual humans are subordinated to the concerns of the group—the greatest good for the greatest number.

The unique value of human beings is evident in their capacity for freedom or personal responsibility as well as reasoning—not to mention the capacity to know and relate to God through God's gracious provision in Christ. But this does not mean that the unborn or the elderly with dementia are less valuable because they are not fully developed or they are physically blocked from realizing this capacity. Aristotle does not have more value than Joe the Plumber because the former towers over the latter in powers of ratiocination.

HUMAN DIGNITY AND BIOETHICS

What does a biblical view of human dignity mean for bioethics? In this section I shall briefly review two areas in bioethics: mental disability such as Down syndrome (here I follow Yong 2007: 155–91) and end-of-life issues. I examine the implications and applications of the concept of human dignity with regard to these two areas. (Since the points I make overlap significantly, I shall not treat these two topics separately.)

Possessing the image of God—along with life itself—is a gift, even if the expression of key human capacities (cognitive, relational, emotional) is blocked due to a physical abnormality. Unlike an Eastern karmic perspective (such a condition is payback for deeds in the previous life), the biblical perspective views such physical or mental disabilities as opportunities for the "works of God" to be "displayed," as in the case of the man born blind (John 9:3). The glory and strength of God may often be evidenced through weakness and suffering (2 Corinthians 12:10). This does not mean that one should receive a birth defect passively or fatalistically (perhaps as a curse). Rather, Jesus urges human engagement and action: "We must work the works of Him who sent Me while it is still day" (John 9:4). Indeed, Jesus himself routinely engaged in physically healing persons who were "born that way." Without giving even a compassionate embrace or lending a merciful hand, our calling another's disability a "blessing" may prove our words a sham.

Our earliest ancestors' fall away from God produced a spiritual death and alienation—a breach in humanity's fellowship with God and with other human beings. This has led to injustice, violence, and alienation between human beings—most often the mentally disabled and unwanted being ostracized from the nondisabled and those "contributing to society." According to Christian theism, Christ's redemptive work on the cross includes all sinners—the physically whole as well as the ill and broken in body. In fact, Christ's *body* that is broken on the cross for humanity (Mark 14:22)—repeatedly presented as broken in the Lord's Supper/Eucharist. Christ became broken so that we might be healed. He was the disfigured, despised "suffering servant" (Isaiah 53:2–3).

Humans are surrounded by limitation, finitude, and contingency. We should not load these themes with negative connotations but rather receive them as the realities of a broken world in which God is "not far from each one of us" (Acts 17:27)—whether we are disabled or not. We can affirm that such disabilities do not catch God by surprise, nor do we need to affirm that they are directly intended by God. God's final goal is the redemption and healing of all in the new heaven and earth, in which every willing human may participate.

The Son of God entered into the world of humanity, experiencing oppression, discrimination, limitation, and even disability. In his post-resurrection state, his body still bore the marks of his injuries (Luke 24:39–40; John 20:24–28). In order to empathize with us, "he had to become like his brothers and sisters in every respect" (Hebrews 2:17, NRSV). According to Luther, God is both hidden and revealed in his suffering. God's glory—indeed, God's greatest achievement—is revealed in stooping to the lowest of all possible humiliations and stigmatizations: dying naked on a cross for all to see. God is exalted in his humiliation—a key theme in John's gospel (Bauckham 1998). The cross provides a framework for embracing the disabled and stigmatized. Jesus became handicapped for our sakes.

Just as Jesus valued the societally devalued or valueless, so we too are to accept and value "the other." If all of us are honest, we can readily see how the disabled serve as reminder of the weaknesses and deficiencies of the nondisabled. They serve as reminder that the nondisabled are not as strong, as "together," as secure, as self-sufficient as they let on.

All humans—disabled and nondisabled—share the gift of life from God (Genesis 2:7). While all may not be able to fully engage in *intersubjective* relations (self-consciously engaged persons), the mentally disabled and nondisabled can still engage in *interpersonal* relationships (Yong 2009: 184–85). In the words of Jesus, "Just as you did it to one of the least of these who are members of my family, you did it to me" (Matthew 25:40, NRSV). Such actions of compassion apply not only within the context of the Christian community but to humans in general: "He who oppresses the poor taunts his Maker" (Proverbs 14:31, NASB).

Earlier we noted Singer's justifying the death of certain "humans" because they are not "persons." For Singer, humans must meet certain criteria for personhood—self-consciousness, rationality, quality of life. Certain *functions*—not some underlying human *nature*—make one's life worth living. The comatose and those with Down syndrome or Alzheimer's disease do not qualify. On the contrary, the image of God is ultimately located in who we *are* as persons, not in what we *do*. Humans share a divine dignity and honor, even if they do not "contribute" to society in some utilitarian manner. Indeed, removing the handicapped and infirmed from our midst would actually damage the nondisabled and diminish their capacity to care and show genuine, full-bodied humanness.

Caring for others with chronic illness or even end-of-life care tends to reveal the strength of our own character. When we sacrificially care for those who cannot care for themselves, particularly over a long period of time, we celebrate a love for life—for theirs and ours. Our care for the elderly and the chronically ill "is a crucial act for witnessing our celebration of their lives and ours" (Hauerwas 1977: 114).

An important Christian implication is learning to suffer with those who suffer—to "weep with those who weep" (Romans 12:15). Rather than shun the suffering, the nonsuffering must become part of the world of those who suffer. After all, those who suffer do not have lesser value than those who do. Commonly, the suffering and the nonsuffering are often isolated, and suffering tends to make a person a stranger to the nonsufferer. After a loved one dies and the funeral is past, onlookers tend to withdraw their warmth and support. What is more, those in the medical field tend to "guard us from those who suffer" (Hauerwas 1977: 26).

However, Christians follow a suffering Savior and are called to identify with the less-fortunate, disempowered, and suffering. In his book on the valuable contribution religion makes to society, agnostic political scientist Guenther Lewy contrasts the mind-set of the naturalist and theist in this regard:

> adherents of [a naturalistic] ethic are not likely to produce a Dorothy Day or a Mother Teresa. Many of these people love humanity but not individual human beings with all their failings and shortcomings. They will be found participating in demonstrations for causes such as nuclear disarmament but not sitting at the bedside of a dying person. An ethic of moral autonomy and individual rights, so important to secular liberals, is incapable of sustaining and nourishing values such as altruism and self-sacrifice. (Lewy 1996: 137)

Jack Kevorkian ("Doctor Death") espoused the view of absolute autonomy. That is, the individual—not God—is sovereign over life or death. A person is "free" to take her life if she believes that life is no longer worth living. Yet to take such a view is ultimately self-undermining. Do we, as J. S. Mill asked, have a right to enslave ourselves? Is not suicide itself the ultimate *anti*-self-expression—the utter extinction of being master over one's body?

HUMAN DIGNITY IN A PLURALISTIC SOCIETY

How does our account of human dignity function—indeed thrive—within a pluralistic society? Does it offer a public vision with a view toward the common good? Does it share common ground with other perspectives? How does one defend or advance our position in a pluralistic society? While I began to lay the groundwork for my position in the first section, the following points follow from the larger worldview picture.

1. To affirm the universal is not to deny the particular.

An initial difficulty about human dignity concerns a tension between the universalists and the communitarians. The former group holds that a single unified concept of human dignity should apply to all cultures, whereas the latter group holds that each culture should have its own unique concept of human dignity. The problem seems to be that universalism appears too abstract to be concretely applied, while communitarianism is too parochial to be cross-culturally relevant. In the face of this tension, the challenge is to coherently advance a universal yet locally applicable concept of human dignity. Fortunately, the disjunction between universalists and communitarians is a false one.

In my view, the legitimate moral horror about Auschwitz and Soviet gulags undermines a pure communitarian position. Likewise, one can legitimately critique a government that hides behind the breastworks of "Asian values," for example, to preserve economic and political structures while individuals' religious rights are trampled and any dissent is crushed. It is true, however, that some Asian nations feel no compunction to follow the West's emphasis on democracy and rights; they prefer economic growth and prosperity, even if this means a more authoritarian government (e.g., Malaysia, Singapore). The community and family take precedent over the individual. Even so, global and widespread religious representation of the United Nations Universal Declaration on Human Rights (UNUDHR) signatories suggests that utterly irreconcilable differences on rights are more perceived than real (Ignatieff 2001: 62–63). As Stackhouse observes, despite metaphysical differences in human worldviews, all languages and nations have terms for and laws against "sacrilege, disrespect, lying, cheating, stealing, murder, rape, [and] exploitation" (Stackhouse 2002). Mary Ann Glendon concurs: "If there are some things so terrible in practice that virtually no one will publicly approve them, and some things so good in practice that virtually no one will oppose them, a common project can move forward without agreement on the reasons for those positions" (Glendon 2001: 78). Or, as Ignatieff writes, "not much more than the basic intuition that what is pain and humiliation for you is bound to be pain and humiliation for me. But this is already something . . . the shared vocabulary from which our arguments can begin" and the pursuit of human rights can take root (Ignatieff 2001: 95).

To illustrate, 1991 Nobel Peace Prize winner Daw San Suu Kyi (a Buddhist) led Burma's (now Myanmar's) National League for Democracy party, and her party received 82 percent of the popular vote in 1990. Despite her Asian context, her impulse to affirm human dignity was universally compelling and widely celebrated in her own country and beyond (Paris 2001). Along these lines, consider Gandhi's defense of the dignity of India's "untouchables"—or other reformers such as Martin Luther King Jr., Nelson Mandela, and Desmond Tutu. They remind us of the universal longing to affirm human dignity and worth, even when government powers or communitarian societies would suppress it.

More than any other time, national constitutions affirm human rights. Whether in principle or on paper, the widespread assumption within foundational national documents is that humans have dignity and worth which cannot be taken away or ignored by governments. This consensus, with or without actual compliance, provides a starting point for dialogue and international pressure on violators to conform to these stated assertions.

Intellectuals challenging human rights as a Western invention in favor of, say, variations of communitarianism (or "Asian values") or postmodern "diversity without universality" undermine important gains for humanity. What's more, their condemnations actually adopt a universalistic position themselves (is ethnocentrism universally wrong, or is *that* merely cultural?). Celebrating plural perspectives as equally legitimate—shades of moral relativism here—runs the risk of our inability to condemn any perspective, including those favoring despotisms and brutal treatment of individuals (Stackhouse 2004).

The general, universal, and (for some) abstract notion of human dignity can be applied in particular cultures and situations. Human dignity entails the universal wrongness of certain practices in particular settings, which requires particular reforms—sexual slavery and other forms of human trafficking, forced arranged marriages, honor killings of raped women, enslavement, arbitrary imprisonment, forced military conscription of young boys, forced child labor, coerced religious conversion, exploitation and oppression of ethnic groups, wife beating. Positively, universal human dignity involves particular applications: the right to convert to another religious perspective or worldview, to form a family, to find a life's vocation or calling, to undergo a fair trial, and so on (Stackhouse 2005: 36–38).

2. Christian ethics reflects a shared humanity and thus a fundamentally human ethic.

Christians affirm that Christian ethics is fundamentally human ethics (Graham 2001). The Christian view holds that all people—regardless of their metaphysical views—have dignity and rights. This includes Buddhists, Taoists, Muslims, and others. So, even though Buddhists and others do not have their *own* legitimate metaphysical foundation for human dignity and rights, Buddhists and others should still be treated with dignity and respect, since, on a Christian view, all people are made in the *imago Dei*, are objects of divine love, and are potential recipients of redemption through Christ's atoning sacrifice. Since humans are made in the divine image, their political and religious groupings can reinforce the implementation of structures conducive to human flourishing.

Clearly, this does not mean that all must become Christian to embrace universal human rights. Because all humans are made in God's image, the affirmation of human dignity arising from this fact is not easily suppressed—especially when one is mistreated or oppressed. So strong is this impulse that many will remain committed to humans' dignity and worth—even they are agnostic about what grounds it. Moreover, taking individual human rights seriously, while bringing honor killings or wife beatings (in certain Muslim contexts) to an end, does not require that a husband and wife must cease being Muslims. Seeking protection of one's rights hardly means becoming a Westerner. Human rights need community support to sustain them, such as educating girls and women (Ignatieff 2001: 68–69, 75).

The Christian affirms that human dignity is not merely individual; it is also communal. In the Incarnation of Christ, the "Word became flesh, and dwelt among us" (John 1:14, NASB)—the individual (representing the triune God) entering into the human community as a human being for the sake of human beings. *Human* dignity is directly tied to the common good. A government that is doing its God-given duty (whether acknowledging God or not) will keep peace and order, which includes promoting human dignity abroad and preventing abuses of dignity at home.

3. Even secularists who deny the connection between God and human rights should acknowledge the historical connection between theistic belief and the preservation of human dignity.

In the 18th century, two notable rights documents were drafted connecting human dignity and the existence of a good, benevolent (albeit deistically tinged) God. The United States Declaration of Independence (1776) asserts: "We hold these truths to be self-evident, that all men are created equal, that they are endowed by their Creator with certain unalienable rights." Though the French Declaration of the Rights of Man and Citizen (1789) does not clearly specify the divine metaphysical foundation for human rights, it affirms these rights "in the presence and under the auspices" of "the Supreme Being." Both of these documents affirm a *theological*—not secular—basis for human dignity. Indeed, the rights language found therein is indebted to philosopher John Locke himself, who affirmed that humans ought not harm another "in his life, health, liberty, or possessions" precisely because humans are the property, workmanship, and servants of *"one omnipotent, and infinitely wise maker"* (Locke 1690: 2.6, italics mine).

Even if secularists deny the existence of a personal God as the basis of human dignity and worth (and thus universal human rights), they should honestly acknowledge the well-documented historical connection between contemporary rights language and theological—not secular—rootedness in the Judeo-Christian tradition (Glendon 2001; Stackhouse and Healey 1996: 486).

Even Jürgen Habermas, one of Europe's leading philosophers and a secularist himself, is clear on Western civilization's indebtedness to its Jewish-Christian heritage for monumental moral gains over the centuries:

> Christianity has functioned for the normative self-understanding of modernity as more than just a precursor or a catalyst. Egalitarian universalism, from which sprang the ideas of freedom and a social solidarity, of an autonomous conduct of life and emancipation, the individual morality of conscience, human rights, and democracy, is the direct heir to the Judaic ethic of justice and the Christian ethic of love. This legacy, substantially unchanged, has been the object of continual critical appropriation and reinterpretation. To this day, there is no alternative to it. And in light of current challenges of a postnational constellation, we continue to draw on the substance of this heritage. Everything else is just idle postmodern talk. (Habermas 2006: 150–51)

Even non-Westerners have come to recognize the remarkable impact of the Christian faith in the West. *Time* magazine's well-respected correspondent

David Aikman reported the summary of one Chinese scholar's lecture to a group of 18 U.S. tourists:

> One of the things we were asked to look into was what accounted for the success, in fact, the pre-eminence of the West all over the world. . . . We studied everything we could from the historical, political, economic, and cultural perspective. At first, we thought it was because you had more powerful guns than we had. Then we thought it was because you had the best political system. Next we focused on your economic system. But in the past twenty years, we have realized that the heart of your culture is your religion: Christianity. That is why the West has been so powerful. The Christian moral foundation of social and cultural life was what made possible the emergence of capitalism and then the successful transition to democratic politics. We don't have any doubt about this. (Aikman 2003: 5)

This lecturer was not some ill-informed crackpot. To the contrary, he represented one of China's premier academic research organizations—the Chinese Academy of Social Sciences. So one need not just take the word of Westerners here.

4. A further affirmation of the Christian faith's universal truth and appeal is evident in the rapid growth of the Christian church in non-Western settings.

With all of our discussion on Eastern and Western perspectives on human rights, we should remember the biblical faith, which has profoundly influenced the West, originated in the East. Thus, Western biblical exegetes have been wise to attend to the Bible's very Eastern features (e.g., Bailey 2008).

Moreover, Philip Jenkins has ably documented the Christian faith's explosion across the globe, far outstripping all of its religious rivals (Jenkins 2007; 2008). Whether in China, Muslim nations, or sub-Saharan Africa, multitudes of believers are being added daily. According to Jenkins, this is the biggest world news story of our era.

We have seen that the triune God presents the strongest metaphysical basis for affirming dignity and rights for all humanity. In addition, the global appeal of the gospel ("good news") of Christ's salvation for all human beings—not simply Westerners—reinforces this theme.

CONCLUSION

Are freedom to worship, freedom of conscience, freedom of speech, the right to a fair trial, freedom from genocide, and the like merely a cultural (tribalistic) Western expression? Or are these rights universal and fundamental to our very humanity? Are our belief in human dignity and the presumed rights that flow from it ultimately rooted in biological and/or cultural evolution? Is human dignity founded on something more substantial—indeed, metaphysical? What if human dignity really is the entailment of a divinely given gift to all humankind? If it is a gift from God rather than from culture, then culture or government must not attempt to take it away.

There is the question of whether certain political systems or cultures are more conducive to sustaining and nourishing these basic rights. The UNUDHR affirms human rights, and a nation may subscribe to this declaration on paper. Yet if the nation is ruled by a thug or tyrant, such rights-talk means very little. Indeed, despotisms and dictatorships lack the moral framework and cultural wherewithal to nurture this God-given human dignity (Witte 2009: 23). We should remember that the UNUDHR was drafted precisely because of the "barbarous acts" of World War II.

Thus, those opposing imperialism and colonialism affirm a certain universality and opposition to tribalism or parochialism. Unless human dignity is genuinely acknowledged to be universal rather than the mere expression of Western values, the world's peoples will ever be vulnerable to the dictatorship of this cultural relativism. Even non-Westerners increasingly acknowledge the appeal of freedom to convert out of the religion of one's birth, to participate in the choice of one's marriage partner, to be protected from honor killings—as opposed to allowing families and cultural institutions to overwhelm and suppress individual choices and appropriate expressions of human dignity (Ignatieff 2001: 66–70). The dangers in bioethics are equally frightening.

The Christian faith—with its emphases on the *imago Dei* and the Incarnation, atonement, and resurrection—is well positioned not only to affirm these universally valid values and to cooperate with those of various non-Christian faiths that affirm them, but to metaphysically ground them as well.

NOTES

Thanks to Stephen Dilley for his helpful comments on an earlier draft of this chapter.

1. Similarly, the French Catholic philosopher Jacques Maritain helped draft the 1948 United Nations Universal Declaration on Human Rights (UNUDHR). This document acknowledges "the inherent dignity" and "the equal and inalienable rights of all members of the human family." Further, it affirms: "All human beings are born free and equal in dignity and rights. They are endowed with reason and conscience and should act towards one another in a spirit of brotherhood." However, the sweeping assertion of the UNUDHR takes much for granted without any clear metaphysical substantiation. What metaphysical foundations could ground such intuitions of human dignity? And what does the UNUDHR mean by "endowed"? By *what* or *whom*? We are not told. Regarding the philosophical discussion behind the drafting of the Declaration, Maritain wrote: "We agree on these rights, providing we are not asked why. With the 'why,' the dispute begins" (Maritain 1951: 77). It is telling that several of the drafters of UNUDHR privately expressed their

particular Judeo-Christian perspective (e.g., Maritain 1964: 439). We should not ignore this fact, which offers a much richer metaphysical context for affirming human dignity than nontheistic worldviews: the existence of a good Creator provides a basis for human beings' moral worth and responsibility— a concept embedded in the phrase "the image of God" (Genesis 1:26–27).

2. The (alleged) necessity of moral truths does not diminish their need for grounding in the character of a personal God. God, who necessarily exists in all possible worlds, is the locus of all necessary moral truths that stand in asymmetrical relation to God's necessity. Consider the necessary truth: "Addition is possible." This is rooted in a more fundamental necessary truth—the existence of numbers, which have certain essential properties. The latter grounds the former, not vice versa; hence, the asymmetrical relationship. Or consider the statement, "Consciousness necessarily exists." This is true only because God—a supremely self-aware being—exists in all possible worlds. Similarly, God's existence also means that objective moral facts are necessary—that is, they exist in all possible worlds precisely because a supremely good God exists in all possible worlds. That is, God's existence is explanatorily prior to these moral facts (Craig 2009: 168–73).

5   Postmodern Perspectives on Human Dignity

Mark Dietrich Tschaepe

> On the whole, then, we must conclude that no philosophy of ethics is possible in the old-fashioned absolute sense of the term.
>
> —William James (1949: 208)

Human dignity is a foundational concept in ethics, especially with regard to the applied fields of medical ethics and bioethics. According to Roberto Andorno, "The emphasis on human dignity is impressive enough to lead scholars to characterize this notion as 'the shaping principle' of international bioethics" (Andorno 2009: 227). It is also one of the more vague concepts, typically occurring in official documents without definition. For instance, the term *dignity* occurs five times in the Universal Declaration of Human Rights (United Nations 1948), but no definitive meaning is attached to the term within the document. Similar to other ethical concepts, there is no clear and distinct definition for human dignity, but we are somehow expected to "know it when we see it."

As a foundational concept, human dignity might be assumed to be an absolute concept—unchanging, necessary, essential. Such an assumption often relies on an overarching, universal source, such as God or Reason.[1] In other words, the narrative of human dignity as a necessary quality of the human person is assumed to fit within a grand narrative upon which the concept depends. The grand narrative of theism or rationalism is the foundation for human dignity. In turn, human dignity is often considered the grounding principle for human rights. Once recognition of human dignity has been stipulated, we are thereby guided, in the broadest manner, with regard to how all humans are to be treated. Various classes of action are forbidden because of human dignity; for instance, "organ sales from living 'donors'; seeking patent rights over human genes; making animal-human chimeras; obliging someone to live in abject poverty; pornography; torture; sex selection by preimplantation genetic diagnosis; death in irremediable physical or psychological suffering; abandonment to senility in a nursing home" (Ashcroft 2005: 679). At the bare minimum, all humans are to be dignified *as* humans; we are to recognize "an unconditional worth that everyone has simply by virtue of being human" (Andorno 2009: 229). If we subscribe to what Jean-François Lyotard claims is the *modern perspective*, we are to deduce what such dignity entails based upon a grand narrative—that is, an ontology to which we abide (Lyotard 1979: xxiii). If, however, we do not take a modern perspective—if we do not adopt a grand narrative—then our concept of human dignity radically changes. In this chapter, I describe the concept of human dignity once we recognize our *postmodern condition*, which leads to a multiplicity of postmodern perspectives.[2] These perspectives undermine the modern perspective of the grand narrative, disrupting the notion of an ontological foundation for human dignity.

Is human dignity even a possible quality of humans from a postmodern perspective? Perhaps the common assumption would be that there is no human dignity once we acknowledge that we have entered what Jean-François Lyotard deemed the *postmodern condition*. I argue that this is not necessarily the case. We need not buy Ruth Macklin's argument that the term *dignity* is meaningless or redundant (Macklin 2003). Human dignity is a concept that can function in multiple ways within a postmodern context. Before exploring three distinct approaches to human dignity from postmodern perspectives, I first want to dispense with what I find to be a common misunderstanding that the postmodernist is necessarily a *nihilist* akin to Bazarov, who, in Turgenev's novel, *Fathers and Sons*, claims to accept only the findings of empirical science, rejecting all traditions, customs, and authority. Postmodernism is not the standpoint from which all standpoints are necessarily done away with, vanquished, or deconstructed in favor of either science or playfulness. Scientific activity is an important feature of the postmodern condition, but all ethics is not necessarily thrown out the window or not taken seriously in favor of mere scientific research and production. Science is *not* accepted from the postmodern perspective as a grand narrative that replaces all other grand narratives. Although the conversation concerning ethics has changed with the postmodern condition, it remains a serious conversation involving those deemed postmodernists.

Here I offer three postmodern perspectives with the hope that the reader will note the relations between these positions. Each perspective addresses at least one of William James's three questions concerning morality: the psychological question (what are the origins of an idea?), the metaphysical question (what is the meaning of an idea?), and the casuistic question (what is the measure of an idea?) (James 1949: 185). The first position I address is that of Jean-François Lyotard, whose work, *The Postmodern Condition* (1979), is considered a definitive postmodern text. I chose to begin with Lyotard because he provides what is perhaps the most easily understood definition of postmodernism, especially as contrasted with modernism. In addition, Lyotard is recognized as one of the leading postmodernist figures in philosophy, and he provides an initial answer to the psychological question of what is the basis of human dignity. The second perspective is that of Richard Rorty. Rorty is a distinctly American postmodernist who reveals the strong connection between the pragmatist tradition and the postmodern condition. His perspective adopts Lyotard's answer to the psychological question but also provides an answer to the metaphysical question of what human dignity is. The third perspective is that of Larry Hickman, who advocates what he terms *post-postmodernism*. Hickman's position not only includes that of the postmodern condition but extends beyond it with the use of an instrumentalist perspective that pertains to the casuistic question of determining the value of a given conception of human dignity. All three perspectives offer interesting and varying approaches to human dignity, and although each relates in being postmodernist, none of the positions is identical to the others.

JEAN-FRANÇOIS LYOTARD

> Postmodernism thus understood is not modernism at its end but in the nascent state, and this state is constant.
>
> —Jean-François Lyotard (1979: 79)

Postmodernism does not refer to an ideological movement within a specific period of time. Rather, to be postmodern, in its most basic form, is to have "incredulity toward metanarratives" (Lyotard 1979: xxiv). The postmodern perspective is one, taking its cue from scientific practice, of experimentation. Lyotard refers to postmodernism as modernism "in the nascent state" because it is the process of *formulating* the rules of discourse and behavior rather than the attempt to discover or deduce rules that are preexisting (Lyotard 1993: 79, 81). To take a postmodern perspective is to be oriented toward the future with the understanding that we are contributing to that future before it has arrived. As Lyotard states: "Post modern would have to be understood according to the paradox of the future (*post*) anterior (*modo*)" (Lyotard 1993: 81). Such a perspective undermines the modern reliance on grand narrative, because, rather than attempting to locate that which is necessary, unchanging, and essential, one who takes the postmodern perspective formulates that which is contingent, changing, and accidental. In the wake of the transition from modernist to postmodernist, ethical concepts such as human dignity become malleable, relative concepts, the meanings of which are constantly reformulated with the perpetual alteration of context and its variables. The grand narrative is erased, and underlying it is a series of local narratives that are in the process of being created. Human dignity is not a singular narrative but rather an element in multiple narratives. Lyotard specifically locates the foundation of ethical narratives in the pragmatics of language; it is the relationship between speakers and their speech that "governs the formation of the figure of the other" (Lyotard 1993: 137). Thus, human dignity begins with the recognition of a person *as a speaker*—that is, one whose *voice* is acknowledged.

Pragmatics is a classification of a type of linguistic relationship—that between speakers as well as between speakers and what is spoken. One should not confuse pragmatics as a foundation for ethics with the grand narrative approach to founding ethical concepts. Pragmatics simply offers a generalized term for localized relationships that intersect, overlap, and develop shared histories. Lyotard's localization of ethical relationships based upon pragmatics is strikingly similar to John Dewey's claim that morality is social. According to Dewey, morality is not social because we *ought* to acknowledge and account for others and our actions with regard to them; the fact of the matter is simply that "others do take account of what we do, and they respond accordingly to our acts" (Dewey 1922: 316). In this regard, Lyotard shares with Dewey the understanding that human dignity is not necessarily something we *ought* to acknowledge in the other; it is something that we allow for when we recognize the voice of the other—the other as interlocutor.

According to Lyotard, what threatens human dignity is the "foreclosure of the Other," what he deems the "postmodern affliction" (Lyotard 1993: 146). This foreclosure is the refusal to allow one to enter into the local narrative: the pragmatic relation is denied. Lyotard equates this with children stating to another child, "We're not playing with you" (Lyotard 1993: 145). Such a negation of an-other—a deletion from the field of discourse—is an imposition upon the local relationship from which ethics is formed; the other is ignored as an-other based upon an underlying prejudice—a refusal—that trumps the local pragmatics of the situation. Refusing the voice of the other threatens human dignity because it undermines the very basis of the ethical relationship: the pragmatics of discourse.[3]

The dignity of the person—granted by the person's *right* to speak, according to Lyotard—is not a natural right; it is not based in the essence of being human. Returning to the pragmatic relationship, such a right is *merited* (Lyotard 1993: 141). From this postmodern perspective, human dignity might be understood as the authority to speak and be heard on the level of local discourse; it is the ability—merited to the speaker by the other speakers—to engage on the level of an active participant instead of as a mere *object* of discourse. Here we find the major difference between what Lyotard defines as the *modern* perspective and the *postmodern* perspective: in the former, I am necessarily an object of the grand narrative; in the latter, I am a subject—a participant—in co-created local narratives.

What Lyotard provides with regard to ethics and human dignity is an answer to what William James referred to as the *psychological* question: the origin of morality (James 1949: 185). For Lyotard, in a fashion similar to that of Emmanuel Levinas on an individual level and to that of John Dewey on a community level, the ethical relationship—the relationship upon which human dignity is based—is engagement with an-other.[4] What Lyotard fails to address, however, is what James referred to as the *metaphysical* question: what *is* human dignity? Lyotard simply localizes and locates the relationship from whence ethics begins. He does not claim that the right to speak *is* human dignity, but only that meriting such a right gives birth to human dignity. Richard Rorty provides the answer to James's second question, not only locating the basis of the ethical relationship but also supplying a postmodern perspective with regard to the question of what human dignity is.

RICHARD RORTY

> I also prefer to abandon expressions like ontological status. Pragmatists do not employ this term.
>
> —Richard Rorty (2007: 32)

The American philosopher Richard Rorty occupies an interesting place with regard to postmodern perspectives. His philosophical position signifies a series of connections between what Lyotard defined as the postmodern condition and the American pragmatists of the late 19th to the mid-20th centuries. Quite infamously, Rorty claimed: "On my view, James and Dewey were not only waiting at the end of the dialectical road which analytic philosophy traveled, but are waiting at the end of the road which, for example, Foucault and Deleuze are currently traveling" (Rorty 1982: xviii).

The pragmatist perspective is a postmodern perspective; it undermines grand narratives in favor of localized problem solving. The pragmatist trades in absolutism and essentialism for fallibilism and meliorism. Rorty nicely captures this shift in perspective from the modernist to the pragmatist/postmodernist:

> the pragmatist answers this question [Is "X is ϕ" true?] by inquiring whether, all things (and especially our purposes in using the terms "X" and "ϕ") considered, "X is ϕ" is a more useful belief to have than its contradictory, or than some belief expressed in different terms altogether."
> (Rorty 1982: xxiii–iv)

Thus, the question of what is human dignity is not an absolute question for Rorty; there is no *essence* to human dignity. Rather, the usefulness of human dignity (even as a phrase) is questioned in contrast to its contradictory or to other beliefs "expressed in different terms altogether" (Rorty 1982: xxiv). For Rorty, the question shifts from the traditional modernist question of "What is our nature?" to the postmodern question "What can we make of ourselves?" (Rorty 1993: 115). Human dignity here becomes future-oriented: if we are to ascribe to a notion of human dignity, how are we to do so, and is doing so beneficial in our situation? How does doing so facilitate our becoming better than we already are?

Rorty argues that to attempt to supply "a purportedly ahistorical human nature" does no useful work (Rorty 1993: 118–19). What is beneficial is the realization that human dignity—what Rorty refers to as part of "human rights culture"—has been historically fostered by our increased receptivity to "hearing sad and sentimental stories" (Rorty 1993: 118–19). This realization facilitates working toward utopian ideals in a fallibilistic manner denied by modernist thinking. Rejecting the grand narrative scheme in favor of the pragmatic approach to human dignity allows us to "concentrate our energies on manipulating sentiments on sentimental education" (Rorty 1993: 122). These sentiments exist on a localized level, not on a "human" level. Rorty locates dignity on the level of the family, the clan, and the tribe, especially in non-Western cultures (Rorty 1993: 125). However, over time these spheres of dignity have expanded and continue to expand, but not because of some underlying recognition of an ontologically prior category of humanity. According to Rorty, this expansion has been quite rapid because of undermining grand narratives that seek "understanding of the nature of rationality or of morality" and being increasingly "moved to action by sad and sentimental stories" (Rorty 1993: 134). Human dignity, from this postmodern perspective, expands beyond the recognition of an-other from the pragmatics of interaction to the emotional sentiment felt between interlocutors.

In one respect, Rorty's perspective seems to resemble that of Ruth Macklin, who states quite simply that dignity is "respect for persons" (Macklin 2003: 1419). Rorty's perspective, however, is a bit more nuanced than Macklin's, at least insofar as he includes an emotional bond between interlocutors. That emotional bond—the feeling of sentiment—is, for Rorty, in what human dignity consists. Introducing Macklin's perspective here does showcase the major difference between the postmodern perspective of Lyotard and that of Rorty. Lyotard considers human dignity as merely that pragmatic recognition of an-other, which we could quite easily consider as respect *for* an-other *as* a speaker. Rorty moves a step beyond this; his perspective entails both the respect for an-other as well as the type of consideration shown to an-other, which is one of emotional sentiment. Such sentiment is beneficial to us because it binds us into a community, which facilitates bringing about our utopian ideals. Our families, our clans, our tribes expand with increased emotional sentiment. We gain more allies in our pursuit of a better world.

Rorty addresses the metaphysical question of what human dignity is (the sentiment we feel for an-other) from a postmodern perspective, but he fails to address adequately what William James referred to as the *casuistic* question with regard to ethics. In terms of human dignity, Rorty does not answer the question: how do we measure the benefits accorded by our conception of human dignity and its relation to our goals? Although Rorty does refer to the importance of achieving utopian ideals, this is far too vague to be adequately addressed. Such ideals do change over time, thus undermining the modern adherence to grand narrative, but such an abstract ideal (or series of ideals) seems entirely too imprecise and slippery to establish any means by which to effectively measure differing accounts of human dignity. Rorty's prescription for preferring one notion of human dignity over another seems something like: "that which brings us closest to our utopian ideals," but this is largely unsatisfactory, being, at best, an abstract goal on a never-reached horizon.

Counter to Rorty's vague perspective, Larry Hickman provides a more satisfactory answer to the casuistic question with his *post*-postmodern perspective. By combining the postmodern localization of narrative with John Dewey's instrumentalism, Hickman provides us with theoretical tools by which to judge one form of human dignity against another within varying contexts.

LARRY HICKMAN

> Where there is not conflict, there is no need for inquiry.
>
> —Larry Hickman (2007: 46)

Reading both the pragmatists of the late19th to the early 20th centuries, and the later postmodernists of the 20th century, Larry Hickman realized similar connections between the two groups that Richard Rorty had realized. In addition, Hickman realized that the pragmatists offered a corrective to what the later postmodernists lacked: "a theory of experimental inquiry that takes its point of departure from real, felt existential affairs" (Hickman 2007: 29). Like Lyotard, pragmatists like William James and John Dewey had rejected the search for absolute truth proposed by modern thinkers, favoring localized narratives that were based upon experience. As indicated above, there are definite affinities between Lyotard's localization of ethics in the pragmatics of communication and Dewey's localization of morality from the interactions within the community. One of the major differences between the two views, according to Hickman, is the "compartmentalizing, hyperrelativistic tendencies latent in most forms of postmodernism" and the "commonality of human experience" conveyed in Dewey (Hickman 2007: 19). Such commonality is not a reversion back to an essentialist perspective but rather the acknowledgment that human beings, in many ways, are incredibly similar. In fact, Dewey and Hickman both argue that human beings are more similar than they are different, which Hickman sees postmodernists, such as Lyotard, as contradicting (Hickman 2007). As Hickman states, "the common features of the human nervous system, having evolved within a common natural environment, provide a platform from which aesthetic diversity can be appreciated and appropriated for productive purposes" (Hickman 2007: 39). In other words, our physicality—our bodies—provides us with more commonality than difference. Human dignity is here not dependent upon the mere recognition of an-other as speaker. Human dignity is based in the fact that we are bodies—human bodies—that tie us together into a biological community and make our social communities more similar than different. With the shift from the pragmatics of communication to the inclusion of the commonality of the body and its environment in addressing the psychological question of human dignity, Hickman provides a solid foundation from which to understand human dignity that does not have the trappings of a grand narrative. From a pragmatic post-postmodernist perspective, such a concept of human dignity becomes a tool among other conceptual tools that may be used in solving ethical problems that emerge within experience.

One of the great strengths of the pragmatic perspective is its commitment to the process of inquiry in solving localized problems. According to Hickman, inquiry provides "an essential component of communication, which can construct pluralistic links across otherwise isolated cultures and disciplines" (Hickman 2007: 19). The pluralism of cultures and disciplines reconnects at points of felt difficulty, in addition to common biological and environmental points, into which disparate groups inquire in search of solutions. For the post-postmodernist, one stage of problem-solving is an instrumentalist tool, which Hickman refers to as "dramatic rehearsal." As he states, "Dramatic rehearsal arises within the context of a sequence of inquiry that itself arises within existential conditions and seeks to find its expression in the same sphere" (Hickman 2002: 33). Through dramatic rehearsal, the post-postmodernist attempts to account for all the foreseeable contingencies and outcomes, rather than simply theorizing based upon the problem presented as if it was in a vacuum. In other words, the inquirer acknowledges the problem and its context as real constraints to possible solutions to the felt difficulty.

Such acknowledgement betrays any adherence to grand narrative and links different inquirers to one another in a shared problem-solving venture. Shared problems lead to a greater unification between persons, often regardless of disparities (Hickman 2007: 46). Whereas Rorty takes his cue from literature with regard to human dignity and its increase via sentimental education, Hickman allies himself more closely with the efficiency both James and Dewey found in scientific experimentation and methodology.[5] From a pragmatist perspective, scientific methodology grants the testability and falsifiability of proposed solutions to problems.[6] Scientific methods do not rely upon grand narrative, especially in an essentialist sense, but they do often result in viable solutions to problems. Hickman reveals the greatest discrepancy between the pragmatist postmodernist and the late-20th-century postmodernist in a description of the perspectives of James and Lyotard.

> [T]here is a significant difference between James's proposal for a working method that privileges no particular result, and the proposal issued by Lyotard and other post-modernists that no particular narrative or standpoint can or should be privileged. James's proposal possesses the virtues of what has been termed "the scientific method" of inquiry (on which it is based): when it is properly applied, it does present a privileged method with respect to knowledge-getting, since it is experimental

and therefore produces practical effects that are objective. It is self-correcting. (Hickman 2007: 34–35)[7]

The ability to experiment and produce solutions to problems that facilitate inquiry is the core of the pragmatic method, which emphasizes "continuity and commonality, thus rejecting the claims of skeptics, racists, and others that the primary features of human life are difference, discontinuity and incommensurability" (Hickman 2007: 37).[8] Human dignity from a post-postmodern perspective thus combines the localizing anti-essentialism of Lyotard's postmodernism with experimentalism as a *valued unifying principle* that grounds our worth based upon our shared biology, environmental circumstances, and felt difficulties. Human dignity is therefore not based in the concept of an atomic, disconnected self; rather, a "self is an individual that exists as an individual only in relation to connections and communities that enable it to do so" (Hickman 2007: 38). The post-postmodernist utilizes the conception of human dignity as a tool of unification and measurement in addressing James's casuistic question of how we are to measure ethical outcomes against one another.

The commonality of human dignity based in the shared circumstances of persons, including their inquiry into felt difficulties, provides a gauge by which to measure ethical outcomes: does a proposed solution actually *solve* the problem at hand?[9] In addition, the outcome is a more viable option if it does not threaten human dignity, which would entail solving one problem at the cost of producing greater difficulties—that is, threatening the very values of community that the initial problem threatened. This would be similar to plugging a tiny hole in a dam, only to cause greater exertion of pressure to produce larger fissures in the structure. In attempting to solve problems, the post-postmodernist operates with a pragmatic theory of truth that pertains directly to those problems, proposed solutions, and probable outcomes. The experimentalism of the post-postmodernist "requires that we regard our ideas as true when they have both an objective basis (i.e., their warrant can be demonstrated before a candid world) and the capacity to resolve some type of objective difficulty (i.e., they are assertible under relevant conditions)" (Hickman 2007: 37). The more warranted a concept is as a tool in solving specific types of difficulties, the more likely it is to become part of our *habits of action*, used in future instances of similar difficulties (Hickman 2007: 46). From the post-postmodern perspective, human dignity is part of our habits of action because it has often been successfully appropriated in addressing past problems. Our solidarity, which is instrumental in solving our problems *as* humans, is also what grants us our worth as humans. This use of human dignity—valuing humans based upon solidarity in problem solving—is both warranted and assertible.

With regard to ethical difficulties, the post-postmodernist undermines an essentialist conception of human dignity in favor of one that is *instrumental*; in many problematic cases, especially those involving medical and bioethical difficulties, human dignity is a warranted concept that both facilitates solving such difficulties and preventing future difficulties. It is not an essentialist trump card concept to be utilized merely to win one side of an ethical argument or to rhetorically sway opinion in one's favor but a tool that can be utilized as part of a solution to an ethical problem.

The power of Hickman's perspective when applied to human dignity and ethics, especially when contrasted to that of Lyotard and Rorty, is in its scientific approach that provides unification through a biological, environmental, and communicative foundation, along with a means by which to determine what ethical solution might best solve a given ethical problem or series of problems. From this post-postmodern perspective, science is not valued over and above human dignity; rather, scientific methods, along with scientific enterprises, are tools for solving human problems. As Roberto Andorno explains, science is "a means for improving the welfare of individuals and society," and "people should not be reduced to mere instruments for the benefits of science" (Andorno 2009: 228). From a pragmatic perspective, people should not be reduced to such instruments, because to do so would undercut the very purpose of scientific experiment: to solve people's problems.

Science—that is, scientific method and experimentation—is utilized by the post-postmodernist because it produces results that often effectively solve our difficulties that initiated inquiry. As Dewey claims in his *Logic: The Theory of Inquiry*, there is no necessary barrier separating the methods of inquiry utilized in most sciences from being utilized in ethical inquiry (Dewey 1938: 488–89). According to Ronald Lindsay, "our moral judgments should continually be tested for adequacy by considering their practical implications. The practical implications of our moral judgments represent the 'evidence' that confirms or refutes their appropriateness" (Lindsay 2008: 38). Both scientific and ethical experimentation involve a community of inquirers attempting to solve shared felt difficulties. In other words, they are engaged in the very practice that signifies their human dignity.

APPLYING POSTMODERN PERSPECTIVES: CHIMERAS AND EUTHANASIA

The following is an examination of two categories of bioethical cases (human-animal chimeras and voluntary euthanasia), applying the above postmodern perspectives to each. Such application is here done *ne multa*, and the analysis is necessarily incomplete due to the cases being abstract and without context. However, the following presents the initial praxis of these perspectives with regard to such cases.

In the case of human-animal chimeras, some theorists have questioned if human dignity is threatened when nonhuman animals are modified with the use of human cellular material (cf. Robert and Baylis 2003). In addressing any specific case of human-animal chimeras, it would benefit us to question what problem or problems are being addressed in, and what possible problems result from, the creation of the specific hybrid species. In addition, we would benefit from asking the Rortian question of whether or not there is, in fact, an ethical issue at all when chimeras are developed.

In the essay "Human-Animal Chimeras: Human Dignity, Moral Status, and Species Prejudice," David DeGrazia focuses specifically on the potential problems that arise with regard to dignity when human stem cells are transplanted into great apes and rodents with the intent of producing human neurons. As DeGrazia states, "The focus on neurons is intended to provoke as fully as possible concerns about the mixing of characteristically human traits with nonhuman traits, since neurons enable the cognition and emotion often thought definitive of humanity" (DeGrazia 2007: 310). I will not here address DeGrazia's specific arguments, but simply utilize the example of transplantation of human stem cells into great apes as a particular case to which I will apply each theorist's conception of postmodernism.

Lyotard's perspective offers very little with regard to solving this particular issue, because he merely points to the pragmatics of communication as the birthplace of ethical relationships. From Lyotard's perspective, the question of human dignity would pertain to whether or not human-animal chimeras were acknowledged as having voices—that is, status as subjects.[10] With regard to human dignity being threatened simply by the implantation of human stem cells into nonhuman animals, Lyotard would likely not see this as an actual issue, because it would be based upon an essentialist account of dignity. There is nothing inherent in the act of transplanting human tissue into a great ape that necessarily involves human dignity. Such a consideration would be culturally relative, based in the pragmatic rules and social mores of a given culture. Without a specific cultural context, determining the inclusion or exclusion of human-animal chimeras as those possessing human dignity is impossible. Even if such a context were here available, Lyotard's perspective only offers the ability to *report* on whether entities have been included as ethical subjects; it does not *diagnose* whether it would be beneficial to include them or their cellular tissue. From what Lyotard states concerning the postmodern condition, one can infer that any consideration of creating human neurons within great apes would bring with it ethical experimentation from which to derive the rules concerning the ascription of human dignity to those great apes that developed human neurons.

The pragmatic/postmodern perspective of Rorty is here more emotive than that of Lyotard. The question of human dignity would rest upon our reaction to such research, if our art and literature advocated an affinity for such research through sentimental education, and how the creation of human-animal chimeras via the implantation of human stem cells into great apes would contribute to our utopian ideals. At the moment, the issue of whether to conduct such research is fairly well divided between people, and Rorty's perspective does not facilitate moving beyond the divide. Such an antagonistic relationship between those who advocate conducting such research and those who do not would seem to be the cause of more problems than solutions, betraying our movement toward utopian ideals, but Rorty does not seem to offer anything pertaining to this issue other than the suggestion that we might facilitate moving out of a stalemate based upon what art and literature we produce.

The emotivism of Rorty's position is replaced by the instrumentalism of the post-postmodern perspective. From a perspective based in Hickman's post-postmodernism, we must determine what problems we are attempting to solve and whether such solutions are warranted and assertible. If the problem can be solved in another way that has a greater benefit with less cost, then why attempt to grow human neurons in nonhuman animals? The proposed chimerical solution would have no merit. Here we would have to address the pragmatic considerations of the problem. Pragmatic variables would include, among others: the existing knowledge, beliefs, and assumptions of those involved as they pertain to the problem at hand, the tools at our disposal for solving the problem, the alternatives, both tried and potential, and the constraints that cause the problem and prevent other solutions.[11] Another consideration would be whether or not human dignity would be applicable as a tool if chimeras were the solution to a given problem. The same issue seems to arise here as with Lyotard: human dignity is only threatened if it is considered absolute and fixed instead of existing on a continuum that would allow such research. On the other hand, the biological similarity between humans and great apes might warrant moral status to the latter, which is similar to the claim DeGrazia makes in his argument.

Because Hickman's perspective is instrumentalist, experimentation is a key component of determining the warranted and assertible course of action. Ethical experimentation may strike one as odd; however, experimentation would involve questioning whether human dignity was, in fact, something that was unnecessarily threatened. In addition, analogous precedence—that is, a collection of similar past dilemmas that have solved or caused problems—functions as an element of experimentation that informs those deliberating on the matter.

One here detects a major difference between the perspectives of Rorty and Hickman. For Rorty, the key to any threat to human dignity would be based in our emotional reaction to the transplantation experiments. For Hickman, the emotional connection is unimportant; such accordance is based upon solving a particular problem without causing greater problems.[12] What all three perspectives have in common is the dismissal of a belief in human dignity as absolute. Even the very idea of what it is to be human is acknowledged by each perspective as constantly changing, which allows for the pragmatic consideration of whether human dignity is threatened by such cases to be less problematic than it might be within a grand narrative scheme.

In a case of voluntary euthanasia, Lyotard's perspective is much more telling than that pertaining to experiments involving the implantation of human cellular tissue; because euthanasia pertains to those who request to die, not seriously considering such a request would be a betrayal of the pragmatic relation *already established*. The person making such a request has already been recognized as having a voice, so the question would arise of why they were being denied a voice—that is, considered as a mere object of ethical discourse rather than what they had previously been considered: an ethical participant. In such a case, the aegis of argument would actually rest upon those claiming euthanasia to be unethical; to reject the request for euthanasia out of hand would be to deny the very ethical relation previously established. Thus, if we take Lyotard's perspective, we would describe denying *consideration* of a request to be euthanized as a denial of human dignity, for one's human dignity rests upon being a participant in ethical narratives, especially those directly involving the person and their requests.

Rorty's sentiment-based perspective is perhaps the most confusing in this case because such a request would likely elicit affective reactions, but the reactions in such situations are typically contradictory. We often are torn between a compassion for the person requesting to die and a compassion for those friends and family members left behind. Sentimental education also runs both ways in this situation, and utopian ideals likely involve the eradication of the need for making such decisions; our utopian ideal is that we will always be healthy, or that health care will be able to mend any condition. Rorty's perspective here is not very helpful, and it seems to leave us where we began with regard to human dignity. We might ask what the ethical dilemma is, or whether such language is useful, but what Rorty would seem to advocate here is a decision based upon emotion, and he offers no device by which to overcome conflicting or contradictory emotions. He acknowledges our affinities but throws up his hands at their conflict, suggesting little more than a democratic process of decision making.

The post-postmodern perspective follows the same approach used in the case of human-animal chimeras. We must approach the problem instrumentally, determining what the problem is and whether or not the decision to euthanize is warranted and assertible. Additionally, this perspective merits being open to experimentation with regard to what alternative solutions might work, even if such experimentation is merely hypothetical and based in probability. In the case of a euthanasia request, the solution *is* assertible, but the major question pertains to its warrant. Are there other measures that are more satisfying to the person making the request? Are they aware of all the tools at their disposal? Has the patient been adequately informed of his or her diagnosis and all of the treatment options? Is/was the patient reasonable enough to grant an advanced directive? In many cases, euthanasia is a decision when there are no beneficent options available.

As with any situation understood from a post-postmodern perspective, the devil is in the details—the specific context of the situation would help determine the pragmatic considerations involved. These would include, but not be limited to, such variables as economic factors, religious beliefs of the parties involved, viable options for treatment, and established precedence within the setting. In this case, human dignity is a useful tool, for its definition can facilitate guiding whether granting a request to be euthanized is dignified. Hickman's perspective actually benefits from that of Lyotard in this situation, because Lyotard's location of human dignity in the pragmatic relation indicates that, at the very least, the request for euthanasia must be considered. To deny such a request would usurp the preestablished ethical relation, thus betraying the pragmatics upon which human dignity is based. From the post-postmodern perspective, the person requesting to be euthanized is the primary pragmatic consideration because the person is dignified. However, one major difference between Hickman's perspective and that of Lyotard is that the post-postmodern perspective attempts to account for a decision's effects upon those parties who help compose the environment of the individual; from the perspective of Lyotard, the individual is only regarded as a subject when treated with what one might consider full individual autonomy—that is, the ability to destroy oneself regardless of the cost to others. In this respect, the post-postmodern perspective is more community conscious than that of the postmodern perspective in its attempts to account for both more considerations than simply the individual's voice.

CONCLUSION

> To say that there is a universal law, and that is a hard, ultimate, unintelligible fact, the why and wherefore of which can never be inquired into, at this a sound logic will revolt, and will pass over at once to a method of philosophizing which does not thus barricade the road of discovery.
>
> —Charles Sanders Peirce (1955: 335)

> Approach to human problems in terms of moral blame and moral approbation, of wickedness or righteousness, is probably the greatest single obstacle now existing to development of competent methods in the field of social subject-matter.
>
> —John Dewey (1938: 489)

A perspective being *postmodern* does not necessitate an eradication, erasure, or annihilation of the concept of human dignity in toto. Here I have presented three varying and intersecting concepts of human dignity from postmodern perspectives. These perspectives have hopefully conveyed that the concept of human dignity is meaningful, even when removed from modern perspectives. Between the three perspectives, Larry Hickman's post-postmodernism provides the more robust approach to the subject of human dignity, addressing all three of William James's questions concerning ethics while partially adopting the perspectives of Lyotard and Rorty. The instrumentalist approach provides tools otherwise lacking in the perspectives of Lyotard and Rorty. This does not negate the importance of the other two perspectives; all three have played integral roles in dialogues concerning methodology, modernity, and morality. In fact, Rorty's analysis, in particular, provides an element within the ethical relationship that Hickman seems to neglect: empathy.[13] The emotional component is important to a postmodern (or post-postmodern) perspective of human dignity, because it galvanizes the recognition of common problems, often suddenly sparking the move from recognition of a common problem to action with regard to that problem. Although intellectuals may not like to admit it—I think this is where Rorty is at his most radical and his most acute—we are often moved to act because we are emotionally affected by something as simple as a person's story, a photograph, a film, or even a person's facial expression.[14] Many times we are affected because we discover that the other is, in fact, much more like us than we had originally suspected. This sentimental aspect of human dignity indicates the profound localization of the ethical relationship that all three thinkers denote.

The three perspectives presented here are admittedly only a sampling of a number of perspectives that could be claimed postmodern, and these perspectives should not merely be judged in reference to one another but also in reference to *modern* perspectives.[15] By no means does the postmodern condition signify the end of dialogue about human dignity. Rather, it signifies a continuation and multiplication of many dialogues; it signifies the continuation of inquiry.

The postmodern perspectives presented here all attempt to accommodate the pluralism within and between societies by undermining what could be considered hegemonic grand narratives and favoring the localized voices of those societies and cultures. Human dignity is a functional concept that can aid in solving ethical problems, and these perspectives pose alternative approaches to both the concept of human dignity and the way in which we attempt to solve those problems. A common thread between these conceptions of human dignity and other perspectives is an underlying idealism. The idealism underlying postmodern perspectives of human dignity is what we might refer to as *instrumental idealism*. As John Dewey stated in *The Quest for Certainty*, "The active power of ideas is a reality, but ideas and idealisms have operative force in concrete experienced situations; their worth has to be tested by the specified consequences of the operation. Idealism is something experimental not abstractly rational" (in Alexander 1986: 50; Dewey 1960: 166–67). From postmodern perspectives, human dignity is an instrument that we use to help achieve our goals in overcoming ethical problems; there is an operative force to the idea of human dignity. In this regard, these perspectives are perhaps not so completely different from other perspectives on human dignity. Such similarity facilitates an ongoing dialogue between the various perspectives, even in light of their profound differences.

NOTES

I would like to thank Rebecca Bamford, Larry Hickman, and the editors of this volume for incredibly insightful comments and suggestions pertaining to this essay.

1. One may also include what C. S. Peirce referred to as the doctrine of necessity, "that every single fact in the universe is precisely determined by law" (Peirce 1955: 324).
2. I have intentionally avoided perspectives that are considered affiliated or derived from postmodern thought, but not postmodern per se, including poststructuralism, deconstruction, and postcolonial thought. In addition, I have chosen to forgo a historical analysis of postmodernism. For those interested in such an analysis, I recommend Jameson (1991).
3. Lyotard does not attempt to fashion the genesis of the ethical relationship as if power is not a factor, which gives his account an honesty lacking in modernist grand narratives that do not acknowledge that power is a factor in the determination of the participants and the types of participation permitted in ethical discourse. Postcolonial thinkers, such as Spivak (2010), specifically address the issue of power and those prohibited from participation.
4. Levinas locates the basis of the ethical relation in our encounter with the face of the Other (cf. Levinas 1969: 194–219).
5. For a convenient explanation of scientific method and experimentation from a pragmatic point of view, cf. Dewey (1997: 150–56).
6. One of the distinguishing marks of scientific methodology, as opposed to nonscientific methodology, for the pragmatists, is the insistence on experimentation over and above what Peirce referred to as the method of tenacity, the method of authority, and the method of reason—that is, the a priori. (cf. Peirce 1949: 7–31).
7. Hickman postulates that Lyotard's postmodernism replaces modern grand narratives with its own narrative. This is a point of disagreement between us, although I will not go into a lengthy discussion of our arguments here. For the purposes of this chapter, it is sufficient to simply note that the experimentalism of James and Dewey fills the lack of methodology that haunts the postmodernism of Lyotard, upon which Hickman and I agree.
8. Against possible objections that post-postmodernism does not account for *out-groups*, Hickman's retort is that the very notion of an out-group is a construct that is much more difficult to defend than the inclusion of individuals in a community. His very basic point is that the pragmatic method, through its focus on problem solving, emphasizes the sameness of individuals as biologically similar individuals living in similar environments who are *problem solvers*. This precedes all difference, leaving the aegis of responsibility on those who want to argue for *exclusion due to difference* rather than on those who want to argue for inclusion due to sameness.
9. For Hickman's own application of his post-postmodernist method to problems arising from genetic screening cases, see his essay, "Pragmatic Resources for Biotechnology" (Hickman 2002).
10. Part of Degrazia's argument actually pertains to the dignity of great apes being threatened. He argues that, like children or the mentally disabled, they

are *borderline persons*. Lyotard would perhaps disagree with meriting such personhood based upon the fact that great apes do not have a voice. However, one could argue in a similar vein as Bruno Latour that we have simply not recognized the great apes as having voices (cf. Latour 2004).

11. From Hickman's perspective, issues raised from a grand narrative, such as Christianity, might be important, but only with respect to the particular context and parties involved. Such narratives would be diminished to one pragmatic consideration among many.

12. If the shoe pinches, we want to stop it from pinching, but not at the cost of cutting off the foot.

13. I am quite sure that Hickman is intensely aware of the importance of what one might call the *aesthetic dimensions* of experiences, but his focus pertains much more to the instrumentality of such experiences than the "feeling of" such experiences. Here I am simply indicating that Rorty's pointing to the sentimentality involved in the ethical relationship—the feeling of connectedness to an-other—is an important factor to such a relationship.

14. Rorty's notion of sentimentality can be understood as what thinkers like Solomon R. Benatar refer to as *moral imagination*, "the ability of individuals and communities to empathise with others" (Benatar 2005).

15. A major difficulty in locating postmodernist perspectives is the fact that most thinkers do not label their thought as postmodernist. Perhaps to do so would be to commit to a grand narrative of sorts, thus undermining one's own postmodernism.

6 Dignity for Skeptics
A Naturalistic View of Human Dignity

Richard McClelland

The concept of human dignity has been under intense discussion in a variety of humanistic and scientific disciplines in recent years. Much of this discussion focuses on the history of the concept, including its ancient, Christian, and Kantian treatments (e.g., Cancik 2002; Wright 2006: 537–48; McCrudden 2008: 656–63). In the Christian tradition, it has been customary to ground the notion of dignity in the doctrine of creation (humans as created "in the image of God"). Kant and his modern *epigoni* (e.g., Zagzebski 2001; Lee and George 2008) ground it in rationality, which is often taken to be distinctive of humans.[1] At the same time, human dignity has played a substantial role in political documents of the late 20th century, especially in state constitutions drawn up after the end of World War II (see Iglesias 2001; McCrudden 2008). In many of these contexts, of course, the tasks of defining human dignity and giving rational grounds for it are avoided, thereby rendering "dignity" a mere placeholder for an undefined and unjustified content. Legal and quasi-legal documents, especially those dealing with human rights, make similar appeals to the foundational value of human dignity, but also without extending clear definitions or justifying grounds for that concept (see summaries in Iglesias 2001: 127–28; cf. Dicke 2002; Frowein 2002; Chaskalon 2002; McCrudden 2008: 665–67). Paradigmatic documents of these kinds are the 1948 United Nations Declaration of Human Rights, Convention against Torture (1987), and the Common Article Three of the Geneva Conventions of 1949. In all of these settings, violations of human dignity are held to be blame-worthy and grounds for punishment (see discussions in Nagan and Atkins 2001, especially pp. 95–102; Duff 2005).

Not surprisingly, ideas of human dignity have also made their way into contemporary bioethical debates (see review in Caulfield and Brownsword 2006). Some bioethicists, perhaps disturbed by the absence of consensual definitions of the leading term or justifying grounds for its use, have complained that dignity is "a useless concept" and have argued that it should be replaced by more substantive notions such as respect for personal autonomy or a catalog of specific human rights. Ruth Macklin (2003) is a case in point. Replies to Macklin's pot-boiler have been astonishingly thin. Killmister, for example, insists that dignity is not a useless placeholder for other ideas, but the best she can do to describe dignity is "the capacity to live by one's standards and principles" (2010: 160). This seems hopelessly broad, for it would not exclude the case of the sincere Nazi whose racist standards and principles require him to murder Jews and other social minorities, thereby simultaneously upholding his own dignity while grossly violating the dignity of others. Such incoherent implications render her proposal vacuous. Schroeder (2008; 2010) has vigorously disputed Macklin's thesis and offers to both ground the notion of dignity and to unite various uses into a single focal meaning. However, she concludes: "The fact that human beings are understood to have dignity must then be a contractual agreement between legitimate representatives and their peoples, transformed into written law" (2010: 124). The importance of social context should not be merely dismissed and will be followed further below. However, Schroeder's final position seems functionally equivalent to Macklin's own deflationary view of dignity. It also has some of the same problems as Killmister's reply, for Schroeder's conception of dignity is compatible with all kinds of social compacts, many of which seem obviously not to promote human dignity in any widely acceptable form (e.g., the real constitution of the old Soviet Union).

Conspicuous by its absence from all of these discussions is any mention of human biology. But herein, in my view, lies rich and fertile ground for understanding human dignity in a thoroughly naturalistic fashion that also possesses considerable conceptual unity and depth. Moreover, such a view of dignity has important connections to both of the two great historical traditions that have most commonly been represented in these contemporary discussions: classical theism and Kantianism. At the same time, however, a naturalistic view of dignity draws some of its strength from the failures of those two traditions. More particularly, the failure of Kantian emphases on rational autonomy as a unique characteristic of humans, and agnostic or atheistic rejection of classical theism, can help to highlight strengths of a biologically based understanding of human dignity. Fundamental to such an understanding is cooperation.

No other animal species known to us cooperates on the scale that humans do. Moreover, it is clear that our capacity for cooperation is essential to the long-term health and survival of our species:

> evolution is constructive because of cooperation. New levels of organization evolve when the competing units on the lower level begin to cooperate. Cooperation allows specialization and thereby promotes biological diversity. Cooperation is the secret behind the open-endedness of the evolutionary process. . . . Thus, we might add "natural cooperation" as a third fundamental principle of evolution beside mutation and natural selection. (Nowak 2006: 1563)

On this view, then, cooperation is a fundamental biological feature of the long-term history of our species. Cooperation comes into its own, especially once populations reach sizes orders of magnitude larger than ancestral hunter-gatherer bands (Dubreuil 2008). We also know that cooperation depends upon reciprocity. It is plausible to hold that reciprocity is also fundamental to at least a wide range of the social or interpersonal uses of the notion of human dignity. This would account for the regularity of appeals to equality in juridical and political contexts engaged with human dignity. For while reciprocity may not entail dignity, dignity entails reciprocity. And reciprocity is a fundamental feature of human biology.[2]

THE BIOLOGY OF HUMAN RECIPROCITY

At least three different forms of reciprocity are widely recognized in the biological and anthropological literature today. The simplest form seems to be mutualism, which has to do with the way two organisms may interact such that each and both of them derive some relatively immediate fitness benefit from that interaction. The bacteria in the guts of ungulates (cows, for example) derive such a benefit as do their hosts, who could not digest vegetable foods without those bacteria. Mutualism is known between humans, of course, but is the least interesting form of reciprocity. Robert Trivers, in a ground-breaking paper in 1971 identified "reciprocal altruism" as more characteristic of humans. Trivers was concerned to provide an evolutionary account for all those behaviors between biologically distant human persons which may benefit one party currently and may postpone returning benefit for long periods of time. More recently, discussion about "strong reciprocity" has taken center stage. This notion presumes both mutuality and reciprocal altruism but points to human practices in which individuals who defect from social norms for reciprocity will be punished by third parties, most of whom are not biologically related to the individuals being punished (Gintis 2000; Fehr and Fischbacher 2004). The debate over what forms of reciprocity must be accounted for in human behavior, and how they are related to one another, continues (e.g., Clutton-Brock 2009 and Tomasello 2009 have argued that mutualism is the best explanation for cooperation). Similarly, arguments about how best to account for the latter two forms of reciprocity in evolutionary terms also continue. It is not my purpose to settle these matters here. It suffices here to point out that the appearance of widespread cooperative behavior in our species is what might be expected on the basis of evolutionary theory.

The first indication that this may be so is that reciprocal behavior appears very early in human ontogeny. This is notably so in the case of the affective development of normal human infants. What we know about this process is that it depends vitally on affective exchanges between infants and their primary caretakers (most often, the biological mother). Synchrony in these exchanges can be tracked by observing very closely the manner in which mother and child look intently at one another or look away from one another (Brazelton et al. 1974; cf. Feldman 2007a, 2007b). While every normal human neonate comes into the world with a suite of basic emotions available to it, everyone has to acquire the capacity to regulate those emotions, and these kinds of intimate reciprocal exchanges between infants and their caretakers is the matrix of that regulatory achievement (Schore 1994; Sroufe 1995; Tronick 2007). Moreover, both later moral development and capacity for thinking almost certainly depends upon these early affective, and non-linguistically mediated, regulatory capacities (Gopnik et al. 1999; Hobson 2004).

Decoding the emotional content of faces is a crucial element in these developments. This depends upon a subcortical system that operates very fast (under 200 milliseconds, with peak activities at 100 milliseconds and 170 milliseconds post stimulus), a system in which the amygdala (a small almond-shaped area of the brain found bilaterally deep in the limbic system) plays a prominent role (Vlamings et al. 2010), together with the cingulate and prefrontal cortices (Etkin et al. 2011). The basic point for us is that these developments depend entirely on thoroughly reciprocal exchanges between infants and their caretakers of affectively laden information. Since these exchanges are also the very matrix from which human personality arises, we may say that reciprocity is built in to human personality (in the normal case, that is; there are many ways in which this process can go awry). Reciprocity, then, is ontogenetically early in its appearance, nonrationally cognitive, and affectively laden. Moreover, it appears to have a distinctive neurobiological basis—which will concern us further below.

We also know that reciprocity is a human universal, appearing in similar forms in all stable human cultures known to us (Henrich et al. 2005). Especially important is the evidence that it is widespread among human hunter-gatherer and hunter-agriculturalist societies (Boehm 1999: 212–17; Gray 2009; Tomasello 2009). For we commonly suppose that such cultures are important guides to how things worked for social groups of our species much earlier in its evolutionary history. Cooperative behavior grounded in reciprocity is also found in some nonhuman animal species besides our own, notably other primates (Boehm 1993, 1999; de Waal 1996, 2008, 2009). Whether such reciprocity rises to the level of full-blown empathy in our close primate relatives and in other animals is much debated.[3] But evidence of reciprocity appearing early and automatically in human ontogeny, evidence for it across human cultures, including hunter-gatherer cultures, and evidence for it among nonhuman animals, especially our close primate relatives—all this is what we would expect to find if reciprocity is indeed the product of evolutionary processes. Such reciprocity also depends on our highly developed capacities for social cognition.

It is now evident from imaging data that social cognition generally takes place in a distributed network engaging elements of the limbic system (a cluster of brain regions deep in the midbrain and sitting roughly on top of the brain stem) and elements of the lateral and prefrontal cortices (Tomlin et al. 2006; Wilson 2006; Rilling et al. 2008; Moll and Schulkin 2009). Social cognition includes the most basic affective capacity of the human animal to promote prosocial behavior, and especially cooperation by means of various kinds of reciprocity, namely empathy (see reviews in Preston and de Waal 2002; Decety and Jackson 2004, 2006). Human empathy appears to involve at least three basic components: an affective response to another person, often one that shares or mimics his or her emotional state; a capacity to take the perspective of the other person; and a regulatory capacity that prevents loss of self-identity and preserves our capacity to distinguish between our own feelings and those of the other person. Neural networks engaged in empathic experience include regions of the limbic system (notably the amygdala), the anterior cingulate cortex, the hippocampus (which is heavily involved in emotional memory), the anterior insular cortex, together with neural projections of these regions into the frontal lobes, especially the prefrontal cortex.[4] Humans can feel empathy for a wide range of targets, including animals of other species. Empathy is also costly, demanding scarce cognitive resources, and sometimes places considerable emotional burdens on the empathic person (Hodges and Klein 2001). Empathy tends very strongly to promote prosocial behavior, notably reciprocal cooperation (Page and Nowak 2002) and inhibits aggression (Schore 1994: chap. 27). However, empathy is not always prosocial, and harmful behaviors can also spring from empathic connection to another person (Batson et al. 1995; Van Lange 2008). Nevertheless, since empathy gives one person a stake in the emotional condition and welfare of another person, it is not surprising that it generally supports prosocial and cooperative behavior, especially the giving and receiving of support, especially where the relationship between the two persons is otherwise emotionally close and positive (de Waal 2008). Moreover, from an evolutionary perspective, it makes little sense to suppose that empathy would be selected for and thus preserved in the long run if it did not tend strongly toward prosocial behavior. Thus, we tend to empathize more strongly with offers made in social exchanges that are perceived to be fair than we do to those that are perceived to be unfair (Singer et al. 2006). Different cortical regions are active during cooperative exchanges than during competitive exchanges (Decety et al. 2004). Both results suggest that the role of empathy in social cognition is closely bound up with perceptions of fairness versus unfairness, sensitivity to which I consider next.

Human brains respond differentially to perceived unfair offers in an economic game, as compared to our neural responses to offers perceived to be unfair (Tabibnia and Leiberman 2007; Tabibnia et al. 2008). It is clear from these findings that we find fair offers in social exchanges more rewarding than we do unfair offers and that this sense of greater reward is realized in the brain's reward system, especially the striatum, a very active region of the limbic system heavily involved in computation of the expected reward value of actions (King-Casas et al. 2005; Delgado et al. 2005). The insular cortex, especially its larger anterior portion, is also heavily implicated in such behaviors (Rilling et al. 2008; Takagishi et al. 2009; Güroğlu et al. 2010). The anterior insula plays a large role in the regulation of emotions, interoceptive experience (our awareness of our own bodily conditions and states), and homeostasis of several subsystems vital to human life (e.g., maintenance of blood pressure within tolerable limits). It has significant neural connections to other parts of the limbic system, including the thalamus, hippocampus, and amygdala, all of which also engage with our emotions. (The amygdala further interacts in complex ways with the reward system, notably the nucleus accumbens; see Stuber et al. 2011.) We will see similar connections in regard to punishment dynamics. The anterior cingulate cortex also plays an important role in this neural network governing our assessments of fairness in social exchanges (Güroğlu et al. 2010). These brain regions also project to the frontal lobes, notably areas of the orbitofrontal cortex and areas of the prefrontal cortex. Reward assessments implicate various neuropeptides and thus significant elements of brain biochemistry as well, notably dopamine, serotonin, oxytocin, and vasopressin, both of the latter two promoting affiliative behavior (Crockett 2009; Crockett et al. 2009; De Dreu et al. 2010). Judgments of fairness and unfairness thus engage all three of the major systems of the human brain: brain stem (reward system), limbic system, and frontal and lateral cortices. The entire system bears a marked bias in favor of fairness, for which reciprocity is a necessary (but not sufficient) condition.

The other side of our sensitivity to fairness and equity in social exchange is our capacity to detect and respond differentially to "cheaters," those persons we perceive to be likely to defect from fair reciprocity. Cosmides and colleagues posited "an evolved neurocognitive system that is functionally specialized for reasoning about social exchange, with a subroutine for detecting cheaters" (2005: 505; Cosmides et al. 2010). Given what we know about human sensitivity to fairness considerations in social exchanges, it seems a reasonable posit and is now widely accepted (Yamagishi et al. 2003; Farrelly and Turnbull 2008; Moll and Schulkin 2009). Automatic detection of cheaters is found in every human culture so far investigated, ranging from hunter-horticulturalists to urbanized market economies (Sugiyama et al. 2002). Such judgments, though made unconsciously, very quickly and without evident stepwise reasoning, are 65 percent to 80 percent accurate in distinguishing defectors from cooperators in situations of social exchange. We are, apparently, canny judges of the likelihood of reciprocity (and associated fairness or unfairness in exchange) among our social interlocutors.[5] It is not surprising, then, that conspicuous systemic failure of reciprocity is an indicator of severe psychopathology.

Given our hypothesis, and the evidence surveyed above, we might further expect that failures of reciprocity, of cooperative behavior, and of empathy would reflect relevant neural dysfunctions and that such failures would be regarded as maladaptive formations of the personality. And this is, indeed, just what we find in antisocial personality disorders (including psychopathy), borderline personality disorder, narcissistic personality disorder, autistic spectrum disorders, and schizophrenia (Decety and Moriguchi 2007). Deficits of empathic functioning are especially characteristic of psychopaths and cases of acquired sociopathy (Blair et al. 2005; Patrick 2006). Borderline personalities regularly show "a profound incapacity to maintain cooperation" in an integrated trust game, and generally lack a well-functioning capacity to repair ruptures of cooperation (Austin et al. 2007; Silbersweig et al. 2007; King-Casas et al. 2008). Autistic persons (Lombardo et al. 2007; Baron-Cohen 2009; Hobson et al. 2009) and schizophrenics (Montag et al. 2007; Bora et al. 2008) also show disruptions of empathic functioning and its associated neural network, together with broad failures of reciprocity and cooperation in social encounters. What the study of these pathological conditions tends to show is the breakdown of the matrix of empathic and cooperative behaviors, along with deficits in the functioning of the associated neurobiological systems (including associated neurochemistry).

For all these reasons, I think that reciprocity (promoting fairness and equity) in social exchanges is a fundamental human expectation, one that is integral to our biological development as a species. It is not surprising, on this account, that reciprocity and equity should appear as fundamental features of notions of human dignity. On this view, dignity may serve as a shorthand expression for biologically based reciprocity in human social exchanges. Something similar applies to the association of dignity with punishment, to which I turn next.

PUNISHMENT AND COOPERATION

Actions that violate dignity may be subject to social and legal sanctions themselves specified in legal protocols or practices. There is, then, a deep connection to be drawn between equality or equity and punishment. And, as I hope to show here, there is also a deep connection between punishment and cooperation among humans and other animals. Understanding this may help to explain the connections contemporary human cultures draw between rights claims and punishment, especially when both are founded on some conception of human dignity.

The role of punishment in promoting cooperative behavior has to do with social norms and our characteristic response to violations of those norms. It is not uncommon to further distinguish norms of cooperation (including moral norms) and norms of conformity (Fiddick 2008; Tomasello 2009: 34). Enforcement of norms often takes the form of conditional prescriptive rules: If you do X, then you must also do Y. Developmental studies show that human children (in a variety of cultural frames) acquire the ability to understand and appropriately apply such rules as early as age three years (Rakoczy, Brosche, et al. 2009; Rakoczy, Warneken, et al. 2009). Moreover, children spontaneously learn to understand and apply such rules specifically in contexts involving social exchange and social agreements (Wyman et al. 2009).

Punishment as norm enforcement is not unknown in the nonhuman animal world (Clutton-Brock 1995, 2009). However, humans practice both second-party enforcement (punishment of you by me for what you did to me) and third-party enforcement (punishment of you by me for what you did to that other person) more widely and more vigorously than any other animal known to us (Haidt 2001). Moreover, as we have already seen, sustained difficulty internalizing norms or following norms is often a mark of severe psychopathology. In his study of hunter-gatherer cultures and nomadic foragers, Boehm discovered a wide range of types of such punishments (Boehm 1999: 64–89; see also Fehr 2004; Panchanathan and Boyd 2004). He argues that such punishments among foragers are mainly motivated by egalitarian concerns and that the most severe of these punishments are almost always aimed at despotic usurpers of power within the social group. But there is a further biological function of punishment, and it arises especially in connection with so-called altruistic punishment.

Altruistic punishments are those that are costly for the punisher to administer and which may not reap any positive return to the punisher, whether in the short term or the long. The ultimate purpose of such punishment is to promote cooperative behavior, especially over large populations of non-kin (e.g., Rockenbach and Milinski 2006; Egas and Riedl 2008; Boyd et al. 2010). It appears that both egalitarian and retributive motives play important parts in our punishing activities but that retribution is the dominant and more primary of these two motives (Carlsmith 2006; Carlsmith and Darley 2008). Moreover, violations of fairness or equity tend to mobilize the retributive emotions of righteous anger and indignation as well as associated (but nonpunitive) emotions such as disgust or contempt (Brebels et al. 2008; Dubreuil 2010). The game-theoretic and other empirical investigations of Fehr and his colleagues seem to me to decisively support the same order of priority: retributive motives are more fundamental than equity or egalitarian motives (Falk et al. 2005; Fehr and Gächter 2002, 2005).[6] Moreover, retribution as an action is driven by retributive emotions, which tend to evaluate their targets much faster and more automatically (though still reliably) than conscious reasoning processes can.

Culture has a strong influence on reciprocity, both positive and negative (Gächter and Hermann 2009). Following Aristotle, we may argue that durable and well-functioning alliances are regularly based on common values. Cultures that promote the equal dignity of their members generate such values. We may expect, then, that durable and well-functioning alliances will be especially common in such cultures. This claim is one that could be investigated empirically in selected cultures. My prediction is that cultures that score high on measures of dignifying their members will also score high on such affiliative measures. Groups that confer dignity on all their members and are assiduous in protecting and honoring the dignity of their members are likely to have a strong internal egalitarian ethos and to be particularly attractive to newcomers who are prepared to respond in kind. Genuine like-mindedness tends thereby to reduplicate itself. We may speak of such like-mindedness (together with its behavioral expressions) as a kind of ethical or axiological reciprocity. It is not surprising, then, that coordinated and like-minded reciprocity should promote and secure cooperation. We might further predict that societies (on whatever scale) that promote dignity will also be more cooperative than those that do not (see Margalit 1998; and, for relevant confirming evidence, Wilkinson and Pickett 2009).

We have good reasons to think that punishment, and especially altruistic punishment, is a natural concomitant of reciprocity and tends to promote cooperative behavior among groups of humans. Moreover, I hope to have showed that some concept of human dignity can fit into this part of the biological framework very well. Indeed, one thing illuminates the other. Practices of punishment and intentions to punish, of course, do not get off the ground in any intentional group without first some signaling of those intentions to the other members. Similarly, coordination of punishment across many members of such groups (which makes punishment promote cooperation more effectively than it would otherwise do) also requires signaling of intentions between group members (Boyd et al. 2010). Moreover, such signals have to be honest ones if they are to be effective. This introduces the third dimension of our biological matrix for dignity: dignity as a communicative signal.

DIGNITY AS A COMMUNICATIVE SIGNAL

We commonly associate dignity with a certain kind of bodily presentation of the self (Schroeder's "dignity of comportment"). Usually we have in mind a combination of elements that include a certain kind of facial expression (one that is calm, confident, even determined or grave), a very erect posture, a slow and stately gait unaccompanied by large movements of the arms or hands, and a certain prosody of the voice (including frequency, pitch, intonation, rhythm, and tempo, the latter generally being slow). This combination constitutes a multimodal signal that has a particular communicative function that can be vital to success in social exchanges and that has been subject to selection pressures throughout our long evolutionary past. This function has mainly to do with disarming (or at least inhibiting) the operation of a very fast perceptual system in humans that automatically categorizes traits of persons on the basis of how they appear. That system is biased toward aggressive responses, and it is important to success in many social encounters that its operation be inhibited. Before visiting some of the details of this hypothesis and its empirical support, a somewhat wider context should be set, for detecting dignity signals depends upon our ability to detect biological motion.

Humans are remarkably adept at detecting biological motion (as opposed to mechanical motion) in their near environment. Such detection depends on a distinctive neural substratum, both in humans and in monkeys (Peelen et al. 2006; Blake and Shiffrar 2007). Under favorable circumstances, the system is about 90 percent accurate, and it is remarkably robust. That is, it performs well when the information available to it is severely distorted or degraded. For example, if we outline the shape of the human body by attaching small LED lights to shoulders, hips, knees, and ankles and view their motion in the dark, humans are adept at recognizing walking, running, and dancing. We can reliably detect gender, direction of movement, emotional expressions communicated by movement, and intentions from such point-light displays and also under further conditions of informational degradation (Casile and Giese 2005; Chang and Troje 2008). The system is also very fast, completing its discrimination efforts in around 200 milliseconds (Hunt and Halper 2008; Reid et al. 2008). The system is normally functioning as early as three months of age (Booth et al. 2002; Pinto 2006). Finally, it appears that the single most important datum for detecting biological motion is the contralateral motion in the horizontal plane of the hands and feet as they pass one another and the vertical midline of the body (Giese et al. 2008). This may explain why gait is so important for reading the affective expressiveness of the body as a whole when it is in motion.

Scientists have long studied our ability to read the affective expressions of the human face, but only recently have they begun to look at the emotional expressiveness of the body as a whole. But, of course, whole body motion is detectable at a much greater distance than is the face alone. Indeed, many social encounters will begin well before the face can clearly be seen, and in some, of course, the face may be obscured. Much is now known about "emotional body language" (EBL) its neural underpinnings, and its biological significance (De Gelder 2009; De Gelder et al. 2010; Sinke et al. 2011). The neurobiological underpinnings of this system involve prominent roles for limbic structures and the prefrontal cortex (De Gelder 2006; Grézes et al. 2007; Urgesi et al. 2007), the same combination of cortical and subcortical regions we have mentioned before in connection with social cognition and affective regulation. Similar body-selective cortical areas are found in our fellow nonhuman primates (Pinsk et al. 2005; Kriegeskorte et al. 2008). Moreover, sensitivity to the configuration of bodily parts in a perception of the body as a whole appears in human children as early as three months, along with the general capacity to detect biological motion, and reaches adult levels of categorization accuracy for some emotions by age five years (Vieillard and Guidetti 2009). Like other functionalities that depend on the same or closely related neural networks, this one is also very fast. Processing of the affective qualities of whole-body gestures or motion can begin in as little as 115 milliseconds post-stimulus and may be completed within 260 milliseconds post stimulus (Meeren et al. 2005; Pourtois et al. 2007; Kret and De Gelder 2010).[7] Categorization of affective bodily expression can occur with exposure to a stimulus as brief as 26 milliseconds and is often completed by 400 milliseconds with up to 96 percent accuracy (Joubert et al. 2007). These time spans are short enough to indicate that processing of emotional body language is automatic, can take place without our conscious awareness of it, and does not require stepwise linguistically mediated reasoning. The system operates reliably even when information from the face is missing or degraded, and when facial expressions are available, the system is biased toward the emotional expression of the whole body. Thus, EBL trumps facial affective expressiveness. It further appears that the EBL-detection system is especially sensitive to large motions by the arms (including the hands) and legs, generally interpreting these movements as expressing aggressive intent (De Gelder 2009). It seems likely, then, that gait could have an important role to play in inhibiting or enhancing aggressive responses in social encounters, especially between strangers. The EBL-detection system thus appears to be hard-wired to the human neonate brain, to appear very early in normal psychosocial development, to be very fast and automatic in its operations, and to be entirely independent of our linguistic abilities.

The emotional expressiveness of the body is further enhanced when the prosody of the voice matches what the rest of the body is communicating. In general, multimodal signals are more effective than monomodal signals (Rowe 1999; Hebets and Papaj 2005). Cross-modal effects of voice prosody (especially its emotional content) and facial expressions are well known (Ethofer et al. 2006; Kreifelts et al. 2007). But it is also the case that body motion and voice prosody, taken together, trump information available from the face and voice together (Van den Stock et al. 2007, 2008). Emotional prosody of the human voice involves a combination of cues, some universal to human cultures and others more specific to cultural location (Elfenbein and Ambady 2002, 2003; Balaguer-Ballester et al. 2009). Here, as in so many cases of emotive expression and its detection, the neurobiological substratum is supplied especially by the amygdala, the insula, striatum, and frontal cortices (van Rijn et al. 2005; Wildgruber et al. 2006).[8] The same set of neural networks appears to be especially sensitive to angry prosody (Grandjean et al. 2005; Quadflieg et al. 2008). Prosodic detection and categorization also gives evidence of being very fast (within 110 milliseconds post stimulus), automatic in its operation, and ontogenetically early, starting around three months of age. It is plausible also that other mechanisms might exist in humans to exploit this system for the purpose of supporting cooperative behavior. My view is that dignity signals are just such a mechanism. They are especially important given the special role of the amygdala in forming first impressions.

First impressions are those very quick and intuitive judgments we tend to make about traits of persons on the basis of how they look to us (see Ambady and Skowronski 2008 for a thorough review). We are especially prone to this in social situations that we perceive to be risky. We may draw conclusions about the trustworthiness (or untrustworthiness) of another person, his or her competence, dominance, or aggressiveness from what little we can see of his or face (Montepare and Dobish 2003; Oosterhof and Todorov 2009). Such quick trait inferences from appearances are fallible but reliable, though markedly less so when confronted by a deceiver. The speed of the system is quite impressive: consistent first impressions can be formed based on whatever information is available in the first 39 milliseconds post stimulus (Bar et al. 2006). The amygdala plays a central role in these trait inferences (Todorov and Engell 2008; Sinke et al. 2010) and is probably responsible for the speed of the system (Willis and Todorov 2006). Investigating the neural activity of the right amygdala during trustworthiness inferences, Said et al. (2008) found an interesting bipolar result: the activity peaked for faces perceived to be maximally untrustworthy (from among those on offer) and again for those perceived to be maximally trustworthy. The lowest levels of amygdalar activity during first impression formation was for neutral faces. In my view, this is a singularly important finding. For it suggests that this very fast, automatic, and effortless process of forming first impressions operates at minimal levels of neural activity only when processing neutral faces. There is a role here for the calm, confident, and grave facial expressions that we commonly associate with dignity. They may have some power to inhibit first impression formation, thereby causing the system to function at minimal levels of amygdalar activity (and associated levels of arousal).

Sinke et al. (2010) found similar patterns of activity in the amygdala for whole bodies, and we know that whole bodies (especially when the face is obscured) are also a common basis for first impression formation. And, as we have already seen, humans are especially sensitive to perceived threats, whether conveyed facially or bodily. So here is the situation when two human strangers first meet, especially when at such a distance from one another that facial details are not readily perceived, but bodily postures, gait, and prosody of the voice can be accurately perceived. Each individual is likely to form very quickly a series of inferences about the intentions, emotional condition, and personality traits of the other. Each individual's first impression system is operating very fast and is largely outside of his or her conscious control. Moreover, the system is biased in the direction of perceiving threats. This makes sense from an adaptive point of view, because it would be in our interest (especially during the early period of our evolution as a species) to form too many false positives (taking situations to be threatening when they are not) than to form too few (which could very quickly be fatal). This is the context within which I propose to think about dignity as a communicative signal. Before considering the general meaning of such a signal, something more should be said about gait and posture.

Scientific investigation of gait is still in its infancy, not least with respect to identifying emotions or other mental states from gait information. However, we know that people can identify others known to them from their gait alone at levels of accuracy well above those expected by mere chance

(Cutting et al. 1977). We also know that it is possible to read emotions off of gait information, again, well above merely chance levels. Such reading appears to depend upon a variety of kinds of information, from configural (the arrangement of body parts in a whole) to kinematic (especially relative to motion and angles of knees, ankles, and hips) as well as temporal and spatial information (Barclay et al. 1978; Atkinson et al. 2007; Bouchrika and Nixon 2008). More important for our purpose, exaggeration of body movements improves recognition accuracy of emotions expressed thereby (Atkinson et al. 2004). When information from gait is combined with facial biometrics, identification gets much faster and more reliable (Liu and Sarkar 2007). Neural processing of gait information is likely to be similar to that for other forms of biological motion in the region of 100 to 180 milliseconds post stimulus (Pavlova et al. 2004, 2006). So what does all this mean for dignity signals? Common lore associates slow and stately gait with dignity or pride. Montepare et al. (1987) attempted with some success to detect pride from gait but found that such detection was substantially less accurate than for sadness or anger. They also found that length of stride and degree of arm swing played roles in such affective gait recognition. This suggests that gait information alone is a signal in need of amplification. Such amplification may well be the function of the other elements associated with dignity: voice prosody, erect posture, limited arm swing, open and relaxed hands, calm and confident facial expression (Atkinson et al. 2004 tends to confirm amplification effects). If so, then dignity of comportment may prove to be a reliable and honest signal by design (Taylor et al. 2000).[9]

The matter of erect posture has been held to be a mark of dignity or justifiable pride since ancient times. In his much later book on the expression of emotions in the faces of men, Charles Darwin observes:

> A proud man exhibits his sense of superiority over others by holding his head and body erect. He is haughty (*haut*), or high, and makes himself appear as large as possible; so that metaphorically he is said to be swollen or puffed up with pride. (1890: 263f)

This suggests that erect posture could well be a bodily marker for something like dignity. Erect posture also means that the large muscles of the chest are not ready for a deep intake of air, preparatory to fight or flight, and the shoulders will be well back, also indicating absence of aggressive intent. An adult human person who exhibits dignity as a combination of facial expression, gait, posture, and affective voice prosody is plausibly taken to be sending a signal to potential social partners that may have some power to inhibit their fast limbic system from forming first impressions.[10]

Such "dignity of comportment" would be especially valuable in encounters between strangers or in other circumstances entailing risk or stress. For in those circumstances we are most likely to adopt a hostile stance toward the other, unless such a signal is received. What, then, does the signal communicate? It seems to me that the bodily dignified individual is communicating two things: (a) that he or she is a cooperator (or is at least prepared to cooperate) and is not present with hostile or threatening intent; and (b) that he or she expects the other person to be the same and is ready to receive cooperative overtures from the other. It may suffice to make encounters with strangers less dangerous for dignity signals merely to retard the action of the amygdalar system, slowing it down. Indeed, part of the stately gait, the sonorous and drawn-out prosody of the voice, and the grave mien may be to entrain slower responses in this very system. Like other multicomponent and multimodal signals, it acts as a functional unit (Hebets and Papaj 2005) and is the product of the operation of selection forces. In the early environment of adaptation for our species, after all, encounters between strangers would have been unusual, since most humans lived in relatively small bands (fewer than 150) of direct biological kin. Whether meeting strange conspecifics or (more rarely) encountering members of other hominid species, humans could find dignity signals useful to prevent unnecessary hostilities that would otherwise have ensued. Dignity as a communicative signal would have had considerable adaptive value in the early environment of human adaptation. It remains to suggest some ways to operationalize my notion of a biological approach to dignity.

EMPIRICAL FOUNDATIONS FOR DIGNITY

It may be possible for us to develop a science of dignity using technology currently available. For example, there is a set of pictures that is widely used to study affective responses to standard and varied human facial expressions or whole scenes (the International Affective Picture System, or IAPS), which could be adapted to represent an array of whole-body affective expressions, and this could give a useful tool for identifying responses to such emotional body language. So far, no attempt has been made to do so, but there does not seem to be any reason in principle why this could not be done (for the IAPS, see Mikels et al. 2005; and for its auditory parallel, the International Affective Digitized System, see Stevenson and James 2008). Should such an adaptation take place, it would seem to be possible to develop a subset of the adapted IAPS to deal with conventional or common expressions of dignity in whole-body language.

That latter move would of course entail our developing an appropriate profile of what constitutes dignified behavior. This would include a profile of a dignified gait. Here the algorithms being developed in the artificial recognition of individuals by their gaits (in computerized security systems, on which see Zhang et al. 2007 and Mu and Tao 2010) might be the foundation for a further and more specific adaptation to produce empirically testable recognition criteria for a dignified walking style (including length of stride, speed of walking, associated arm swing, and the characteristic open but relaxed hand configuration). Similarly, there seems to be nothing to prevent us from observing paradigmatically dignified erect posture, together with associated disposition of the head relative to both the trunk and the direction of movement, to see if we can discover commonalities (or more likely a range of commonalities) that might belong to such a profile. Much the same goes for what we might classify as a dignified facial expression, or range of such expressions. In a recent study of communication of emotion through qualities of the voice, Gobl and Chasaide (2003) picked out emotive qualities, including confident, unafraid, friendly, interested, and neutral, all of which might have a bearing on analogous facial expressions relevant to dignity of comportment (see similar results in Banse and Scherer 1996).

I have argued that prosody of the voice is plausibly taken to be an amplifier of larger whole-body emotive language. Prosodic processing is dissociable from processing of the lexical qualities of the voice (i.e., its semantic contents) and exerts a stronger influence on listeners' judgments than does lexical content of speech (Bänziger and Scherer 2005; Wildgruber et al. 2006). Scherer et al. (2001) showed that inferences of emotional state from vocal cues (here "sentences" made up out of randomly selected meaningless syllables), across nine culturally different countries, reached average accuracy rates of 66 percent. One affective quality that was particularly accurately perceived was the neutral condition expressed by German actors to German-speaking subjects (88 percent accurate). They conclude that such results confirm phylogenetic continuity of emotional sounds. Scherer (2003) found that voices are less accurate representations of affective states than are faces, but that recognition of affective qualities conveyed by voices was 55 percent to 65 percent accurate (faces were up to 75 percent accurate), a rate five to six times greater than that expected on the basis of mere chance. Recognition of affective condition from voice prosody is sensitive to the time course of utterance and to any social interactions that may be part of the communicative scene, both of which are also relevant to study of dignity (Cowie and Cornelius 2003). Scherer's model took into account expression of emotion (encoding), transmission of emotion, and recognition of emotion (decoding). Such a model responds well to what studies of animal signals generally have emphasized: the importance to the meaning of signals of the psychology of those who receive them (see Grandjean et al. 2006). We know also that discrimination of pitch variability is an important aspect of prosody decoding (Pihan 2006; Pihan et al. 2008). Technical capacities for such work could be transferred to the study of dignity if we could isolate the pitch and intonation contours of that sonorous tone of voice that is commonly taken to be an aspect of dignity of comportment. There seems to be no reason in principle why this could not be done. There thus seems to me little standing in our way to develop measurable and empirically testable profile of dignity as a communicative signal with regard to facial expression, posture, gait, and prosody of the voice.

Finally, there does not seem to be any reason why such investigations should not extend to the study of relevant neurobiological and other physiological infrastructures. There is already evidence that affective prosody of the voice, for example, is lateralized to the right hemisphere, with some assistance from left hemispheric structures for such matters as the segmentation of sounds and the temporal course of utterances (Pihan 2006; Wildgruber et al. 2006; Pihan et al. 2008; Wiethoff et al. 2008; Wildgruber et al. 2009). EEG studies, coordinated with various brain imaging technologies, and studies of important affective neurotransmitters could all combine to provide powerful information about regions of interest in the brain engaged in recognition of dignity signals (for a very suggestive EEG study in this connection, see Bostanov and Kotchoubey 2004).[11] Indeed, it is not difficult to predict that corticolimbic networks such as we have already discussed will figure prominently in that account. That, in turn, could illuminate problems recognizing dignity in various forms of psychopathology (especially autism, personality disorders, psychopathy), notably in those pathologies that show severe to moderate deficits in empathy and associated failures of reciprocity. Study of the effects of brain lesions, especially to the right temporal gyrus, caused by strokes, accidents, or disease, might also be drawn into this orbit and add confirmation to the results of brain imaging and neurochemical investigations.

Should a workable and empirical profile of dignity behavior become available, it should also be possible to investigate physiological correlatives of the detection of dignity signals, as hypothesized in this chapter. Here the work of Stephen Porges and his colleagues is of particular importance. They have shown a deep connection between physiological states and qualities of human social behavior, especially with regard to prosocial communicative behaviors. The myelinated vagus nerve (cranial nerve X) functions as a tightly regulated vagal brake on the heart (systematically altering the rate of change of heart rates) in social contexts that are perceived to be safe (Movius and Allen 2005; Porges 2011: chap. 7). When applied, the vagal brake slows heart rate to promote a calm behavioral state, which in turn fosters social interaction. Vagal activity can be monitored closely (via ambulatory electrocardiograph) by measuring the amplitude of the naturally occurring spontaneous fluctuation of heart rate occurring at respiratory frequencies (respiratory sinus arrhythmia, or RSA). This has to do with the relative lengthening or shortening of the interval between heart beats. Large decreases in RSA are associated with greater sociability, while increasing RSA promotes alternative patterns of behavior, either sympathetic nervous system mediated fight-flight behaviors or parasympathetic nervous system mediated immobilization behaviors (freezing or feigning death: see Porges 2011: chaps. 1–3; and cf. Field and Diego 2008). Both practical and statistical methods for measuring RSA with great accuracy already exist (Denver et al. 2007; Heilman et al. 2007; Porges 2007). RSA activity is also correlated with production of cortisol, a stress hormone (Doussard-Roosevelt et al. 2003), and measurement of cortisol is a well-established technique in the study of human stress responses.[12]

The same cranial nerve complex that regulates heart rates also regulates various other head muscles, including those of the middle ear, mouth, larynx, and pharynx. These muscles control looking, listening, vocalizing, and facial gesturing. Of particular interest here is the middle ear. The stapedius muscle acts to stiffen the chain of small bones in the middle ear when excited. Such activity is essential for dampening out the low-frequency sounds that flood the human environment and make it harder for us to distinguish human voices (and their associated prosody) from surrounding noise (Porges 2011: chap. 12). It is possible to monitor this activity by measuring the degree of responsiveness of the tympanic membrane. The combination of RSA, cortisol production, and tympanic membrane compliance could provide a powerful tool for investigating how agents respond to dignity signals as those have been conceived here. If my view is even approximately correct, then receipt of dignity signals should result in decreased RSA, decreased cortisol levels, and greater tympanic membrane compliance. Experimental determinations of these measures could, in turn, suggest modifications of our behavioral profile for dignity.[13]

It thus appears to me to be possible for us to develop a rigorous, empirical, and scientific treatment of the notion of human dignity, particularly in its role as a communicative signal. Both current technology and existing research programs are available to create a template for such an investigation. Moreover, there seem to be several advantages to conceiving human dignity along such lines.

CONCLUSION: OUT OF THE SHADOWS

On the view of it offered here, human dignity is primarily a function of the human social networks in which it is practiced and conferred. It is not some mysterious property of individual human beings but rather a shorthand term for a complex dynamical reciprocity that subserves the primary biological aims of promoting social cohesion and cooperation. Such cohesion and cooperation include practices of altruistic punishment (whether driven by retributive or egalitarian motives). As I have shown, punishment is also an essential functional element of human dignity, as we may now conclude, on biological grounds. Dignity also functions as a multimodal signal to conspecifics engaged in such social networks, a signal that also has the power to promote cohesion and cooperation in large part by disarming or inhibiting the function of a fast and automatic neural system for threat detection. As a function of our sociobiological makeup, dignity is subject to evolutionary pressures and has arisen out of our long evolutionary history as a species. Since the attempt to ground and justify dignity ascriptions in traditionally conceived rationality seems to fail tests of empirical adequacy for human moral agency generally, and since the attempt to ground and justify dignity in classical theism is unusable from any naturalistic point of view, such a scientific approach to dignity comes into its own.

One advantage of this approach is that it brings dignity out of the shadows of obscure, confused, and poorly evidenced folkways about human uniqueness, human moral agency, and human rationality, and gives it a firm empirical basis. Of course, it remains to be seen whether the science of dignity that I have pointed to in this essay can be further developed as a viable research program. It also remains to be seen whether this approach can be extended to include dignity-promoting social arrangements (including state constitutions) and dignity-promoting institutions. We have preliminary indications that both of these developments are in the offing. But there is a great deal of work yet to be done. Nevertheless, such an empirical approach has the distinct advantage of making our understanding of dignity revisable in the light of further evidence available to any competent observer. There is also nothing to prevent theists and Kantians from using this approach to dignity, even where their metaphysics are ineluctably at odds with that of naturalistic defenders of dignity.

The biological approach to dignity also rescues that concept from the skepticism with which this essay began. Far from being a "useless concept," on this view, dignity is almost irreplaceable in a wider understanding of the social and biological roots of human moral agency, human moral systems, and the social structures and processes that arise from those roots. Indeed, on this view of it, we have ample reasons to press the skeptics in another direction. We can also by these means supply a well-developed and empirically defensible understanding of those biological roots as a justification for the widespread invocation of human dignity as a basis for state constitutions, for international laws (e.g., forbidding torture), and schedules of human rights as well as for other kinds of social institutions. Haidt and his colleagues have argued compellingly that the concerns with fairness/justice and rights that are central to much of human morality have a distinctively different and independent basis from that of the concerns with social cohesion, loyalty, and other aspects of "coalitional psychology" that are central to other elements of human moral systems.[14] The notion of dignity may function as a bridging concept that links these two broad streams of human morality; for dignity can be a basis for robust concepts of individual human rights (see the argument in Gewirth 1996: 6–20) and also for social cohesion, as suggested above. It is difficult to imagine a better testimony to the success of a theoretical point of view.

From the outset of this discussion, I have also tried to suggest that cooperation is in our interests. It is, indeed, one of the hallmarks of our species, for no other known animal species cooperates over as long a temporal duration or over as large a spatial distribution as we do. No long-term or large-scale human project succeeds without very widespread cooperative behavior among conspecifics. It is this feature of human action that gave rise to the whole issue of what might be the biological underpinnings of cooperation in the first place. And if we pose the very sensible question "why should I cooperate?" the only good answer is "because it is in your interest." This need not be a matter of egoism (ethical or otherwise) but merely of genuine self-interest. And the promotion of self-interest is compatible with the promotion of the interests of others—that is, with altruism. It is only a confusion to assert or to infer that pursuit of self-interest is *eo ipso* selfish. It is, of course, true that cooperation may prove in some cases not to be in our interest (this problem lies at the heart of so-called altruistic punishment, for example). Indeed, engagement with conspecifics in large-scale or long-term cooperative enterprises is almost bound to result in many individuals not satisfying (much less maximizing) their self-interest. It remains the case that belonging to a species that is adept at cooperative behavior is in our interest. Similarly, it is in the interest of the individual bees in a hive that they possess stingers, even if it means on a given occasion that an individual bee may lose its life by using its stinger to defend the hive against attacking predators. Even free riders in a broadly cooperative social context promote their own interests by belonging to such a species. (Their problem is to ensure that there are not so many free riders as to cause cooperation to collapse altogether.) That dignity might promote cooperation, then, implies a powerful appeal to self-interest. At the same time, it appeals to the biological roots of social and emotional attachment and thereby the other track in human morality of social cohesion concerns.[15] It thus has some power to conciliate self-interest and altruistic social cohesion norms and practices.

In closing, I want to underscore the long-term value of cooperation. According to our best understanding, for the longest part of our evolutionary history (perhaps reaching as far back as four and half million years) members of our species lived in small bands of biologically related hunter-gatherers. It is in that context that the demands for cooperation and altruism, for social equity, and fairness in exchange gained their deepest foothold in our psychological phylogeny. Eventually, around 15,000 years ago, we settled down and began to practice agriculture and to build cities. At that point, we made a relatively successful transition to living in much larger units, up to three orders of magnitude higher than we had previously known.[16] This adaptive change depended on the invention of cities as a distinctively human form of social life. At around the time of the industrial revolution, a couple of centuries ago, we made a further transition to living in groups on the scale of tens to hundreds of millions, in the form of what eventually became the modern nation-state. Here, too, we proved to be adept at maintaining a set of broadly cooperative social structures, though the other side of our nature as the most aggressive species on the planet also took its toll in large-scale intraspecies violence. I take it that, at the present time, as we have finally become a genuinely global species, we face a significant challenge to our cooperative nature: can we adapt successfully to forms of social life (including institutions) that engage billions of our conspecifics (and at least some other nonhuman animals)? Our success in making transitions of this kind in the past suggests that we can rise to meet this challenge also. But it has to be said that the book is still open, and that we probably do not yet possess the institutions and other social processes and structures that will be needed to make the transition successfully. What has become painfully clear is that the future of our species on this planet probably depends upon our becoming a genuinely global cooperative species. It may be that a deeper understanding of the biology of dignity may make a small contribution to the likelihood of our success in making this vital transition.

NOTES

1. This complex history is explored by David Calhoun in his contributions to this volume: "Prospects for Human Dignity after Darwin" and "Human Exceptionalism and the *Imago Dei*: The Tradition of Human Dignity." Other historical studies of note are Nordenfelt 2004, and Killmister 2010.
2. What follows is a shortened version of McClelland 2011. The longer article contains substantially greater detail and further documentation. My thanks to Ray Russ and the Institute for Mind and Behavior for permission to publish in this chapter a condensed version of the following: Richard McClelland (2011), 'A Naturalistic View of Human Dignity,' *Journal of Mind and Behavior*, 32, 5–48, © Institute of Mind and Behavior.
3. Frans de Waal is perhaps the most vigorous champion of this view. See also Langford et al. (2006), Pierce (2008), and Johnson and Pierce (2008) for apparent empathy in rodents; Bates et al. (2008) for empathy in elephants. For a more skeptical position, see Cheny and Seyfarth (2007). There is a very balanced discussion in Tomasello (2009).
4. Some neuroscientists have argued that the mirror neuron system is involved in empathy (Gallese 2001; Preston and de Waal 2002; Decety and Jackson 2004; Gazzola et al. 2006; Singer and Lamm 2009; and Zaki et al. 2009). The issue can remain moot here.
5. Recent work by Bell and his colleagues indicates that we have enhanced memory functions with respect to persons perceived to be cheaters in social exchange (see Bell and Buchner 2009, 2010, 2011; Bell et al. 2010; Buchner et al. 2009).
6. Evidence that punishment, when aimed at cooperators, decreases levels of cooperation (Hermann et al. 2008; Rand et al. 2009; Wu et al. 2009) is based on faulty assumptions and loses its significance in the presence of better models (Boyd et al. 2010).
7. Our attention is drawn especially quickly to scenes or objects that arouse fear or represent a threat (including whole body postures: see Bannerman et al. 2008; Pichon et al. 2008). Perceptions of bodies appear also to be further specialized in so far as one neural circuit is specialized for body parts (Downing et al. 2001; Astafiev et al. 2004) while another configures those parts into wholes (Pourtois et al. 2007; Taylor et al. 2007). Kriegeskorte et al. (2008) present evidence for strongly analogous systems in monkeys.
8. Persons with autism commonly show difficulties in perceiving and categorizing prosodic emotional expressions, and such deficits are associated with dysfunction of just these circuits, especially in the right hemisphere (see Kujala et al. 2005; 2007; Järvinen-Pasley et al. 2008).

9. The reliability of animal signals is highly relevant to the study of dignity as a communicative signal among humans but cannot be further explored here (see Espmark et al. 2000; and Searcy and Nowicki 2005 for reviews of this complex issue). Both consistency (Wolf et al. 2011) and commitment (Rusbult and Agnew 2010) in social behavior are also relevant to establishing reliability of signals.

10. Body posture is also known to have affective regulatory effects (see Riskind and Gotay 1982; Duclos et al. 1989; Stepper and Strack 1993; Schnall and Laird 2003), which lends further support to the idea of posture as a communicative signal.

11. Of the several neurotransmitters likely to be involved here, dopamine is of special importance in a system that determines emotional salience of events, objects, and so on, and thus their motivational force (see Berridge 2007; Berridge and Aldridge 2008; and Zhang et al. 2009 for reviews). I have reviewed the reward system more generally in McClelland 2010: 108–10.

12. For extensive reviews of the stress system, see Schulkin 1999, 2004. For the evolution of that system and its characteristic patterns of response to stress, see now Giudice et al. 2011.

13. Autistic persons show distinctive differences in RSA from normal controls (Van Hecke et al. 2009; Bal et al. 2010). Austin et al. (2007) show that a group of patients with borderline personality disorder tended to produce physiological states supporting fight-flight behavior (in keeping with the impulsive aggression characteristic of the disorder), while a control group tended to produce a physiological state supporting social engagement. Such natural experiments tend not only to support Porges's polyvagal theory of autonomic functioning but also suggest further ways in which hypotheses about dignity as a signal could be tested.

14. Haidt 2009; Haidt and Joseph 2004; Haidt and Graham 2007, 2009. Cf. also Fiske 1991, 1992. The term "coalitional psychology" is from Kurzban and Leary 2001. The subject of coalitions and their underlying biology is too large to tackle here, but see Tooby et al. 2006 and Cimino and Delton 2010; and for the importance of signaling for coalition formation and development, see Hagen and Bryant 2003, Hess and Hagen 2006, Johnson et al. 2008, Barrett and Rendall 2010, and Price 2011. I have currently in preparation a study of character as a communicative signal in coalitional dynamics.

15. Attachment theory grew out a John Bowlby's assimilation to his psychotherapeutic practice of the ethological writings of Konrad Lorenz, a development that in many respects already anticipates much that appears in this essay. For reviews of attachment theory, see Karen 1998, Cassidy and Shaver 2008, and Schore and Schore 2008.

16. This schema and its associated argument are drawn from Boehm 1993, Van Vugt 2008, and Van Vugt et al. 2008.

Part II

The Politics, Law, and Science of Human Dignity

7 International Policy and a Universal Conception of Human Dignity

Roberto Andorno

This chapter offers a brief account of the policy statements issued by inter-governmental bodies that appeal to human dignity and focuses on those specifically dealing with bioethics. At the same time, it discusses the cultural challenges that arise from the adoption of universal or transcultural understandings of human dignity and responds to the objection that human dignity and human rights standards are mere products of Western culture and are therefore inapplicable to other regions of the world.

HUMAN DIGNITY: A KEY CONCEPT OF INTERNATIONAL HUMAN RIGHTS LAW

Although the notion of human dignity has a very long history in philosophy, it has reemerged with great vigor after the Second World War as an *international legal and political concept* that aims to stress the need for unconditional respect for every human being in the most different areas of social life. It was indeed in response to the horrors of that tragic period of history that the international community felt it necessary to emphasize the idea that every individual has inherent worth and accompanying rights in order to prevent "barbarous acts which have outraged the conscience of mankind" from ever happening again (Preamble of the Universal Declaration of Human Rights 1948, hereafter UDHR). Indeed, simultaneously with the end of the Second World War, the Member States of the newly created United Nations reaffirmed their "faith . . . in the dignity and worth of the human person" (Preamble of the United Nations Charter 1945). Subsequently, the UDHR served as the cornerstone of the new international human rights system, which was grounded on the "recognition of the inherent dignity and of the equal and inalienable rights of all members of the human family" (UDHR, Preamble).

From the very beginning, the Declaration puts forward that "all human beings are born free and equal in dignity and rights" (Article 1). Human dignity is clearly the bedrock of all the rights and freedoms set forth in the Declaration and, in particular, the basis for the prohibition of all forms of discrimination (Article 2); of slavery (Article 4); of torture; and any cruel, inhuman, or degrading treatment or punishment (Article 5). In this respect, the tireless human rights and public health advocate Jonathan Mann did not hesitate to characterize the statement made in Article 1 as a "seismic shift in human consciousness," which is "so profound that, paradoxically, its importance may not be fully realized" (1998: 31).

There is no doubt that the emphasis on human dignity that dominates the ethical and political discourse since 1945 can be to a large extent explained by the horror caused by the revelations that prisoners of concentration camps, including children, were used by Nazi physicians as subjects of brutal experiments. In this regard, the American bioethicist Robert Baker asserts that the UDHR was in part informed by the discoveries of these abuses, which led in 1947 to the development of the Nuremberg Code by the trial that condemned the Nazi physicians. Baker claims that "the details revealed daily at Nuremberg gave content to the rights recognized by Articles 4 through 20 of the Declaration" (Baker 2001: 241–52).

Thus, both modern medical ethics and international human rights law emerged simultaneously from the same tragic events and are conceptually much closer than usually assumed. Similarly, George J. Annas (2005: 160) points out that "World War II was the crucible in which both human rights and bioethics were forged, and they have been related by blood ever since."

Even if the Nuremberg Code does not explicitly include the expression "human dignity," it is clear that this notion lies at the background of its 10 principles, which are presented as *nonnegotiable*. This is very significant as it puts in evidence that the idea of an unconditional human worthiness was in the mind of the judges that formulated the Code, and that, in their view, the rules governing medical research should be based on categorical, and not on merely utilitarian, grounds. In this respect, Jay Katz (1992: 227) writes that:

> The Nuremberg Code is a remarkable document. Never before in the history of human experimentation, and never since, has any code or any regulation of research declared in such relentless and uncompromising a fashion that the psychological integrity of research subjects must be protected *absolutely*. (emphasis added)

As mentioned above, simultaneously with the development of the Nuremberg Code, the international community made an unprecedented political move with the adoption of the Universal Declaration of Human Rights, which would become the cornerstone of the international human rights system. A brief excursus into the drafting history of this founding instrument will be helpful to clarify how its intellectual authors conceived the role of human dignity.

There has been long debate among scholars about the relative importance of John Humphrey and René Cassin—the Canadian and the French representatives, respectively—in the drafting of the Declaration. Even if there is no doubt that Humphrey played a key role in the first draft of the document, it seems well that the incorporation of the concept of human dignity was due not to him but to Cassin, who introduced corrections to Humphrey's text. The inclusion of dignity was initially controversial, even if it had already been incorporated into the Preamble of the UN Charter. Humphrey himself considered that the reference to dignity did not add anything to his draft and that its incorporation as Article 1 of the Declaration was merely rhetoric (McCrudden 2008: 677). For others, however, it was a vital attempt to articulate their understanding of the basis on which human rights could be said to exist. In her detailed account of the history of the UDHR, Mary Ann Glendon (2002: 146) recalls how, when the South African representative questioned the use of the term *dignity*, Eleanor Roosevelt, who was chairing the commission that drafted the Declaration, argued that it was included "in order to emphasize that every human being is worthy of respect . . . it was meant to explain why human beings have rights to begin with."

Consistent with this understanding, the two great support pillars to the Declaration—the International Covenant on Civil and Political Rights (ICCPR) and the International Covenant on Economic, Social and Cultural Rights (ICESCR), both of 1966—solemnly affirm that human rights "*derive* from the inherent dignity of the human person." This statement is extremely important because it puts in evidence that, regardless of academic discussion on the proper role and value of human dignity, international law regards this notion as the *source* of human rights: people are recognized as rights bearers *because* their lives and their flourishing as persons are viewed as having intrinsic worth.

Actually, to talk about persons having rights would not make any sense unless moral value is previously attached to the very core of human personhood (Black 2000: 131). But dignity does not only *precede* human rights as their source; it also *follows* them as it embodies their raison d'être. Indeed, the recognition of basic rights does not have ultimately any other aim than that of securing the conditions of a minimally good life for all people (Nickel 1987). Therefore, it can be said that rights are simultaneously *grounded on dignity* and *aim to promote dignity*.

If human dignity (and not merely contingent agreement) is the foundation of human rights, then it is understandable that compliance with human rights norms is not discretionary but mandatory and therefore cannot be simply ignored or disregarded by any authority. Certainly, the practical efficacy of promoting and protecting human rights is significantly aided by their legal recognition by individual states. But the ultimate validity of human rights is characteristically thought of as not conditional upon such recognition (Nickel 1987), and this is so because they are grounded on the inherent dignity of every human being. In other words, legal norms do not create individuals' rights from nothing; human rights are not the capricious invention of domestic lawmakers or of the international community, which could legitimately revoke them in a change of humor. Rather, both individual states and the international community are morally obliged to recognize that all people have basic rights (i.e., that they have equally valid claims to basic goods), because they derive from the dignity that is inherent in every human being. The explicit use of the verb *to recognize* in the UDHR, which denotes the formal acknowledgment of something that already exists, is very illuminating in this respect.

The notion of human dignity has become so central to the United Nations' conception of human rights that the UN General Assembly provided in 1986, in its guidelines for the development of new human rights instruments, that such instruments should be "of fundamental character and derive from the inherent dignity and worth of the human person" (Resolution 41/120 of 4 December 1986). Since then, all major international human rights texts— such as, for instance, the conventions on the Rights of the Child (1989), the Rights of Migrant Workers (1990), Protection from Enforced Disappearance (2006), and the Rights of Persons with Disabilities (2007)—include references to human dignity.

It is true that human dignity is never clearly defined in international law. Such a thing would be as difficult as trying to define freedom, welfare, solidarity, or any other key social value. In any case, this lack of definition does not entail that dignity is a merely formal or empty concept or a purely rhetorical notion. It is not because it is too poor but because it is too rich that it cannot be encapsulated into a very precise definition. In reality, its core meaning is quite clear and simple and embodies a very basic requirement of *justice* toward people. Such requirement means, in the words of Rawls, that "each person possesses an inviolability founded on justice that even the welfare of society as a whole cannot override" (1973: 3). This latter statement is very revealing of the inescapable need for the idea of intrinsic human worthiness as the precondition of any society seeking to realize justice among its members. Even a contractualist account of equal respect, like the Rawlsian one, which intends to ground society's norms on a hypothetical agreement between individuals (and not on any substantive moral reason) can ultimately not avoid referring to human dignity. The truth is that consensus alone is not enough to justify the moral norms governing a human community, both locally and globally. The basic principles guiding every just society are not right just because its members have agreed on them, but it is the other way around: they have agreed on them because such principles are right. Torture, slavery, or any other practices that seriously disregard human dignity are not wrong because they are prohibited, but they are prohibited because they are wrong.

Why are there certain things that cannot be done to anyone, under any circumstances? Why is this moral requirement previous to any domestic law or international agreement? The answer to these questions is very simple: every human individual is a *person*, not an *object*; everyone is an end in him- or herself, and never a mere means to another's end. The widely shared conviction that human beings deserve to be treated with respect is not merely subjective and arbitrary but is grounded on an indisputable fact: human beings are entities capable, *as a kind*, of understanding, self-understanding, loving, self-determining by judging and choosing, expressing themselves by means of art, and so on. These extraordinary abilities that characterize human beings as a kind and qualitatively distinguish them from all other known living beings (even if those capacities are not currently present in all human individuals, or not in all to the same degree) make of every human being something absolutely unique, precious, and irreplaceable. In other words, those typically human features make of every human individual a *self* (i.e., a person), an entity with a certain absoluteness of being.

It is not difficult to see that, as the above-mentioned characteristics are inherent in the human condition, they preexist to any social consensus or legal regulation. This means that states do not *create* personhood from nothing but simply *recognize* it. Moreover, they are morally *obliged* to grant human beings with such recognition. This is precisely what is meant by the Universal Declaration of Human Rights when it provides that "everyone has the right to recognition everywhere as a person before the law" (Article 6).

It should also be noted that, contrary to what is sometimes maintained in bioethical circles, the idea of intrinsic human dignity does not rely on a speciesist claim—that is, on the merely *biological* fact of belonging to the species *Homo sapiens*. If there were other entities in the universe besides human beings having, as a kind, the above-mentioned capacities, they would also have intrinsic dignity (Sulmasy 2007).

This helps to explain why it is a conceptual mistake to claim that human dignity is a mere slogan that means no more than "respect for autonomy" and therefore could simply be eliminated without any loss of content (Macklin 2003). This criticism overlooks the fact that respect for persons' autonomy is just a *consequence* of human dignity, not dignity itself. If people would not have any inherent value, then there would be no reason to respect their autonomous decisions. In other words, the (usually implicit) ultimate ground for respecting people's autonomy (as well as any other basic human good or interest) is the intrinsic worthiness that we attach to every human being.

HUMAN DIGNITY IN THE INTERGOVERNMENTAL INSTRUMENTS RELATING TO BIOETHICS

An Overarching Principle of Bioethics/Biolaw

As mentioned above, the need to put some limits to medical research involving human subjects played a decisive role in the renewed importance of the idea of human dignity. In this regard, it is very revealing that the only provision of the ICCPR that directly deals with bioethical issues is the one relating to medical research. According to Article 7, "no one shall be subjected without his free consent to medical or scientific experimentation." Twenty years after the Nuremberg Code, this article is still like an echo of that historical trial decision that condemned the Nazi physicians. More importantly, it enshrines for the first time in history the requirement of free consent for medical research in an international *binding* instrument.

Between the end of the 1970s and the end of the 1990s, the recourse to human dignity in relation to medicine went beyond the field of medical research and began to be applied to very different practices, especially those that operate at the edges of life, such as abortion, embryo research, preimplantation genetic diagnosis, futile treatments, assisted suicide, and euthanasia. In this varied context, it is not surprising that the term dignity was sometimes used to support different and even opposed views. Simultaneously, the concept began also to be employed to criticize what was regarded as new forms of commodification of the human body, such as organ selling and surrogate motherhood.

Gradually, the notion of human dignity acquired the status of an *overarching principle*, that is, of an ultimate and general standard that is called to guide the normative regulation of the whole biomedical field (Andorno 2009). This broad and multifaceted function of dignity is visible in the intergovernmental instruments adopted since the end of the 1990s such as the UNESCO Universal Declaration on Bioethics and Human Rights and, at the European level, the Convention on Human Rights and Biomedicine of 1997 (also known as the Oviedo Convention).

The UNESCO Universal Declaration on Bioethics and Human Rights is unequivocal in this regard, as the promotion of respect for human dignity embodies not only the key purpose of the document (Article 2.c) but also the first principle governing the whole field of biomedicine (Article 3), the main reason why discrimination—including, for instance, genetic discrimination—must be prohibited (Article 11), the framework within which cultural diversity is to be respected (Article 12), and the highest interpretive principle of all the provisions of the Declaration (Article 28).

The Oviedo Convention is another good example of the key role that human dignity is beginning to play in this field. The Explanatory Report to the Convention states that "the concept of human dignity . . . constitutes the essential value to be upheld. It is at the basis of most of the values emphasized in the Convention" (Paragraph 9). Recalling the history of the European document, one of the members of the drafting group recognizes that "it was soon decided that the concepts of dignity, identity and integrity of human beings/individuals should be both the basis and the umbrella for all other principles and notions that were to be included in the Convention" (Kits Nieuwenkamp 2000: 329). The Preamble refers three times to dignity: first, when it recognizes "the importance of ensuring the dignity of the human being"; second, when it recalls that "the misuse of biology and medicine may lead to acts endangering human dignity"; third, when it expresses the resolution of taking the necessary measures "to safeguard human dignity and the fundamental rights and freedoms of the individual with regard to the application of biology and medicine." More importantly, the *purpose* itself of the Convention is defined in Article 1 by appealing to the notion of human dignity:

> Parties to this Convention shall protect the dignity and identity of all human beings and guarantee everyone, without discrimination, respect for their integrity and other rights and fundamental freedoms with regard to the application of biology and medicine.

Human Dignity and the Preservation of Humankind's Identity

At the end of the 1990s, the notion of human dignity began also to be used with a secondary (or derivative) meaning that goes far beyond the worthiness of individuals and aims to stress the need to protect the integrity and identity of *humankind as such* against some potential biotechnological developments such as reproductive cloning and germline interventions. These two procedures appear indeed to jeopardize basic features of the human species and our understanding of what it means to be *human*, as well as the interest of children in having an open future.

In the case of reproductive cloning, the root of the ethical problem is the total loss of *biparentality* (i.e., the fact that every individual is conceived by the fusion of genetic information of two individuals, a male or father, and a female or mother). While biparentality is a key feature of advanced animals, asexual reproduction can only be found in the most primitive living beings. It is indeed hard to see how the promotion of asexual reproduction in humans could represent a progress for humankind. Rather, it seems well that it would be the most dramatic regression in history that the human species could ever suffer. But even leaving aside the purely biological perspective, it should be noted that the element of chance that characterizes the combination—always original—of the genetic material of two progenitors is not morally indifferent. Rather, on the contrary, it has great ethical relevance for the individuals themselves as it is a precondition for a full and free discovery of one's *self* (Jonas 1985: 182–94). Indeed, the cloned children are to some extent deprived of such a freedom, as the model of their sole progenitor is—at least tacitly—imposed upon them by the mere fact that they have been deliberately conceived as a genetic copy of the original individual.

The basic objection to germline interventions for enhancement purposes is similar to the one leveled against reproductive cloning. Human genetic engineering would put at risk people's *freedom* from deliberate predetermination of their genetic makeup by third persons and, in the long run, the principle of *equality* between generations. Such intergenerational freedom closely depends upon the condition that each individual's features are more due to *chance* than to *choice*, to *contingency* than to *human design*. This is definitely a paradox, because the idea of a human's mastery of nature has always been regarded—at least in the Western world—as an expression of the special place of humans on earth. But when that mastery reaches the basic features of the human condition itself, then it becomes problematic and even self-contradictory since it entails reducing future people to the condition of an *object* of the subjectivity of present people; human beings are at the same time subjects and objects, and the increase of human subjectivity leads strangely to an increase of human's objectification (Bayertz 1996: 88). Thus, it can be reasonably held that contingency in human reproduction (from which depends the non-predetermination of future people) is a value in itself that needs to be protected (Habermas 2003). This latter is an entirely novel idea in the history of moral philosophy and stems from the awareness that the increasing biotechnological control over our own nature leads in reality not to a greater control over ourselves but over *those who will succeed us*. And the exercise of such unprecedented power is in open conflict with the idea of intergenerational justice, which presupposes that the interests of future generations do not have less moral standing than the interests of present individuals.

It is important to note that the human rights framework is powerless to face these new tremendous challenges to the identity of the human species, because rights only belong to *existing individuals*, not to humankind as such or to future generations. In this regard, it is interesting to recall that during the intergovernmental discussions that took place between 1994 and 1997 at UNESCO aimed at producing an international instrument for the protection of future generations, all references to the supposed rights of future generations were removed from the initial draft after the ad hoc commission of legal experts concluded that there are no such rights, since this concept always relates to *existing* persons. The instrument that was finally adopted on 11 November 1997—the Declaration on the Responsibilities of Present Generations Towards Future Generations—avoids any mention of the rights of future generations.

Therefore, it can be said that the claims sometimes made that there is a right not to be conceived as a genetic copy of another person or a right to inherit nonmanipulated genetic information are more rhetorical statements than conceptually consistent arguments. Indeed, how could people who do not exist, who have not even been conceived, be *today* entitled to any rights? Humankind as such, including future generations, can probably be an *object* of obligations of the present generation but certainly not a *holder* of rights (Mathieu 2000: 43). This explains why the notion of human dignity begins to be regarded as a kind of last barrier against the deliberate alteration of the basic features that characterize our common human condition (Annas et al. 2002).

Three intergovernmental instruments illustrate this appeal to a broad understanding of human dignity to preserve the integrity of the human species: the UNESCO Declaration on the Human Genome and Human Rights of 1997, the Additional Protocol to the European Biomedicine Convention Concerning Human Cloning of 1998, and the UN Declaration on Human Cloning of 2005.

The UNESCO Declaration on the Human Genome and Human Rights of 1997 refers no fewer than 15 times to human dignity. Its most original feature is included in Article 1, which provides that the human genome is "the heritage of humanity." This notion is inspired by the concept of "common heritage of humanity," which aims to preserve the world's natural and cultural resources for the benefit of humankind as a whole. According to the drafters of the Declaration, when applied to the human genome, this expression means, first, that genetic research engages the responsibility of the whole of humanity and that its results should benefit present and future generations (Gros Espiell 1999: 3); second, that the international community has a duty to preserve the integrity of the human species from improper manipulations that may endanger it (Kutukdjian 1999: 33). It is also interesting to note that the 1997 Declaration directly appeals to the notion of human dignity (not to human rights) to condemn both reproductive cloning and germline interventions (Articles 11 and 24, respectively).

The above-mentioned European Convention on Human Rights and Biomedicine did not address the cloning issue, because the final version of the document was adopted in November 1996, a few months before the announcement of the birth of the cloned sheep Dolly, which launched a worldwide debate on this matter. In response to this gap, the Member States of the Council of Europe decided to urgently adopt a specific Additional Protocol to the Convention to prohibit human cloning, which was finalized in January 1998. The substantive reasons offered by the Protocol for the ban on human cloning are the following: (a) it constitutes an "instrumentalisation of human beings," which is "contrary to human dignity" (Preamble); (b) it poses "serious difficulties of a medical, psychological and social nature" for the individuals involved (Preamble); (c) it is "a threat to human identity," because "it would give up the indispensable protection against the predetermination of the human genetic constitution by a third party" (Paragraph 3); (d) it reduces human freedom because it is preferable "to keep the essentially random nature of the composition of their own genes" instead of having a "predetermined genetic make up" (Paragraph 3). Regarding germline interventions, Article 13 of the Convention prohibits them on the grounds that "they may endanger not only the individual but the species itself," according to the Explanatory Report to the Convention (Paragraph 89).

Another intergovernmental instrument that is heavily drawn on a broad notion of human dignity is the UN Declaration on Human Cloning of 2005, by which Member States were called upon to adopt legal measures "to prohibit all forms of human cloning inasmuch as they are incompatible with human dignity and the protection of human life" (Paragraph d). It is interesting to see that the Declaration, which is only a two page-document, includes no fewer than five explicit references to human dignity.

It is noteworthy that all of the above-mentioned instruments relating to bioethics (like all major international human rights instruments) avoid providing either a definition of human dignity or an explanation of the grounds for forbidding some practices that are regarded as contrary to dignity. This lack of theoretical grounding is sometimes a source of misunderstandings on the part of non-legal scholars, who see it as a failure of such policy documents. For instance, criticizing the ban on human reproductive cloning included in the UNESCO Declaration on the Human Genome and Human Rights, the philosopher John Harris writes: "the document contains not a single argument in support of this claim, nor any indication as to just what is meant by human dignity" (2008: 305). This kind of criticism misunderstands the nature of international policy documents, which are not philosophical treatises aimed at discerning the truth so much as political statements resulting from compromise (Baylis 2008: 323–39). In other words, intergovernmental instruments dealing with bioethics (or with other issues) should not be assessed with purely academic criteria, because they belong to another realm of human action; they are not mainly a product of academic work but rather a kind of compromise between a theoretical conceptualization made by experts and what is practically achievable given the political choices of governments (Andorno 2007). After all, the law has an eminently practical, not theoretical, purpose; it aims to crystallize the existing consensus about a particular policy option to promote the common good, not to define the ontology of things nor to provide a theoretical justification of the choices made.

IS A UNIVERSAL UNDERSTANDING OF HUMAN DIGNITY COMPATIBLE WITH CULTURAL DIVERSITY?

Since the end of the Second World War, a universal understanding of human dignity has gone hand in hand with the recognition of the universality of human rights. Before 1945, the language of rights merely existed on the country-specific level. Despite the unquestionable political and intellectual impact of the 18th-century declarations of *natural* rights such as the American and French declarations, the dominant tendency was that of a merely contractual legal account of rights that resulted in positive law. With the creation of the United Nations in 1945 and the adoption of the UDHR and other subsequent international human rights instruments, the contractual account of rights was to some extent replaced with the idea of rights as *universal moral standards necessary for human flourishing.*

As noted above, human dignity is presented in this new context as the *foundational notion of rights*—that is, as the ultimate rationale for the recognition of equal rights to every human being regardless of ethnic origin, nationality, social status, religion, sex, age, or any other particular feature. Human dignity and human rights are regarded as universal by definition, since they are attached to every human being qua human. If human rights were not universal, they could not be called human rights at all. Moreover, human dignity and human rights could not even been *thought* if they were not understood as universal claims but as particular to some groups of people.

Certainly, respect for human dignity and human rights rests upon the belief in the existence of a truly universal moral community comprising all human beings; it assumes that there are some moral truths that transcend boundaries between countries and cultures. Such respect embodies the conviction that "there are some things that should not be done to anybody, anywhere" (Midgley 1999: 160), or, in the words of Dworkin (1994: 236), that people must "*never* be treated in a way that denies the distinct importance of their own lives." This categorical nature of human dignity raises the question of whether it is *compatible* with respect for cultural diversity or whether both values are necessarily opposed.

My view on this is that human dignity and cultural diversity are not only compatible but that dignity is the necessary precondition for the prima facie moral duty to respect the cultural specificities of each society. If every individual would not have any inherent (i.e., universal) worthiness, how then could the particularities of the community he or she belongs to be worthy of respect? If a universal understanding of human dignity were wrong *because of its universal character*, then how could the principle of respect for other cultures (which is itself a universal principle) be justified?

At this point it is helpful to remember that, contrary to what is sometimes asserted, cultural diversity and the existence of different moral standards are not at all new phenomena but are old as humanity. Ancient philosophers like Plato and Aristotle were perfectly aware of the existence of different and even opposed moral codes. Moreover, their search for universal ethical criteria emerged precisely as a consequence of the awareness of cultural diversity; different moral views acted on them as a strong incentive to think about the objectivity of morality and to try to discover the objective goods that may guide moral behavior (Spaemann 1991: 14).

At the same time, it should be emphasized that universal principles (i.e., human dignity and human rights) do not need to be understood as rigid ones but rather as standards that are flexible enough to be compatible with respect for the specificities of each culture. As a matter of fact, the human rights system allows some local variations not in the substance but in the *form* in which particular rights are implemented (Donnelly 1989: 109–42). For instance, the right to marry and found a family is a universal human right. However, as Donnelly exemplifies, rural Thai people might be expected to give greater weight to the views and interests of their families in decisions to marry than do urban Norwegian individuals. This difference, which is simply due to cultural diversity, is perfectly legitimate and not necessarily inconsistent with the universality of human rights.

Of course, it may happen that human rights and *certain* practices that could be seen as cultural traditions of a particular society are in conflict. In such cases, human rights must prevail. Practices such as lapidation (stoning) for adultery, female genital mutilation, child labor, the so-called honor killing of women who are regarded as having brought dishonor upon the family, and discrimination against people of lower castes, even if accepted by large part of a particular community, are seriously incompatible with most basic human rights and therefore do not deserve to be given due regard on the ground that they reflect the cultural specificities of that society. This is clear not only in international human rights law in general but also in the specific field of bioethics. The UNESCO Universal Declaration on Bioethics and Human Rights is a good example of this; while recognizing in Article 12 that cultural diversity should be given "due regard," the Declaration makes it clear that its respect is subjected to the condition that cultural specificities are not "contrary to human dignity, human rights and fundamental freedoms."

Thus, *cultural diversity* is not synonymous with *cultural relativism*. Cultural diversity is a *fact*; cultural relativism is a *theory* about morality. Cultural diversity is a value in itself as it puts in evidence how societies develop differently across time and space depending on their specific historical circumstances, environmental factors, traditions, religion, and so on. All these specificities lead to significant variations in the way societies organize themselves and in the way they conceptualize ethical norms and values. The cultural particularities of every society are worthy of respect, because they make part of its own identity and self-understanding. Far from necessarily being a source of conflict or tension between societies or ethnic groups, they are a precious asset of humankind that deserves to be preserved against the risks of a cultural homogenization.

On the other hand, cultural relativism claims that *morality is merely a product of culture*, and for this reason there are no objective moral truths, but only truths relative to each specific cultural setting. According to this view, a practice may be good in one cultural setting and bad in another, but there are no practices that are intrinsically good or intrinsically bad; thus, *there are no universal truths in ethics*.

Yet cultural relativism is not as plausible as it may appear at first sight. To show this, it is helpful to consider the syllogism that, as James Rachels has rightly noted, is at the background of cultural relativistic positions (1993):

Premise 1:   Different cultures have different moral standards.
Premise 2:   All standards cannot be simultaneously right.
Conclusion: Therefore, there is no objective moral truth in morality.

It is not hard to see that the conclusion does not logically follow from the premises. Even if premise 1 is true (or true to some extent), the conclusion might be false because it deals with totally different matters. Premise 1 concerns what people *think or believe* (in some societies, people may think that degrading treatment of individuals of lower castes is right; in other societies, people may think that that practice is wrong). But the conclusion attempts to say something not about what people may think or believe but about *what really is the case.* In other words, from the mere fact that people may disagree about the rightness or wrongness of a particular behavior (premise 1), it does not follow that there is no objective truth in morality (conclusion); it could be that some practices are objectively wrong and that the views supporting them are simply mistaken.

IS THE PROMOTION OF HUMAN DIGNITY AND HUMAN RIGHTS A FORM OF WESTERN CULTURAL IMPERIALISM?

One of the most common objections to the very idea of human dignity and human rights applying universally is that they embody a Western liberal-individualistic perspective and are therefore alien to other cultures. Attempting to impose respect for human rights standards on non-Western countries would be tantamount to cultural imperialism. This argument has also been made in the specific field of bioethics to object to the human rights approach adopted by intergovernmental bodies. For instance, commenting on the UNESCO Universal Declaration on Human Rights and Bioethics, some scholars criticize the document on the ground that "human dignity and human rights, both strong features of European enlightenment philosophy, pervades this Declaration" (Schuklenk and Landman 2005).

Although the philosophical controversy between universalists and relativists is too complex to be adequately covered in this chapter, some responses to the above-mentioned objection are immediately available. First, it is true that the current notion of human rights has its immediate origins in the insights of the European Enlightenment philosophers and in the political revolutions of the end of the 18th century—notably, the American and French Revolutions. However, this historical circumstance is not sufficient to discard the idea that people have inherent dignity and equal rights, just as the fact that Mozart or Bach were Europeans is not a good enough reason to deny the extraordinary beauty of their works. The relevant question is whether the notion that every human being has inherent dignity and rights makes sense and deserves to be promoted, no matter where this idea was conceptually developed. It can even be claimed that the current widespread conviction, well enshrined in international law, that people have basic rights simply by virtue of their humanity, and irrespective of ethnic origin, sex, nationality, religion, or social and economic status, is one of the major achievements of modern civilization, much more important than any scientific or technical development. In a few words, merely pointing to moral diversity and the presumed integrity of individual cultures does not, by itself, provide a philosophical justification for cultural relativism or a sufficient critique of universalism.

It is indeed paradoxical that the most severe criticisms of the universality of human rights come from Western scholars. In this regard, the Indian-born economist and philosopher Amartya Sen (1998: 40–43) points out that those criticisms are often based on a misconception of non-Western (largely Asian) societies, as if people in these countries had little or no interest in their rights and were only concerned with issues of social order and discipline. In confirmation of Sen's remark, it is interesting to mention that the main weakness that some Asian bioethicists see in the 2005 UNESCO Universal Declaration on Bioethics and Human Rights is not the adoption of a human rights approach but precisely the opposite: the fact of not having emphasized enough the universality of human dignity and human rights! (Jing-Bao 2005; Asai and Oe 2005).

In addition, it should not be forgotten that, after all, international human rights law has been developed along the last six decades by representatives of the most diverse countries and cultures and therefore it is hard to claim that it intends to impose one *cultural* standard. Rather, it can be said that it seeks to promote a minimum legal standard of protection for all people in our common world. As such, universal human rights can be reasonably seen as the "hard-won consensus of the international community" and not as the cultural imperialism of any particular region or set of traditions (Ayton-Shenker 1995: 2).

The previous remarks do not intend to ignore the fact that many Western nations have placed an excessive emphasis on rights and freedoms for the *individual*, sometimes to the detriment of family and community values, which are of paramount importance to most non-Western (mainly Asian and African) societies. However, it would be equally fair to say that international law has made substantial efforts over the last decades to be more attuned to the communal and collective basis of many non-Western countries. This was done, in particular, through the development of the "second generation of rights" that are included in the above-mentioned International Covenant on Economic, Social and Cultural Rights of 1966, such as the rights to education, fair remuneration, healthy working conditions, health care, and the protection of the family and children. This tendency toward a broader understanding of human rights has been even further developed with the "third generation of human rights"—the so-called rights of solidarity, which include the rights to development, peace, self-determination, and a healthy environment. This is to say that, although human rights remain philosophically grounded within an individualist moral doctrine, there can be no doubt that serious attempts have been made by the international community to adequately apply them to more communally oriented societies (Andorno 2009).

In any case, the truth is that today the objections to the universality of human rights have lost much of their practical significance because virtually all states accept the authority of international human rights law. The six core international human rights treaties (on civil and political rights, economic, social, and cultural rights, racial discrimination, women, torture, and children) have an average of 166 ratifying states, which represents a truly impressive 85 percent ratification rate (Volodin 2009: 16–24).

CONCLUSION

Since 1948, the notion of human dignity operates as a central organizing principle of the international human rights system. It not only plays a *foundational* role by offering the ultimate reason that people have rights but also *identifies* those practices that are unacceptable as they entail treating people in a way that denies the distinct importance of their lives or instrumentalizes them in a way that implies using them as mere means to another's ends.

Regarding bioethical issues, respect for the dignity of every individual has acquired in the last decade the status of an overarching principle that is called to guide the whole biomedical field. By the end of the 1990s, it began also to be used with a secondary or extended meaning, which relates to the integrity and identity of humankind as such and to the promotion of intergenerational justice.

Certainly, there is not always unanimity between countries (and within each country) about the concrete implications of human dignity, especially regarding those medical practices that operate at the edges of life. However, divergent results in the most controversial issues may not necessarily mean that a universal conception of dignity does not exist but suggest only that a universal understanding of dignity does not exist *at the margins* (McCrudden 2008: 711).

This chapter has attempted to show, first, that human dignity has an increasingly important role in international policy instruments, especially in those relating to bioethics; and, second, that human dignity and human rights, which are by definition universal, are not in conflict with respect for cultural diversity. This latter conclusion is certainly also valid in the field of bioethics. Indeed, the circumstance that bioethical issues are closely linked to the deepest sociocultural and religious values of every society is not an obstacle to the formulation of universal principles. Quite the contrary. Precisely because bioethics is close to the most cherished aspirations of people, and since people are essentially the same in the United States and in Guinea, in France and in Japan, it is not that difficult to identify some minimal standards that are valid worldwide. Human dignity plays in this regard a unifying role by reminding us that there are certain things that should not be done to anybody, anywhere (negative requirement) and that all human beings are entitled to some basic goods (positive requirement). From this perspective, human dignity is not only the ultimate conceptual ground for the recognition of equal and inalienable rights of all members of the human family, but also the most valuable bridge between cultures that we have.

8 Human Dignity and the Law

O. Carter Snead

Human dignity is a ubiquitous concept on the European legal and political landscape. Thirty-seven constitutions adopted since 1945 explicitly invoke it as a grounding norm. The *Basic Law (Grundgesetz) of Germany*, for example, holds that "Human dignity is inviolable. To respect and protect it is the duty of all state authority." Human dignity likewise serves as the core normative principle of myriad international instruments, including the seminal and much-revered Universal Declaration of Human Rights, which declares the recognition "of the inherent dignity and of the equal and inalienable rights of all members of the human family" to be "the foundation of freedom, justice, and peace in the world" (Schulman 2008: 12–13). In addition to anchoring the apparatus of human rights erected by European constitutions and key declarations and conventions, human dignity is likewise the pole star of intergovernmental instruments relating to bioethics. To take but three examples among many, UNESCO's Universal Declaration on Bioethics and Human Rights, the Council of Europe's Convention for the Protection of Human Rights and Dignity of the Human Being with regard to the Application of Biology and Medicine: Convention of Human Rights and Biomedicine (also known as the Oviedo Convention), and the United Nations Declaration on Human Cloning all cite the protection and promotion of human dignity as their foundational purpose and key animating rationale.

It is clear that human dignity is a live legal concept in European and international circles. But what about in the United States? More specifically, does human dignity have purchase as a legal concept in the domain of American *public* bioethics—in the *governance* of science, medicine, and biotechnology in the name of ethical goods? This chapter offers an analysis of this question, focusing on three concrete areas of pressing concern: abortion, embryonic stem cell research, and end-of-life decision making. For each topic, there will follow a discussion of the uses of human dignity as a legal or political principle in the United States as well as an analysis of what, if any, analogous normative concepts are invoked in its stead. The chapter will conclude with a brief reflection on how (if at all) human dignity should be further integrated into American public bioethics.

Before proceeding to the case studies, it will be necessary first to discuss briefly the limited scope of the domain of inquiry (namely, American public bioethics), the notion of a legal concept, and the variable and contested definition of human dignity.

Defining the Domain of Inquiry: American Public Bioethics[1]

It is well beyond the present inquiry to offer an analysis of human dignity in U.S. law and public policy writ large. To do so responsibly would require a systematic examination of a massive array of sources of law—the U.S. Constitution; 50 state constitutions; state and federal statutes, regulations, and administrative materials as well as the judicial decisions interpreting these authorities. It would also require exploration of a vast and diverse expanse of legal disciplines, both from a substantive and procedural perspective. To take only a handful of examples, it would require rigorous inquiry into the spheres of criminal law and procedure (including especially the law pertaining to the justifications for and the administration of punishment; the law governing searches, seizures, and interrogations by state actors), tort law, civil rights law (including disability rights, prison conditions, and elder law), poverty law, family law, and the like.

This chapter focuses on the far narrower, yet crucially important, domain of American public bioethics. What are the contours of this domain? Bioethics emerged in the United States as a field of scholarly reflection in the 1960s.[2] The field concerns itself with fundamental questions, including what it means to be human, the nature and value of human life (and death), the ends of medicine, and the purpose of science. It began with a series of conferences convened to discuss the tensions between the humanistic and scientific dimensions of medical practice wrought by extraordinary advances in biomedical science and biotechnology.[3] Shortly thereafter, several centers were founded to explore bioethical questions in a sustained and rigorous way.[4] At roughly the same time as these physicians, life scientists, theologians, legal scholars, and social scientists were considering such questions at academic conferences and in newly founded centers, congressmen on Capitol Hill turned their attention to the *public* dimension of bioethics and took up the issue of governance in this domain. Senator Walter Mondale convened hearings in 1968 in connection with his proposal to create a President's Commission on Health Science and Society. This Commission would recommend policies on organ transplantation, genetic engineering, behavior control, human subjects protections, and the financing of research. Mondale's initial efforts foundered, but in 1973 Senator Edward Kennedy convened hearings to discuss proposed research on living fetuses slated for abortion as well as the discriminatory and abusive treatment of human subjects in scientific research (such as those that had occurred in Tuskegee, Alabama). After much debate, these hearings culminated in the passage of the National Research Act.[5] Among other things, this Act created the National Commission for the Protection of Human Subjects of Biomedical and Behavioral Research. The Commission was charged "to identify basic ethical principles that should underlie the conduct of biomedical and behavioral research involving human subjects and develop guidelines that should be followed in such research" and to conduct a "comprehensive study of the ethical, legal, and social implications of advances in biomedical research." The Act directed the Secretary of Health, Education and Welfare (now known as the Secretary of Health and Human Services) to implement the National Commission's advice within a stated period of time or to show cause why such action was not taken.

With the passage of this statutory mandate directing the National Commission to "do bioethics" in an enforceable way, public bioethics in America was born. Since then, public bioethics has been a permanent and active feature of the work of the political branches of government. At the federal level alone, the executive and legislative branches have taken many actions in response to ethical issues raised by advances in biomedical science and biotechnology. Numerous federal commissions on bioethics have been convened to offer advice. Administrative agencies such as the Department of Health and Human Services, the National Institutes of Health (NIH), and even the Justice Department have propounded regulations that touch and concern bioethical matters. Such issues have included human subjects protections,[6] the federal funding of embryo and fetal research,[7] gene therapy research,[8] conscience protections for health care providers,[9] and physician-assisted suicide.[10]

Similarly, Congress has enacted several laws concerning abortion and fetal personhood,[11] research involving embryos and fetuses,[12] conscience protections for health care providers,[13] the patenting of human embryos,[14] organ transplantation,[15] and end-of-life decision making.[16] Congress has debated (though not enacted) many more bills in this domain relating to issues such as abortion,[17] human cloning,[18] and funding for nonembryonic sources of pluripotent cells (stem cells).[19]

Even the federal judiciary has contributed to the development of public bioethics. The U.S. Supreme Court has reserved the bulk of the abortion question to itself in the seminal case of *Roe v. Wade* and its progeny.[20] It has also issued opinions in landmark cases relating to end-of-life decision making,[21] physician-assisted suicide,[22] and the patenting of living organisms.[23]

Governors, state legislatures, and state courts have been similarly active in public bioethics. Given their nearly plenary authority to act in the name of the health, safety, welfare, and morals of citizens, it should not be surprising that the states have been fertile soil for this type of governance.[24] To take only a few examples, states have been involved in the regulation of abortion (to the extent permitted by Supreme Court precedent),[25] the use of assisted reproductive technologies,[26] embryo research,[27] human cloning,[28] research involving human subjects,[29] physician-assisted suicide,[30] end-of-life decision making,[31] and the definition of death.[32]

What Is a Legal Concept?

Before proceeding to the question of the role of human dignity as a legal concept in American public bioethics, it is necessary to reflect briefly on the notion of legal concept itself. There are many ways in which a normative principle can enter the domain of law and public policy. The most obvious point of entry is by explicit invocation in the positive law enacted and enforced by the political branches of government. For example, the Fifth and Fourteenth Amendments of the U.S. Constitution introduce the principle of due process, which must be provided by the government before depriving any person of life, liberty, or property. The Fourth Amendment clearly cites "reasonableness" as a normative legal principle that constrains government officials in their conduct of searches and seizures. Such concepts comprise the currency of constitutional jurisprudence (and, by extension, litigation). Legal concepts also emerge in decisional law set forth in judicial opinions. For example, the law of torts features the "reasonably prudent person" standard as the duty of care owed by all, the breach of which constitutes negligence. Likewise, the still-controversial right to privacy is an unenumerated yet enforceable norm arising from the interpretation of certain provisions of the U.S. Constitution by Supreme Court Justices.

In the foregoing examples, the legal concepts growing out of the positive and decisional law are *directly* applied to the questions at hand. Did the defendant act as a *reasonably prudent person* would have under the circumstances? Was the police officer's search of a dwelling *reasonable*? Does a statute improperly encroach on a constitutionally protected interest in *privacy*? But this is not the only way that legal concepts shape law and public policy. In many instances, normative concepts are invoked not as doctrinal principles to be applied directly but rather in an indirect, rhetorical manner to add moral force to legal arguments made by or to officials in the judicial, legislative, or executive branches of government. As will be discussed below, in the context of American public bioethics, human dignity is deployed most often in this latter, indirect manner to add normative weight to well-established, explicitly enumerated, and directly applicable legal principles such as justice, equality, and liberty.

Which Sense of Dignity?

Other chapters in this volume will explore more fully the vexed philosophical debate about the meaning of human dignity. But for present purposes, it is necessary to identify briefly some of the principal competing definitions of this term in order to better understand its usage in American public bioethics. A useful way to divide the manifold approaches to human dignity is to distinguish those conceptions that treat dignity as a *contingent* standard of valuation from those that regard it as an *intrinsic* attribute of human beings.

Those who construe human dignity as a *contingent* property ascribe it only to those individuals whose actions (or active capacities for certain actions) satisfy a predetermined set of criteria. As Daniel Sulmasy and Adam Schulman, among others, have noted, in classical Antiquity, *dignity* was a term reserved for those exceptional persons who exhibited human excellence at the highest levels. It was a synonym for "worthiness for honor and esteem"[33] (Schulman 2008: 6). Relatedly, this contingent notion of dignity has also been applied in an aristocratic sense, wherein special esteem or worth is accorded to individuals (or not) by virtue of social standing or rank rather actual achievement. A more modern species of contingent human dignity has been defended by bioethicist Ruth Macklin. According to Macklin, human dignity is properly attributable to "a person whose actions, thoughts and concerns are worthy of intrinsic respect because they have been chosen, organized and guided in a way that makes sense from a distinctly individual point of view" (Davis 2008: 20). Thus, Macklin equates human dignity with the exercise of autonomous choice. Indeed, she has famously argued that "dignity is a useless concept," because it is redundant of the more important and precise ethical principle, "respect for autonomy" (Macklin 2003: 1419). Under both the classical rendering of dignity-as-excellence and Macklin's conception of dignity-as-autonomy, certain actions—even certain kinds of lives—are dignified or not (or dignified to greater or lesser degrees) depending on how the individual conducts him- or herself. Dignity is an attribute that ebbs and flows—it can be acquired or lost. For example, under Macklin's approach, human dignity can be diminished by restraints on autonomous choice wrought by disease, disability, or coercion (whether in the form of restrictive laws or private violence). Lives characterized by radical dependence and vulnerability would, according to this view, be undignified. Similarly, people who lack the cognitive capacity for free choice—young children, the mentally disabled, and individuals suffering from dementia— do not possess human dignity in any measure. Conversely, those individuals who have the strength, intelligence, and means to exercise robust free choice possess human dignity in abundance.

By contrast, human dignity has been alternatively conceived as an *intrinsic*, inalienable, absolute property possessed by individuals simply by virtue of their status as human beings. This understanding of dignity holds that all human beings—regardless of their characteristics (such as size, strength, wealth, independence, or social standing), actions, or circumstances—are equal in basic worth and thus entitled to at least a minimal standard of moral respect and care. Here, dignity is understood to be a pre-political quality that is not conferred by others and cannot be taken away. This conception is sometimes associated with the Judeo-Christian principle of *imago Dei*—that all people are made in the image and likeness of God and should be treated accordingly. Another rendering of intrinsic human dignity is attributed to the moral philosophy of Immanuel Kant, who famously declared that all people possess inalienable dignity because of their rational autonomy and thus must always be treated as ends in themselves and never as mere means to an end. (Though as Adam Schulman has noted, it is not entirely clear whether and how the Kantian conception of dignity applies to those human beings whose capacity for rational choice is absent or radically diminished.) Many international human rights instruments, including the much-revered Universal Declaration of Human Rights likewise seem to embrace an intrinsic/absolutist conception of human dignity. In the wake of the atrocities of the Holocaust, the drafters of such documents tried to recover the notion that all human beings have inherent value that must be respected always and everywhere, without exception. Under this intrinsic conception of human dignity, the individual is inviolable and entitled to moral concern as well as the protection of the law. The intrinsic conception of human dignity encompasses even the weakest and most vulnerable among us. This stands in stark contrast to various accounts of contingent dignity discussed above, wherein dignity is attributable in largest measure to the strongest members of the human community.

The foregoing is merely a rough sketch of the variable approaches to human dignity and could surely be refined, deepened, and improved. But for present purposes, the broad conceptual distinction of human dignity as *contingent* versus *intrinsic* is sufficient to negotiate the usages of the term in American public bioethics.

CASE 1: ABORTION

Abortion is arguably the most contentious and inflamed public question in the United States. The lines of disagreement are well familiar. The principal arguments in favor of abortion rights (sometimes made individually, sometimes in conjunction) are, first, that the fetus is not a person (and thus not entitled the moral concern and protections owed to a postnatal human being) and, second, that the pregnant woman's interest in bodily autonomy gives her the sole right to choose to either carry the fetus to birth or terminate the pregnancy (and thus its life).[34] Those opposed to abortion rights respond that because the fetus is an innocent living member of the human species, it is gravely unjust to intentionally kill it, absent the most compelling justification (for example, to save the life of the mother).

The U.S. Supreme Court has constitutionalized the question of abortion, reserving to itself virtually sole authority to reconcile the competing claims of mother and the state (acting on behalf of the unborn child or in the name of some other state interest).[35] Thus, most (though not all) legally relevant authorities on abortion are U.S. Supreme Court precedents.

What role does the concept of human dignity play in the U.S. law of abortion? The chief use of the concept has not been as a legal notion but rather as a normative principle to bolster the argument *both* for the pregnant woman's constitutionally protected liberty interest in obtaining an abortion as well as on behalf of the right to life of the unborn child. The contingent construction of dignity is usually invoked as a rhetorical amplifier for legal arguments for abortion, whereas the intrinsic understanding of dignity is most often deployed on behalf of the unborn. But for all instances of human dignity in the law of abortion, it is used as a normative principle invoked rhetorically to shore up a more potent, recognizable legal concept—the liberty interest in choosing abortion versus the right to life. Dignity in the context of abortion is not a legal principle with direct, independent potency.

Contingent Human Dignity and Abortion Rights

There are many examples of judges and lawmakers citing human dignity in its contingent sense—more specifically, dignity-as-autonomy—in their legal arguments in favor abortion rights. These proponents of abortion rights argue that laws against abortion violate human dignity, because they restrict the freedom of a woman to make intimate decisions relevant to her bodily well-being and her future. Such arguments reflect a commitment to *contingent* dignity, because they do not ascribe equal worth to all living members of the human species but only to those individuals who have met a certain threshold criterion (e.g., birth).[36] This still-controversial passage (attributed to Justice Kennedy) from the Court's 1992 decision *Planned Parenthood v. Casey* is illustrative:

> These matters, involving the most intimate and personal choices a person may make in a lifetime, choices central to personal dignity and autonomy, are central to the liberty protected by the Fourteenth Amendment. At the heart of liberty is the right to define one's own concept of existence, of meaning, of the universe, and of the mystery of human life. Beliefs about these matters could not define the attributes of personhood were they formed under compulsion of the State.

In his partial concurrence and dissent, Justice Stevens likewise invokes human dignity to support the liberty interest in abortion: "The authority to make such traumatic and yet empowering decisions is an element of basic human dignity. . . . Part of the constitutional liberty to choose is the equal dignity to which each of us is entitled." Nearly 15 year later, Justice Ginsberg, in her dissent in *Gonzales v. Carhart* (upholding the constitutionality of the 2003 Partial Birth Abortion Ban Act) reprised the notion that the freedom to choose abortion is central to a woman's "dignity and autonomy."

Intrinsic Human Dignity and the Right to Life

By contrast, Justices and lawmakers alternatively appeal to an *intrinsic* conception of dignity—one that applies regardless of an individual human being's age, location, size, condition of dependency, usefulness, or value as judged by other—in their legal arguments defending the unborn child's right to life. There are many illustrative examples in this regard. In his bitter dissent to the majority opinion in *Stenberg v. Carhart* (which struck down nearly 30 state bans on partial-birth abortion), Justice Kennedy invokes human dignity in this way (in sharp contrast with his reliance on contingent dignity in *Casey*, as noted above):

> A State may take measures to ensure the medical profession and its members are viewed as healers, sustained by a compassionate and rigorous ethic and cognizant of the dignity and value of human life, even life which cannot survive without the assistance of others.

Similarly, in the legislative debate over partial-birth abortion, supporters of a federal ban invoked intrinsic human dignity as a normative grounding good. Congresswoman Ileana Ros-Lehtinen explicitly declared the procedure in question to be a grave threat to intrinsic human dignity:

> Partial-birth abortion is a gruesome and inhumane procedure in which the child is forcibly pulled from the mother, with only the head remaining inside the cervical canal. The head of the child is then punctured at the base of the skull, and the brain is removed with a powerful vacuum. This is a barbaric act that is a grave attack against human dignity and justice, and it must be banned. Life is a gift, and it must be embraced and respected at all stages.[37]

Analogues to Human Dignity in the Law of Abortion

There are, in the abortion context, well-recognized, directly applicable legal principles that seem to be analogous to human dignity, both in its contingent and intrinsic iterations. Two such concepts—liberty and privacy—do much of the same work as the contingent conception of dignity-as-autonomy. The principle of liberty enumerated in the due process clause of the Fourteenth Amendment is the normative lynchpin of the right to abortion. Prior to *Planned Parenthood v. Casey*, the unenumerated right to privacy deemed implicit in the same clause (pursuant to the still-controversial substantive due process doctrine) served this role. Scholars such as Reva Siegal, and officials including Justice Ginsberg have renewed appeals to another U.S. legal concept—equality (explicitly enumerated in the Fourteenth Amendment)—to justify a broad right to abortion. They argue that access to abortion is necessary for women to achieve full and equal participation in the social and economic life of the nation (Siegal 2007: 815–16).

What, if any, U.S. legal principle is analogous to the intrinsic conception of human dignity? It would seem that equality—a widely recognized American principle with direct and concrete legal impact—comes the closest. More than perhaps any other political official in recent memory, President

George W. Bush invoked this principle in defense of nascent human life in the context of abortion. He rejected the notion that the moral status, dignity, or "personhood" of the unborn was something earned, accrued, or conferred based on the needs or wants of others. He held to the view that human equality is truly an intrinsic, pre-political attribute—one that does not depend on accidental characteristics such as gestational stage, condition of dependency, or the value judgments of others. For him, the notion that the moral status of the fetus should increase as it grows stronger, less vulnerable, and less biologically dependent on its mother effectively inverted the fundamental ethical obligation of the strong to care for the weak. The idea that one human being is entitled to kill another because he or she adjudged that life to be unwanted, burdensome to others, not worth living, or an obstacle to one's full participation in social and economic life[38] was, for President Bush, contrary to the principle of equality on which the nation was founded.

The official actions of the Bush administration with respect to the issue of abortion were calculated to vindicate this robust principle of equality to the extent permitted by the prevailing legal regime.[39] Such actions included the issuance of executive orders and memoranda restricting funding of organizations that promote abortion overseas, issuance of veto threats to shape and block legislation, promotion and signing of bills to protect and encourage respect for unborn children (including the Partial Birth Abortion Ban Act of 2003, the Unborn Victims of Violence Act, and the Born-Alive Infants Protection Act), and allocation of federal funding to maternal group homes and crisis pregnancy centers to encourage and support women to carry their pregnancies to term.[40]

CASE 2: EMBRYONIC STEM CELL RESEARCH

The moral, legal, and public policy dispute over embryonic stem cell research (and related matters, such as human cloning) has been the most prominent issue in American public bioethics for over a decade. Since the derivation of human embryonic stem cells in 1998 at the University of Wisconsin, the issue has been debated and discussed by scholars, politicians, members of the popular media, and the public at large. It has been a key element of political campaigns as well as of the activities of the political branches of government at the state and federal level.

The primary question raised by the practice of embryonic stem cell research is whether it is morally defensible to disaggregate (and thus destroy) living human embryos in order to derive pluripotent cells for purposes of basic research that may someday yield regenerative therapies.[41] The embryos used in this kind of research are typically donated by individuals or couples who conceived them by in vitro fertilization (IVF) in the context of receiving assisted reproduction treatment but who no longer need or want them for such a purpose. There are reports of some researchers creating embryos by in vitro fertilization solely for use (and destruction) in research (Stolberg 2001: A1). Theoretically, embryos for use in stem cell research could also be created by somatic cell nuclear transfer (that is, human cloning for bio-medical research, or so-called therapeutic cloning), though efforts to derive pluripotent cells from cloned human embryos have not yet succeeded.[42] The scientific aspirations for embryonic stem cell research are manifold, including the goals of understanding the mechanisms of early human development, to test and develop pharmaceuticals, and, ultimately, to devise new regenerative therapies. According to prominent researchers in this field, realizing these aspirations will require the creation of a bank of embryonic stem cell lines large enough to be sufficiently diverse both for the creation of models to study all relevant diseases or injuries that might admit of regenerative cell-based therapy, and for purposes of immunocompatibility (should such thera-pies be developed). This program will thus require the use and destruction of millions of human embryos. The most comprehensive study, conducted by the RAND Corporation in 2003, estimated that there are 400,000 or so embryos in cryopreservation, only 2.8 percent of which have been for-mally designated for donation (Hoffman 2003: 1068). Given the scarcity of donated IVF embryos for this purpose, creating embryos solely for the sake of research (by IVF or cloning) is a necessity (Lanza and Rosenthal 2004: 92).

The moral permissibility of embryonic stem cell research depends ulti-mately on the status of the human embryo that is necessarily destroyed in this process. As in the context of abortion, the deliberation and actions of public officials sometimes feature appeals to human dignity—not so much as a direct legal principle but rather as a normative conception meant to bolster the legal or political decision taken. Those public officials who regard the use and destruction of living human beings at the embryonic stage of development as a grave injustice frequently invoke human dig-nity in its intrinsic form. Such officials cite the inalienable and equal right of every human being and the concomitant duty of the law to protect all members of the human family from others who wish to instrumentalize and destroy them.

Examples of this use of human dignity in public argument abound. In vetoing a congressional effort to liberalize his stem cell research funding policy (which authorized funding only for those forms of stem cell research that did not create material incentives for future destruction of living human embryos[43]), President Bush noted that, "as science brings us ever closer to unlocking the secrets of human biology, it also offers temptations to manip-ulate human life and violate human dignity."[44] In opposing embryonic stem cell research, Congresswoman Virginia Foxx invoked human dignity even more explicitly:

> Those embryos are human beings and should not be treated as research subjects. We would never kill to harvest body parts because of the principle of human dignity. We do not even do this with our most heinous criminals.

We do not treat them as things. We treat them with dignity until the time that they die. . . . All human beings have profound human dignity.[45]

On the other side of the issue, some public defenders of embryonic stem cell research likewise invoke human dignity—the contingent human dignity of patients. They argue that limits on embryonic stem cell research funding are an obstacle to bringing about the conditions necessary for a dignified existence for such patients. Two members of President Bush's Council on Bioethics, Drs. Janet Rowley and Michael Gazzaniga included such appeals in their personal statements in two separate Council reports. Rowley wrote: "We talk about protecting human dignity. We should strive to help patients with serious illnesses that could potentially be treated with embryonic stem cells to live as fulfilling and dignified lives as is humanly possible" (President's Council on Bioethics 2008: 264). Similarly, Gazzaniga wrote: "The Koreans have found a way to let biomedical cloning go forward with all of its spectacular promise for restoring human dignity to the seriously diseased and infirmed patients of the world while at the same time not in any way creating a social atmosphere to use such advances for baby making. What could be better?" (President's Council on Bioethics 2004b: 239). Here, Rowley and Gazzaniga seem to regard suffering and dependency that could be ameliorated as a kind of *in*dignity. By suggesting that dignity can be "restored" through the treatment of disease, they imply that disease and disability themselves diminish human dignity in the first instance. Put another way, living with disease and disability is a less dignified form of life than living in a healthier state. This seems a rough species of the contingent iteration of dignity-as-excellence—Rowley and Gazzaniga attribute dignity in greater or lesser degree according to the biological flourishing of living "persons" (whom they both, like many proponents of abortion rights and embryo destructive research, have defined elsewhere [e.g., in their personal statements in several President's Council on Bioethics reports] in exclusionary fashion to encompass certain postnatal human beings possessing a preferred set of active capacities relating to higher cognitive function).

As in the context of abortion, it is clear that in the legal and political debate involving embryonic stem cell research, human dignity is not a concept with direct application and independent potency. It is, rather, a normative claim invoked in the public square by lawmakers or executive branch officials to draw support for official action of one kind or another.

Are there analogous legal concepts in the American context that do similar work, perhaps more directly? There have been some efforts to invoke the familiar notion of liberty—as applied to researchers who wish to pursue work that requires the use and destruction of embryonic human beings. Similarly, there have been some efforts to extend the notion of liberty to encompass the freedom to choose and access treatments according to the patient's wishes. This latter project is, at the moment, inapplicable to embryonic stem cell research, as there are no therapies currently available based on this approach. But both projects to extend liberty into this domain are undertheorized and have not borne any fruit to date.

By contrast, as in the abortion context, the traditional American conception of equality has been deployed in the name of protecting embryonic human life in a way that approximates human dignity in its intrinsic iteration. Here again, President George W. Bush is the most prominent promoter of this notion. In resolving the question of the moral status of the embryo, President Bush appealed both to his radical conception of human equality and the findings of modern embryology. The relevant science confirmed that the five- to six-day-old human embryo used and destroyed in stem cell research is a complete, living, self-directing, integrated, whole individual[46] member of the human species, who, given the proper environment will (if all goes well) move itself along the trajectory of human biological development from embryo, to fetus, to neonate, to child, to adolescent, to adult (Bush 2001: WK12). The biological status of embryos as human organisms did not, however, settle the question of their moral status. For this judgment, President Bush reflected on the notion of human equality as a principle of classical liberalism underlying the nation's founding. He concluded that the only coherent (non-self-destroying) understanding of human equality is one that encompasses all human beings without discrimination on the basis of accidental characteristics such as age, size, condition of dependency or vulnerability, circumstances, or the esteem of others. Accordingly, President Bush concluded that the intentional use and destruction of embryos in stem cell research is gravely unjust. Furthermore, he took the position that the intentional creation of embryos (by IVF or cloning) for use and destruction in research is, a fortiori, morally unacceptable.[47]

In making this judgment, President Bush implicitly rejected the arguments of those who assert that the human embryo is not entitled to a high degree of moral respect because it lacks certain preferred capacities or characteristics.[48] This, in President Bush's mind, was tantamount to the most unjust and invidious kind of discrimination. He likewise rejected the more limited argument in favor of using and destroying donated embryos from fertility clinics because they are destined to be discarded and destroyed in any event. President Bush's understanding of equality dictated that living human beings should not be treated as raw materials to be exploited and destroyed for biomedical research purposes simply because someone else has made the decision that their lives were no longer useful and thus should be terminated (Meilaender 2002: 25). And his devotion to the principle of radical equality and, in his words, respect for the "matchless worth" of every individual, led him to reject a straightforward utilitarian argument that assumed the personhood of the embryo but nevertheless justified its use in research simply by virtue of the hoped-for lifesaving promise of the therapies that might emerge from it. For example, "[Embryonic stem] cell technology stands to benefit everyone. . . . It is this property that may make it reasonable to kill some embryos to conduct ES cell research even if the embryo is a person"[49]

(Savulescu 2002: 529). President Bush's funding policies and his other official actions were clearly animated by this robust sense of equality—a concept closely analogous to intrinsic human dignity.

On March 9, 2009, President Obama rescinded all of President Bush's previous executive actions regarding funding for stem cell research and affirmatively directed the NIH to fund all embryonic stem cell research that was "responsible, scientifically worthy . . . to the extent permitted by law."[50] He gave the NIH 120 days to provide more concrete guidelines. In July of that year, the NIH adopted a policy of federal funding for research involving cell lines derived from embryos originally conceived by IVF patients for reproductive purposes but now no longer wanted for such purposes. The NIH guidelines restrict funding to these kinds of cell lines on the grounds that there is, as yet, no social consensus on the morality of creating embryos solely for the sake of research (either by IVF or somatic cell nuclear transfer, also known as human cloning).[51] Additionally, the NIH guidelines forbid federal funding of research in which human embryonic stem cells are combined with nonhuman primate blastocysts and research protocols in which human embryonic stem cells might contribute to the germline of nonhuman animals. The final version of the NIH guidelines explicitly articulate the animating principles for the policy: belief in the potential of the research to reveal knowledge about human development and perhaps regenerative therapies and the embryo donor's right to informed consent. Neither President Obama nor the NIH guidelines have discussed the moral status of the human embryo.

CASE 3: END-OF-LIFE DECISION MAKING

The U.S. legal context of end-of-life decision making, involving both the termination of life-sustaining measures and physician-assisted suicide, prominently features appeals to human dignity on all sides of the debate. As in the areas of law discussed above, human dignity is not a juridical concept with direct impact (except, perhaps, in one or two state constitutions). Rather, it is a normative claim made to bolster appeals to more recognizably American and directly applicable legal concepts such as liberty and equality.

Unlike abortion, the governance of end-of-life decision making has not been fully constitutionalized by the Supreme Court, and thus states have wide latitude to enact laws according to their own judgment about the relevant goods to be pursued and harms to be avoided. That said, the U.S. Supreme Court has recognized a constitutionally protected liberty interest in refusing unwanted medical treatment and appears to have extended this liberty to those patients who are no longer capable of expressing their wishes.[52] At the same time, the Court allows states to impose the highest standard of proof in civil cases—the clear and convincing evidence standard—on those seeking to prove that the now-incapacitated patient would have wanted to discontinue life-sustaining care under the precise circumstances presented. Within this framework, most states have, in the name of autonomy, adopted laws that grant near-total deference to the preferences of patients regarding end-of-life decision making (provided such preferences can be established by the requisite standard of proof). In those instances where the preferences of the now-incapacitated patient are unknowable, most state laws direct the decision maker to choose a course of treatment (or nontreatment) in the "best interests" of the patient. All ambiguities are resolved in favor of continuing life-sustaining measures.

Regarding physician-assisted suicide, the Supreme Court has ruled that the Constitution does not prevent states from adopting their own preferred laws regarding this practice.[53] Only two states—Washington and Oregon—have formally legalized assisted suicide.

Defenders of a robust right to terminate life-sustaining measures, and even the right to physician-assisted suicide regularly invoke the notion of "death with dignity." In making their arguments for the right to die, such proponents appear to rely on a contingent notion of dignity-as-autonomy. That is, they assert that respect for dignity demands that patients be free to make their own decisions in this profound existential context. Free choice is thus equated with dignity. At the same time, proponents of a right to die argue that a life characterized by agonizing pain or radical dependence on others is, in an important sense, *un*dignified—an unworthy state of affairs that individuals should have the right to choose against. Thus, the right to die promotes dignity-as-autonomy in two key ways. First, it protects the freedom to choose—a necessary condition of dignity. And, second, it safeguards the freedom to choose against living a life of dependence and incapacity—conditions that make dignity impossible. This view is reflected in the dissenting opinion of Justices Brennan in *Cruzan* when he describes the patient's cognitive impairment:

> Medical technology has effectively created a twilight zone of suspended animation where death commences while life, in some form, continues. Some patients, however, want no part of a life sustained only by medical technology. Instead they prefer a plan of medical treatment that allows nature to take its course and permits them to die with dignity. . . . Missouri's protection of life in a form abstracted from the living is not commonplace; it is aberrant.

Similar views were expressed by Justice O'Connor in her concurrence in *Washington v. Glucksberg*:

> This freedom embraces not merely a person's right to refuse a particular kind of unwanted treatment, but also her interest in dignity, and in determining the character of the memories that will survive after her death. . . . Avoiding intolerable pain and the indignity of living one's

final days incapacitated and in agony is certainly "at the heart of [the] liberty . . . to define one's own concept of existence, of meaning, of the universe, and of the mystery of human life."

Finally, in his *Glucksberg* concurrence, Justice Breyer defined death with dignity as "personal control over the manner of death, professional medical assistance, and the avoidance of unnecessary and severe physical-suffering-combined."

On the other side of the debate are voices that raise concerns about the mistreatment, neglect, and abandonment of those patients whose lives are deemed not worth living by others. They raise the worry that life-sustaining measures will be discontinued for those patients who have not and cannot express their wishes based on the judgment of able-bodied and able-minded third parties who do not adequately value life in a diminished state. Such claims are often made in the name of the dignity of such patients—their intrinsic and inalienable worth regardless of their physical incapacities. Such appeals were made in the now-iconic case of Theresa Marie Schiavo. Schiavo was a profoundly cognitively disabled woman with no written instrument declaring her preferences for medical treatment. She was not dying but did require delivery of nutrition and hydration by means of a feeding tube. Her husband (Michael Schiavo) and parents (the Schindlers) violently disagreed as to what course of treatment to pursue. Mr. Schiavo argued that it was her wish under such circumstances to discontinue artificial nutrition and hydration so that she could be "allowed to die." The Schindlers asserted that she had expressed no such wish, and had indeed expressed the contrary view on several occasions. The case ended up in the Florida state courts, which purported to apply guardianship laws that aimed to discern and implement her actual wishes if possible and, failing that, to act in her best interests. The law directed the courts to resolve any evidentiary ambiguities in favor of continuing life-sustaining treatment. The burden was thus on Mr. Schiavo to prove by "clear and convincing evidence" (the highest standard of proof in the civil context) that she would have wanted to withdraw artificial nutrition and hydration under these precise circumstances.

On the basis of four oral statements allegedly made by Mrs. Schiavo (in various informal settings many years prior) recounted by Mr. Schiavo, his brother, and sister-in-law, the Florida trial court held that the standard had been met (Snead 2005a: 60, Snead 2005c: 395). The Schindlers appealed repeatedly over a period of several years pursuant to a number of different theories but ultimately did not prevail. Because of persistent doubts about the soundness of the Florida courts' rulings, the Florida legislature passed Terri's Law, allowing the governor to issue a temporary stay of the order to terminate Mrs. Schiavo's artificial nutrition and hydration and to appoint a guardian ad litem to review the case and to make recommendation to the relevant Florida state judicial authority. The Florida Supreme Court problematically struck down the law as an unconstitutional violation of the principle of separation of powers (Snead 2005a: 88–89).

At the urging of the Schindlers as well as of prominent members of the disability rights community, civil rights leaders (including the Rev. Jesse Jackson), consumer advocates (including Ralph Nader), religious advocates for the sanctity of life (including prominent Catholic public figures), and politicians of both parties (including social conservatives such as Sam Brownback and social liberals such as Tom Harkin, Joe Lieberman, and a substantial percentage of the Congressional Black Caucus), Congress authorized a federal intervention, granting authority to the federal district court in Florida to review de novo any claims of civil rights violations arising from the proceedings. Not a single U.S. senator objected to this intervention. President Bush signed the bill into law. Ultimately, the federal courts declined the Schindler family's petitions for relief under the new law, and Mrs. Schiavo died shortly thereafter.

Those who sided with the Schindlers argued that human dignity, rightly understood, required a more searching inquiry regarding her actual intentions. And barring clear and convincing evidence of her desire to discontinue artificial nutrition and hydration, these advocates (including those congressmen who passed the special federal law on her behalf) argued that there was a duty to care for her as one would any other human being. On the other side, of course, were those who were persuaded that her preferences had been clearly established, and it was a ghoulish violation of her human dignity to continue artificial nutrition and hydration.

As in the contexts of abortion and embryonic stem cell research, human dignity is invoked in the public domain of end-of-life decision making as a normative concept to lend additional support to more recognized and potent American legal concepts. What are these concepts? First, and most obvious, is the concept of liberty, which the Supreme Court has recognized as providing some measure of protection for the right to choose to discontinue unwanted life-sustaining care. This is analogous to the contingent notion of dignity-as-autonomy discussed above. In those circumstances where preferences cannot be reliably discerned, the law in most states directs that a decision to be taken in the best interests of the patient—an open-ended notion that could support the invocation of either intrinsic or contingent dignity.

Additionally, the principle of equality is an American legal concept invoked in this end-of-life context. Disability rights activists such as the group named Not Dead Yet argued that the decision to terminate Schiavo's artificial nutrition and hydration was a form of deadly discrimination based on a prejudice against the disabled and an unfair judgment about the quality of life of patients suffering from cognitive disabilities. In his signing statement on the Schiavo law, President Bush echoed a similar sentiment regarding equality:

> The case of Terri Schiavo raises complex issues. Yet in instances like this one, where there are serious questions and substantial doubts, our society, our laws, and our courts should have a presumption in favor of life. Those who live at the mercy of others deserve our special care and

concern. It should be our goal as a nation to build a culture of life, where all Americans are valued, welcomed, and protected—and that culture of life must extend to individuals with disabilities.[54]

It seems from the foregoing statement that President Bush's equality principle, discussed above in the domains of abortion and embryonic stem cell research, likewise animated his approach to end-of-life matters. His justification for signing into law the congressional intervention was to explore "serious questions and substantial doubts" about the fair adjudication of the Schiavo matter. He was concerned that in cases involving end-of-life decision making for patients who are living in a severely diminished state, surrogate decision makers would be strongly tempted to choose termination of artificial nutrition and hydration because of their subjective appraisal of the patient's quality of life rather than in the name of the patient's actual wishes. Such discrimination on the part of surrogate decision makers would be deeply inconsistent with President Bush's principle that every life should be accorded equal moral worth.

Similarly, those who oppose physician-assisted suicide couch such concerns in equality. They worry that a regime of legalized assisted suicide opened the door to coercion of patients by their physicians, family members, or even society at large to choose to end their lives to discontinue the burdens that they imposed on others. Some are also concerned that legalized physician-assisted suicide is merely a transitional measure on the way to legalized euthanasia, which presents far greater temptations and opportunities for abuse (for example, unconsented mercy killing) based on discriminatory quality-of-life judgments.

CONCLUSION

The foregoing discussion has explored some of the manifold applications of human dignity in American public bioethics. As in the scholarly field of bioethics, it is a term that is used in widely divergent ways, ranging from a term for recognizing exceptional excellence or a proxy for autonomy to an intrinsic characteristic entitling all human beings to at least a bare minimum of respect and protection. As a concept in American public bioethics, it has had a limited direct impact. It has not, for example, been recognized by the Supreme Court as a constitutionally guaranteed principle to the degree that liberty, equality, or even privacy have. It has not been enshrined in positive law as a doctrinal principle applicable to regulate conduct. But it is, however, often invoked by courts and officials in the political branches alike to amplify the moral claims made pursuant to more commonly recognized principles of U.S. law, such as equality and liberty.

Moreover, there appear to be some usages of these traditional U.S. legal principles that serve the same purpose as the variable meanings of human dignity. Liberty (and before that, privacy) do much of the analytic and moral work of dignity-as-autonomy in the contexts of abortion and end-of-life decision making. More recently, equality has been invoked on behalf of the rights of the woman seeking an abortion. In this way, equality has been deployed in the same manner as dignity-as-autonomy. At the same time, equality has been wielded in the name of the rights of the unborn child threatened by abortion, the embryonic human beings sought for use and destruction in stem cell research, and on behalf of the severely disabled whose lives might be deemed not worth living by those in authority.

Thus, one might think that the analogues of human dignity described above render dignity as a redundant (if not "useless") concept for American public bioethics. I am inclined to disagree. Liberty and even privacy are situational goods, and there is (rightly, in my view) a fairly well-calibrated body of law across contexts that helps government officials balance the state's interests against the individual's right to liberty or privacy. Neither right is absolute, and both admit of time, place, and manner restrictions, at least. Equality, while important, is a relational principle: the duty to treat all individuals equally might be discharged by treating everyone equally badly. By contrast, dignity—particularly in its intrinsic iteration—stands as an inviolable threshold rooted in who people are as members of the human family. It is distinctive from liberty, privacy, and even equality in that it makes claims on us all to treat others in a manner befitting a human person. As Avishai Margalit (2002: 114) has written, "dignity, unlike social honor, is not a positional good. It is supposed to be accorded to everybody, even to the one who is nobody, by the most universal common denominator of being human." By my lights, this is a much-needed concept for American public bioethics. How to strengthen human dignity in this area of U.S. law and how to work out its precise applications are complicated questions for another time. But this is a project well worth undertaking.

NOTES

1. For a further treatment of this topic, see Snead 2010.
2. As Gilbert Meilaender explains, "Albert Jonsen dates the 'birth of bioethics' from the year 1962, when Shana Alexander's article describing the Seattle dialysis selection committee appeared in *Life* magazine. Elsewhere, Jonsen describes 1965–75 as the 'formative decade' for bioethics in this country. David Rothman, in what is the first history of the bioethics movement, dates its beginning with the 1966 publication of Henry Beecher's articles exposing abuses in human experimentation" (Meilaender 1995: 1). The origin of the term *bioethics* is contested, though its first usage appeared in 1970. It has been attributed both to Sargent Shriver (original funder of the Georgetown Kennedy Institute of Ethics) and Van Renesslear Potter (research oncologist from the University of Wisconsin). Whereas Shriver used the term to denote the ethical analysis of the development and application of biomedical science, Potter seemingly meant something more capacious, encompassing the relationship between man, his environment, and the civilized world. Shriver's

definition more closely approximates the meaning of the term as it is used in America (Jonsen 1998: 27).

3. Such conferences included "Great Issues of Conscience in Modern Medicine" held at Dartmouth College in 1960, "Man and His Future," held by the Ciba Foundation in London in 1962, the Nobel Laureate Series at Gustavus Adolphus College, which included "Genetics and the Future of Man" (held in 1965, featuring a presentation by William Shockley on eugenics and a rebuttal by Paul Ramsey), and "The Human Mind" (in 1967, featuring a presentation by James Gustafson) (Jonsen 1998: 13–15). S. Marsh Tenney (then dean of the Dartmouth Medical School) noted at one of the very first such events, "Although [medicine's] foundations have become more rational, its practice—the welding of science and humanism—is said to have become more remote and indifferent to human values, and once again medicine has been forced to remind itself that it is often the human factors that are determinant" (Jonsen 1998: 13).

4. Such institutions included the Hastings Center (opened in 1970 to study ethical issues relating to death and dying, behavioral control, genetic engineering and counseling, and population control), the Society for Health and Human Values (opened in 1968 in response to concerns about an undue emphasis on mechanistic explanations in medical education), and the Kennedy Institute of Ethics at Georgetown University (opened in 1971 to study issues in reproduction and ethics) (Jonsen 1998: 26–27).

5. National Research Act, Pub. L. No. 93–348, 88 Stat. 342 (1974).

6. 45 C.F.R. § 46 (2008) (known as the Common Rule).

7. National Institutes of Health Guidelines for Human Stem Cell Research, 74 Fed. Reg. 32,170–02 (July 7, 2009).

8. For a comprehensive discussion of the law, ethics, and science of this matter, see *Reproduction and Responsibility* (President's Council on Bioethics 2004b).

9. 45 C.F.R. § 88.1 (2009); Rescission of the Regulation Entitled "Ensuring That Department of Health and Human Services Funds Do Not Support Coercive or Discriminatory Policies or Practices in Violation of Federal Law," 74 Fed. Reg. 10,207–01 (proposed Mar. 10, 2009) (to be codified at 45 C.F.R. pt. 88).

10. *Gonzales v. Oregon*, 546 U.S. 243, 275–76 (2006) (invalidating Attorney General's efforts to ban physician-assisted suicide by administrative rule).

11. Born Alive Infant's Protection Act, 1 U.S.C. § 8 (2006); Unborn Victims of Violence Act of 2004 (Laci and Conner's Act), 18 U.S.C. § 1841, 10 U.S.C. § 919a (recognizing child in utero as legal victim if he or she is injured during commission of crime); Partial Birth Abortion Ban Act of 2003, 18 U.S.C. § 1531 (2006) (banning form of late-term abortion); Hyde Amendment, Pub. L. No. 94–439 § 209, 90 Stat. 1434 (1976) (restricting federal funding of abortion).

12. Fetus Farming Prohibition Act of 2006, Pub. L. No. 109–242, § 2, 120 Stat. 570, 570–571 (2006); Dickey Amendment, Pub. L. No. 104–99, § 128, 110 Stat. 26, 34 (1996) (prohibiting Department of Health and Human Services from using appropriated funds for creation of human embryos for research purposes or for research in which human embryos are destroyed).

13. Consolidated Appropriations Act, 2005, Pub. L. No. 108–447, § 508(d)(1)-(2), 118 Stat. 2809, 3163 (2004).

14. Consolidated Appropriations Act, 2004, Pub. L. No. 108–199, § 634, 118 Stat. 3, 101.

15. National Organ Transplant Act, Pub. L. No. 98–507, 98 Stat. 2339 (1984) (outlawing sale of human organs).

16. An Act for the Parents of Theresa Marie Schiavo, Pub. L. No. 109–3, 119 Stat. 15 (2005); Patient Self-Determination Act, Pub. L. No. 101–508, §§ 4206, 4571, 104 Stat. 1388–115, 1388–204 (1990).

17. Child Interstate Abortion Notification Act, H.R. 1063, 110th Cong. (2007) (preventing transportation of minors in circumvention of certain laws relating to abortion).
18. Human Cloning Prohibition Act of 2007, S. 1036, 110th Cong. (2007) (proposing to prohibit human cloning); A Bill to Prohibit Human Cloning and Protect Stem Cell Research, S. 812, 110th Cong. (2007) (same).
19. The HOPE Act, S. 30, 110th Cong. (2007) (promoting stem cell research from nonembryonic sources, such as amniotic fluid).
20. *Gonzales v. Carhart*, 550 U.S. 124 (2007) (upholding the Partial Birth Abortion Ban Act of 2003); *Stenberg v. Carhart*, 530 U.S. 914 (2000) (striking down Nebraska law (and all state laws) criminalizing partial-birth abortion); *Planned Parenthood v. Casey*, 505 U.S. 833 (1992) (purporting to reaffirm the core holding of *Roe v. Wade* but changing the constitutional status of the right to abortion (from a fundamental right to protected liberty interest) and abandoning Roe's trimester framework in favor of a pre- versus post-viability dichotomy of regulation (before fetal viability, the state may not "unduly burden" a woman's right to choose abortion, whereas post-viability, the state may restrict abortion provided it includes exceptions for the woman's life and health (as defined in *Doe v. Bolton*)); *Doe v. Bolton*, 410 U.S. 179 (1973) (holding that health includes physical, emotional, psychological, and familial factors, as determined by abortion provider); *Roe v. Wade*, 410 U.S. 113 (1973) (declaring abortion to be a fundamental right and purporting to provide a framework establishing the contours of state regulation of abortion.).
21. *Cruzan v. Missouri*, 497 U.S. 261, 281–82 (1990).
22. *Vacco v. Quill*, 521 U.S. 793, 807–08 (1997); *Washington v. Glucksberg*, 521 U.S. 702, 735 (1997).
23. *Diamond v. Chakrabarty*, 447 U.S. 303, 309–10 (1980).
24. *Poe v. Ullman*, 367 U.S. 497, 539 (1961) ("In reviewing state legislation, whether considered to be in the exercise of the State's police powers, or in provision for the health, safety, morals or welfare of its people, it is clear that what is concerned are 'the powers of government inherent in every sovereignty.' Only to the extent that the Constitution so requires may this Court interfere with the exercise of this plenary power of government." (citation omitted)).
25. Fla. Stat. Ann. § 390.01114 (West 2009) (Parental Notice of Abortion Act) (requiring parental notification for women under age of 18 prior to abortion); 2009 Kan. Sess. Laws Ch. 28 (requiring informed consent of woman); N. D. Cent. Code § 14–02.1–01 (2008) (North Dakota Abortion Control Act) (requiring informed consent and mandatory waiting period).
26. Cal. Health & Safety Code § 125325 (West 2010) (governing procedures for solicitation and donation of oocytes for assisted reproductive technology); Fla. Stat. Ann. § 742.11 (West 2009) (creating presumption that any child conceived by artificial insemination is child of husband and wife); La. Rev. Stat. Ann. 9:126 (stating that any fertilized human ovum is not property of physicians or donors); Va. Code Ann. § 20–156 (2009) (regarding legal status of children of assisted conception); Wash. Rev. Code Ann. §§ 26.26.700–26.26.740 (West 2009) (defining rights of donors and children created through assisted reproduction under Uniform Parentage Act).
27. Mich. Comp. Laws Ann. § 333.2685 (West 2009) (restricting nontherapeutic research on live human embryos, fetuses, or neonates); N.J. Stat. Ann. § 26:2Z-2 (West 2009) (stating that it is policy of state to permit research on human embryonic stem cells); 18 Pa. Cons. Stat. Ann. § 3216 (West 2009) (criminalizing experimentation on unborn child).

28. rk. Code Ann. § 20–16–1002 (West 2009) (making human cloning Class C felony); Cal. Health and Safety Code § 24185 (prohibiting human cloning to produce children); Mich. Comp. Laws Ann. §§ 333.16274, 333.16275, 750.430a (West 2009); N.J. Stat. Ann. § 26:2Z-2 (prescribing administrative and civil penalties to those who engage in human cloning, as defined by the statute); N.D. Cent. Code § 12.1–39–02 (2008) (making human cloning Class C felony); R.I. Gen. Laws § 23–16.4–2 (2008) (prohibiting somatic cell nuclear transfer); Va. Code Ann. §§ 32.1–162.21, 32.1–162.22 (West 2009) (imposing civil penalties to those who engage in cloning).

29. *Grimes v. Kennedy Krieger Inst.*, 782 A.2d 807 (Md. 2001) (holding that parent cannot consent to participation of child or other person under legal disability in nontherapeutic research or studies in which there is any risk of injury or damage to health of subject).

30. Fla. Stat. Ann. § 765.309 (West 2009) (pointing out that nothing in code should construe to authorize "mercy killing or euthanasia"); N.Y. Penal Law § 120.30 (McKinney 2009) (making promotion of suicide attempt felony); Oregon Death with Dignity Act, Or. Rev. Stat. Ann. §§ 127.800–127.897 (West 2009) (allowing terminally ill people to end their lives through voluntary self-administration of lethal medications, expressly prescribed by physician for that purpose); Washington Death with Dignity Act, Wash. Rev. Code Ann. §§ 70.245.010–70.245.904 (West 2009) (allowing terminally ill adults seeking to end their life to request lethal doses of medication from physicians).

31. 2003 Fla. Laws ch. 418 ; *Bush v. Schiavo*, 885 So. 2d 321 (Fla. 2004) (overturning stay issued by legislature to prevent continued withholding of nutrition and hydration from Theresa Marie Schiavo); *In re* Conroy, 486 A.2d 1209 (N.J. 1985) (setting forth guidelines with respect to life-sustaining treatment); *In re Quinlan*, 355 A.2d 647 (N.J. 1976) (holding that decision by daughter to permit noncognitive, vegetative existence to terminate by natural forces was valuable incident of her right to privacy, which could be asserted on her behalf by her guardian); Florida Governor's Office, Exec. Order No. 03–201 (Oct. 21, 2003).

32. Uniform Determination of Death Act (1980); Uniform Law Commissioners, *Uniform Determination of Death Act*, July 21, 2004, http://www.nccusl.org/Update/uniformact_factsheets/uniformacts-fs-udda.asp. Forty-three states have adopted the Uniform Determination of Death Act, or substantially similar legislation.

33. Schulman also notes, however, that the Stoics thought that even though dignity was synonymous with human excellence, everyone was capable of achieving this status through exercise of reason.

34. There are, obviously, many different permutations of this argument. One variation is a developmental approach to moral status, which accords increasing moral worth to the fetus (and its interests as against the claims of the mother) as it progresses through the gestational stages. Other arguments focus on the dependency of the fetus on the mother for bodily support and weigh its claim to life more heavily in comparison with the mother's autonomy rights as it becomes more biologically independent (that is, viable). As Gilbert Meilaender has observed, these arguments about personhood and bodily support, though analytically distinct, are deeply intertwined (Meilaender 1995: 114).

35. *Gonzales v. Carhart*, 550 U.S. 124 (2007) (upholding the Partial Birth Abortion Ban Act of 2003); *Stenberg v. Carhart*, 530 U.S. 914 (2000) (striking down Nebraska law (and all state laws) criminalizing partial-birth abortion); *Planned Parenthood v. Casey*, 505 U.S. 833 (1992) (purporting to reaffirm the core holding of *Roe v. Wade* but changing the constitutional status of the right to abortion (from a fundamental right to protected

liberty interest) and abandoning Roe's trimester framework in favor of a pre- versus post-viability dichotomy of regulation (before fetal viability, the state may not "unduly burden" a woman's right to choose abortion, whereas post-viability, the state may restrict abortion provided it includes exceptions for the woman's life and health (as defined in *Doe v. Bolton*)); *Doe v. Bolton*, 410 U.S. 179 (1973) (holding that health includes physical, emotional, psychological, and familial factors, as determined by abortion provider); *Roe v. Wade*, 410 U.S. 113 (1973) (declaring abortion to be a fundamental right and purporting to provide a framework establishing the contours of state regulation of abortion).

36. Some commentators and public officials might claim that this construction of human dignity is intrinsic and objective, albeit limited only to those individuals who meet the criteria of "personhood." That is, they might assert that human dignity applies equally and intrinsically to all "persons," but this designation does not apply to all living members of the human species. To qualify for the moral status of personhood, a human being must meet a set of criteria predetermined and judged by others. Many criteria have been suggested for this purpose. As John Finnis recently surveyed in the context of prenatal human life, such criteria may include "implantation, development of the primitive streak, brain life, sentience, quickening, viability outside the womb, actual birth, actual birth unless it was an induced abortion, formation of desires, formation of concepts, formation of self-consciousness, [and/or] valuing your own existence." See John Finnis (2010) 'The Other F-Word,' *Public Discourse: Ethics, Law, and the Common Good*, 20 October. Available HTTP: http://www.thepublicdiscourse.com/2010/10/1849 (accessed 20 July 2012). This understanding of personhood entails the possibility of subpersonal, prepersonal, and postpersonal human beings who do not possess human dignity. While under this framework, human dignity might be understood as intrinsic, it nevertheless only applies to a class of living human beings to whom personhood is *attributed*. Thus, by extension, the designation of human dignity is under this framework more properly understood as an attributed rather than intrinsic characteristic. That is, the necessary condition for possessing dignity is attributed.

37. 81 Cong. Rec. H4939 (daily ed. June 4, 2003) (statement of Rep. Ros-Lehtinen).

38. *Gonzales v. Carhart*, 127 S. Ct.1610, 1641 (Ginsburg, J. dissenting) (arguing that the right to abortion protects "a woman's autonomy to determine her life's course, and thus to enjoy equal citizenship stature").

39. It bears noting that since *Roe v. Wade*, the Supreme Court has effectively reserved to itself the principal responsibility to define the contours of the law of abortion. The abortion precedents of the Supreme Court facing the Bush administration upon its arrival in 2001 strongly privileged the pregnant woman's interests over those of the fetus, allowing, for example, a woman to terminate her pregnancy at any gestational stage whenever she and her abortion provider concluded that it was in the interests of her health—defined capaciously to encompass virtually all aspects of well-being, such as "physical, emotional, psychological, [and] familial" concerns. *Doe v. Bolton*, 410 U.S. 179, 192 (1972). The breadth of the health exception has been confirmed by the fact that several federal courts have invalidated limits on abortion because they lacked exceptions for "serious non-temporary threat[s] to a pregnant woman's mental health" (*Women's Medical Professional Corp. v. Voinovich*, 130 F.3d 187, 209 (6th Cir. 1997)). Thus, for much of the Bush presidency, the political branches of government were only able to regulate the manner in which abortions were procured (for example, through the enactment of waiting periods, informed consent requirements, parental

involvement laws, and the like). In April 2007, the Court upheld the first restriction on a particular abortion procedure in its history.

40. For a full discussion of the Bush administration's approach to public bioethics and the role of human equality in this context, see Snead 2009.

41. Pluripotent cells are unique and valuable because they are undifferentiated (meaning that they have the capacity to become any kind of tissue in the body) and, in principle, self-renewing (that is, they can reproduce themselves indefinitely without losing their pluripotency). They can be derived from the inner-cell mass of the early embryo (embryonic stem cells), the gonadal ridge of the early fetus (embryonic germ cells), and perhaps from a variety of other sources, including amniotic fluid, bone marrow, and adipose cells. Recent developments suggest that adult cells can be reprogrammed to pluripotency through the introduction of certain genetic factors (Takahashi 2007: 861).

42. For extended discussion of the science, ethics, and public policy of embryonic stem cell research and human cloning, see President's Council on Bioethics 2004a and President's Council on Bioethics 2002. Also see Snead 2005b.

43. Such forms included research on adult and alternative sources of pluripotent cells as well as research involving only those embryonic stem cell lines derived before the announcement of the policy on August 9, 2001.

44. Veto Letter from President George W. Bush to Congress (July 19, 2006).

45. 66 Cong. Rec. H3552 (daily ed. May 18, 2005) (statement of Rep. Foxx).

46. President Bush also implicitly rejected the argument that because the embryos used in stem cell research are capable of "twinning," they are not yet stable individuals, and thus not entitled to substantial moral respect. Twinning is the process by which cells that become disarticulated from the embryo sometimes, through a process of restitution and regulation, resolve themselves into a new, whole organism. It is believed that twinning occurs in very few cases—monozygotic twins are rare, accounting for only 1 in 240 births. President Bush may have been moved by the argument that, as a biological matter, "indivisibility" is not regarded as a necessary criterion for individuation in an organism. Other species are clearly classified as individual organisms, despite their capacity for the biological equivalent of twinning (for example, flatworms). Rather, organisms are defined according to the level of integration and organization of their constituent parts. Human embryos show highly integrated organization, specialization, and differentiation well before the blastocyst phase of development (that is, when they are used in stem cell research). Accordingly, there is strong support for the proposition that a blastocyst is clearly an individuated organism, that is, a whole, individual member of the human species (Guenin 2006: 463). Moreover, opponents of the twinning argument cite recent research (showing a dramatic increase in incidence of monozygotic twinning after preimplantation genetic diagnosis [PGD]) to support the notion that monozygotic twinning is caused by an extrinsic disruption (for example, blastomere biopsy, as performed in PGD), and is not an intrinsic quality of the early embryo.

47. President Bush has also expressed several other ethical concerns about human cloning for biomedical research, including the following worries: that its practice makes reproductive cloning inevitable (as the only remaining step for that procedure is transfer of the cloned embryo to a woman's uterus), it represents an unprecedented step toward more refined techniques of engineering human organisms with a preselected genetic constitution, and the massive number of ova required to conduct cloning research creates dangerous incentives to exploit women, particularly poor women, as sources.

48. A prominent proposed characteristic for this purpose is the "primitive streak"— a biological structure that marks the location of the vertebral column and

indicates the anterior-posterior axis of the organism (though recent evidence suggests that polarity may be established much earlier, perhaps by the locus of penetration of the egg by the sperm). The primitive streak also marks the moment after which twinning is no longer possible. Other suggested capacities marking personhood include the nervous system, the brain, and more mature human somatic form. For a review of these arguments and rejoinders to them, see President's Council on Bioethics 2004a.

49. One might take issue with the claim on its own terms in light of the speculative nature of the promise of embryonic stem cell research compared to the certainty of the destruction of the embryonic human life on which it depends. Also, the possibility that other nonembryonic sources of pluripotent cells (e.g., adult stem cells or reprogrammed adult cells) might yield similar therapies further complicates the utilitarian calculus in this context.

50. Removing Barriers to Responsible Scientific Research Involving Human Stem Cells, Exec. Order No. 13,505, 74 Fed. Reg. 10,667, 10,677 (Mar. 9, 2009).

51. See National Institute of Health on Human Stem Cell Research, 74 Fed. Reg. 32170–02 (July 7, 2009).

52. *Cruzan v. Missouri,* 497 U.S. 261 (1990).

53. *Washington v. Glucksberg,* 521 U.S. 702 (1997); and *Vacco v. Quill,* 521 U.S. 793 (1997).

54. Statement on the Terri Schiavo Case, 41 Weekly Comp. Pres. Doc. 458 (Mar. 17, 2005).

9   Prospects for Human Dignity after Darwin

David H. Calhoun

Today, more than a century after Darwin's death, we still have not come to terms with Darwinism's mind-boggling implications.

—Daniel Dennett,
Darwin's Dangerous Idea

The concept of human dignity developed in classical paganism and in the Christian tradition as the assertion of human value based on two interwoven threads: human distinctiveness, particularly with respect to rational function, and the claim that humans bear a special relationship to the divine, what in the Jewish and Christian traditions has come to be called the *imago Dei*.[1] Charles Darwin's theory of evolution by natural selection, and the broader Darwinian revolution[2] that it spawned, fundamentally challenged human dignity. Against human distinctiveness and the human-brute dichotomy,[3] Darwinism advanced the notion of common ancestry and the related claim that human beings and animals differ by degree and not by kind (Darwin 1882: pt. 1, esp. chap. 1, 3–4, 6; pt. 3, chap. 21; see also T. H. Huxley 1959). Against the view that human beings reflect divine glory, Darwinism emphasized human kinship with animals and replaced divine creation with natural processes as an explanation for biological complexity (Desmond and Moore 1991: 503–8, 511, 515–16; Brooke 2009a: 199). The doctrine of special creation of species by immediate divine activity, which in the 19th century was a conceptual bulwark for the unique spiritual status of human beings, was the focal target of the "long argument" of *On the Origin of Species* and was further criticized in *The Descent of Man* (Darwin 1859: 459, 484–85; Darwin 1876: 424–25; Darwin 1875: vol. 2, 425–26; Darwin 1882: 607).[4]

Darwin was perceptive enough to understand that the implications of evolutionary theory went far beyond biology. As early as the 1830s, as Darwin was beginning to explore the theory of transmutation of species, he foresaw that in the face of evolutionary ideas, the "whole fabric" of human exceptionalism and transcendence "totters & falls" (Darwin 1987: 263 [Notebook C 76–77]).[5] This view of the revolutionary implications of evolution persisted throughout Darwin's life. Shortly after the publication of the *Origin*, more than 20 years after the research notebook comment,

Darwin remarked to his friend and scientific mentor Charles Lyell: "I am sorry to say that I have no 'consolatory view' on the dignity of man. I am content that man will probably advance, and care not much whether we are looked at as mere savages in a remotely distant future" (F. Darwin 1887: vol. 2, 262; Desmond and Moore 1991: 505).

At the same time, Darwin's constitutional dislike for controversy and conflict led him to minimize in publications and public pronouncements direct contradictions between evolutionary theory and the traditional anthropology and theology that grounded human dignity. Darwin's theological ambivalence is notorious: while he emphatically rejected the interventionist special creation associated with Paley's design argument (Paley 1809; Darwin 1958: 59, 87), he left open the possibility that natural laws are established by God, and famously used theistic language in the *Origin* to describe the origins of life (Darwin 1859: 188–89, 488–89; Darwin 1876: 422, 429; Desmond and Moore 1991: 477; Dilley 2011).[6] Taking all of the evidence into account, it is probably most accurate to say that his theological views fluctuated between the extremes of "undogmatic atheism" and "tentative theism," with the default middle stage being agnosticism (Brown 1986: 27).[7]

Similarly, it is possible that Darwin thought that natural selection could coexist with a conception of human exceptionalism sufficient to justify human dignity. Darwin spoke in passing in the *Descent* of the "dignity of manhood" as a state to which humans had at some point in the evolutionary process attained. Further, appealing to social instincts, social approbation, and "improved intellectual faculties" as the basis for moral judgment, Darwin affirmed human attainment of moral autonomy, even Kantian self-legislation, as fully reasonable on evolutionary assumptions, thereby allowing for a moral conception of human dignity (Darwin 1882: 46, 109–10). On this reading, Darwinian evolution deeply challenged human dignity but left it as a conceptual possibility.

Darwin's ambivalence about the compatibility of evolution with the theology or anthropology necessary to underwrite a concept of dignity is deeply complicated by a further issue: the complex intertwining of evolutionary biology with revolutionary philosophical doctrines beyond species mutability and common ancestry (see Himmelfarb 1968: 239; Brooke 2009a: esp. 197–201; Secord 2000).[8] Whether these philosophical theses were implications from Darwinian premises is less important than the fact that many— including Darwin himself—saw them as logical possibilities, if not natural outcomes, of evolutionary thought. For example, Darwin's approach to biology presupposed "methodological naturalism" (see Ruse 2005: 45–46): the view that natural phenomena should be explained by reference to other natural phenomena and that, in general, nonnatural explanatory accounts should be avoided. However, a focus on naturalistic methods of explanation shades rather easily into a stipulated metaphysical materialism, the view that all reality is material. Similarly, what begins as a commitment to avoid supernatural explanations for biology can easily become succor for the "intellectually fulfilled atheist" (Dawkins 1986: 6). Appeal to the view that nature is ordered and intelligible is taken as license for physical determinism. Celebration of the power of natural science for better understanding our world becomes scientism, the notion that science alone or best can find and justify truth claims about the world (Kass 2007: 37; see also Dupre 2001). While many of these philosophical ideas are found in Darwin in germinal form,[9] they were detected as probabilities latent in biological evolution by both supporters and critics of Darwin[10] and are ubiquitous in the writings of neo-Darwinists.[11] Darwinian evolutionary biology is thus often taken to legitimize what we might call "philosophical Darwinism."[12]

Of course, such putative implications of evolution are *philosophical* doctrines that are neither proven nor necessitated by biological theses such as mutable species and common ancestry. While they are philosophically plausible, reasonable and even compelling counterarguments can be raised against each of them. It is beyond doubt, however, that philosophical claims associated with biological evolution have in fact heightened the challenge posed by Darwinism to traditional ideas, including the idea of human dignity.

What, then, are the implications of Darwinist evolutionary biology for a concept of human dignity? There is widespread agreement that Darwinism fundamentally challenges traditional philosophical anthropology and the ethical norms that are grounded in it (Arnhart 1988: 173–74; see also Rachels 1990: 171; Rue 1994; Dennett 1995: 18; Dennett 2003; Brooke 2009a: 197). Human self-image likewise would be vulnerable to Darwinian ideas (Freud 1920: 246–47; see also Freud 1974: 139–40). Specifically, the basic Darwinian theses of biological gradualism, common descent, and naturalistic explanation raise significant problems for the traditional concept of dignity by challenging human exceptionalism and the special creation of human beings. Philosophical Darwinism, the association of Darwinist biology with philosophical claims of materialism, atheism, determinism, and scientism, is still more threatening to dignity, for it embeds humans squarely in mechanical processes of purely material particles operating without any supernatural governor. Understood either way, "Darwinism leads inevitably to the abandonment of the idea of human dignity. . . . Darwinism undermines human dignity by taking away its support" (Rachels 1990: 171).[13]

The Darwinian challenge to human dignity assumes even greater importance in the context of modern medical ethics. Since the mid-1900s, the concept of human dignity has assumed increasing importance as a benchmark for decisions in medical practice (see President's Council on Bioethics 2008).[14] Darwinian evolution is the regnant framework for the biological sciences, including medicine; Dobzhansky's dictum that "nothing makes sense in biology except in the light of evolution" (Dobzhansky 1973: 125–29; see also Simpson 1966: 472–73) is undeniable for modern medicine and medical ethics. An obvious question follows: is an outlook shaped by Darwinian evolution *compatible* with a robust ethics of dignity?

I answer this question with a two-part thesis: (1) the most obvious proposals for reviving human dignity post Darwin are practically or conceptually flawed; and (2) it is nevertheless possible to construct a robust ethic of dignity consistent with the principles of Darwinian biology based on human functional exceptionalism and the notion that humans uniquely image God, the continuously creative first cause of nature.[15] In short, while Darwinism has played a role in the collapse of the concept of human dignity from the late 19th century to the mid-20th century, and even though the 20th-century revival of the concept of dignity has proceeded apart from and even in defiance of Darwinian challenges, Darwinism and dignity need not be regarded as mutually contradictory.[16] The triumph of the Darwinian evolutionary outlook in modern biological sciences does not require abandonment of the powerful tradition of human dignity.

STRATEGIES FOR HUMAN DIGNITY AFTER DARWIN: EVASION AND RECONCEPTION

Larry Arnhart has argued that Darwinist anthropology precipitates a crisis of value: insofar as it undermines transcendent human goals and purposes, Darwinism inevitably presents us with a "biological nihilism" that must be answered (Arnhart 1988: 173–74; see also Rachels 1990: 171; Rue 1994; Dennett 1995: 18; Provine 1988: 70). Because human beings are not mere mechanisms, nihilism is an unlivable and untenable outlook.[17] So the collapse of value brought about by Darwinism is a step toward "the substitution of a different sort of ethic" (Rachels 1990: 171). Whether one celebrates or bemoans the Darwinian revolution, the consequent collapse of value must be addressed.

Contemporary literature on biology, human value, and human behavior addresses the question of ethics in a number of distinct ways, most of which, however, ignore the question of human dignity (see, e.g., Wilson 1978; Arnhart 1998; Wright 1994).[18] Indeed, many recent evolutionary treatments of ethics and human behavior admit from the outset that dignity is dead as a post-Darwinian ethical concept (Skinner 1971; Rachels 1990). Given the problem of grounding dignity theologically in a context of increasing religious pluralism, the best strategy appears to be to focus on the human exceptionalism thread of the traditional concept of dignity.[19] Even if neo-Darwinists have little to say about human dignity as an ethical value, do they offer sufficiently robust conceptions of human exceptionalism to reground dignity? If not, do they offer accounts that allow some place for dignity as an ethical criterion? The literature here is voluminous and complex enough that all I can hope to do is indicate different lines of response to the Darwinian crisis of dignity and try to offer some assessment of their conceptual vitality. I find two general approaches to the question of dignity in a Darwinist context: *evasion*, which attempts to avoid the Darwinian challenge to human dignity, and *reconception,* which seeks to reground dignity and value in some way other than human exceptionalism or the *imago Dei.*

Evasion: Rejectionism, Dualism, Segregationism

Of course, the most obvious strategy of evasion to preserve human dignity in the face of Darwinism is to reject Darwinism in whole or in part. If Darwinian evolution is false, then there is no crisis of biological nihilism, and all is well for human exceptionalism and human dignity. While I cannot here untangle the evolution-creation debate, I can offer two reasons for bypassing the rejectionist response to Darwinism, one methodological and one more substantive.

The task of the present argument is to determine the possibility of a dignity-based ethic within a Darwinian evolutionary context. A derivative task is to determine the implications of Darwinism as they have spread through culture and ethics. For that reason, the rejectionist response bypasses the task at hand. Even if there are conceptual, theoretical, or scientific problems with Darwinian evolution, we have to take stock of the intellectual and cultural effects of Darwinism.

Substantively, I share with many rejectionists opposition to the conflation of scientific and philosophical Darwinism (e.g., Plantinga 1997a). The power of the scientific account of descent with modification by means of natural selection and the presence of shared elements of biological morphology across nature are not evidence for the philosophical positions of materialism, atheism, determinism, or scientism. Further, reasonable critics of widely varying views have objected to the globalizing and reductionistic tendencies of many proponents of evolutionary theory (see, e.g., Nagel 1997; Dupre 2001; Miller 1999; Haught 2010). However, critics of modern evolutionary theory must also come to terms with the role that evolutionary theory plays as an organizing principle for biology, biochemistry, paleontology, biogeography, and related disciplines (Wilson 1998). While it is possible that evolution will turn out to be the phlogiston theory of the 19th and 20th centuries, that possibility is exceedingly, and increasingly, slim.[20] Consequently, rejectionism is not a very promising strategy for defending human dignity.

An evasive strategy closely related to rejectionism is post-Darwinian dualism, a halfway approach that grants that evolution is a completely adequate account for biological development, including the human *body,* but that simultaneously excepts the human mind or soul. The exception involves some sort of dualism, in a broad sense, which claims that a dimension of the human person—the soul or mind—is different in kind from physical nature and therefore not subject to natural processes such as evolutionary selection. This approach preserves human exceptionalism by the Cartesian strategy: a physical world of mechanisms from which the human mind or soul is distinct. Many of the traditionalist biologists in Darwin's time who were committed to biological transmutation found this approach appealing, for they worried that evolution would bestialize human beings; Darwin's geologist friend and mentor Lyell is perhaps the best example.[21] Of course, versions of the dualist evasion strategy have appealed to many people since Lyell. Some would undoubtedly suppose this to be the only and best approach for Christians and other theists to preserve the possibility of survival of the person after death: a Platonic-Augustinian view of soul melded with a modern scientific view of the body (Ruse 2001: 80; see also Cooper 2000; cf. Green 2008).[22]

The dualist strategy faces several significant problems, including all of the standard philosophical and conceptual problems of any dualist system: mind-body causal interaction, problems about the relationship between spatial bodies and presumably nonspatial minds or souls, and so forth (Ryle 1949: 11–24; Hasker 1999: chap. 6). In addition to these basic problems, a post-Darwinist dualism encounters further difficulties. Dualism locates rational function in the distinct mind or soul, and yet contemporary empirically informed accounts of human beings show that cognition and other mental functions are closely related to physiological conditions and clearly impaired by damage or alteration of those physiological conditions, say, by disease or drugs. The notion of an independent mind or soul has great difficulty explaining these normal experiential phenomena.[23] Most important for our purposes, dualism allows the sort of exceptionalism that can serve as a basis for dignity, but only as a philosophical (and perhaps theological) postulate independent of biology. This remains an evasive strategy, because it works on the assumption that contemporary biology is largely irrelevant to our understanding of human nature, and consequently to our conception of dignity.

A third strategy for evading the challenge of Darwinian evolution to principles of ethics and dignity is simply to divide biology and values into different conceptual spheres. If biological facts are distinct from ethical principles such as justice and human dignity, then innovations in one field will not affect adversely the other. This view is intuitively appealing, especially given the cultural prevalence of the science-religion war narrative. If religion just leaves science alone and vice versa, or science leaves ethics alone and vice versa, everyone can be happy.[24]

This irenic objective is the basis for the famous formulation of the segregation strategy of Stephen Jay Gould, called the "nonoverlapping magisteria" or NOMA proposal (Gould 1997). Gould argues that science and religion are different domains of knowing or magisteria. So long as the proper borders are maintained between these areas, conflict can be easily avoided. The *facts* of evolutionary biology—that humans share common ancestry with nonhuman living things, that living things differ by degree and not by kind—are on this view construed as ethically neutral; problematic challenge to dignity evaded.[25]

While intuitively appealing on many levels, the NOMA proposal faces significant problems. There are clear cases where facts have value implications. Still worse, value-laden facts are particularly prevalent in science-religion issues, the narrow target of NOMA. The same events are often subject to treatment and analysis by both religion and science: the ultimate origins of the universe, competing explanations for purportedly miraculous events, and the factual bases of significant religious doctrines, such as the Resurrection of Jesus (Dawkins 1998; Kass 2007: 39–40). The emergence of the value of dignity in classical paganism and the Jewish-Christian tradition shows that values rest on a network of claims about human beings, nature, God, and reality.[26] Further, while philosophers have fled from is-ought associations in the wake of Hume, and while scientists have fled from teleological assertions about nature since Darwin due to objections to implied supernatural agency or design, the notion that value can be grounded in normative claims about natures of things has an august philosophical pedigree tracing back to Plato and Aristotle (Plato *Republic* 1.352d–353e; Aristotle *Nicomachean Ethics* 1.71097b24–1098a17). Contemporary ethical naturalists of various stripes have revived Aristotelian arguments that good should be understood in terms of the capacities and desires of human beings (Arnhart 1998: 71–73; Dennett 1995: 468).

In the end, Gouldian fact-value segregation is condescending to religious adherents and anyone with thoughtfully held ethical views, because it empties such beliefs of objective content and construes the basis of religious or moral beliefs as a groundless nonrational fideism. It allows a place for moral value, but only by stripping those values of any meaningful reference to the objectively existing world (Haught 2003: 8).[27] Ostensibly protecting religious and moral values by segregating them into a distinct sphere does them no service and provides an inadequate basis for human dignity.

Strategies that seek to evade the values challenge of biological nihilism are therefore unsatisfactory.

Reconception: Darwinian Holism, Darwinian Existentialism, Darwinian Progressivism

A second general approach, reconception, seeks to reground human dignity in some way other than human exceptionalism. One such value reconception traces its lineage directly to Darwin himself. He notes in the closing pages of the *Origin* that individual species, including long-extinct ones, "become ennobled" if we see them as part of a continuous organic whole (Darwin 1859: 488–89). He extends this idea at the close of the book, proclaiming, almost poetically, that "there is grandeur" in the holistic picture of organisms vitally alive as a result of natural mechanisms (Darwin 1859: 489–90; see also Darwin 1987: 228–29 [Notebook B 232]).

Darwinian ethical holism—also called "ecocentrism" and related to the modern view labeled "biocentrism"[28]—has spawned an enormous literature, ranging from scientific to philosophical to popular treatments, and branching out in environmental ethics, animal ethics, evolutionary theology, and religious naturalism.[29] Despite their differences, these approaches take common ancestry as their point of departure and from that ground affirm the value of the whole of nature, rich with vitality and diversity, and sufficiently fecund to produce genuine novelty (Rolston 1988: 197; Haught 2008: 49–60). The commonalities of different versions of Darwinian ethical holism can be captured in two key principles: (1) denial of the claim that human beings alone are morally valuable, a stance critically labeled "speciesism," and (2) affirmation of the intrinsic value of nature and/or natural organisms, particularly as they are capable of realizing the ends appropriate to their nature (P. W. Taylor 1986: 129–56; Rolston 1988: 67; Rachels 1990: 173–74, 181–94, 208–9).[30]

Darwinian holism clearly provides a framework for ethical systems, but it is not at all clear that it can underwrite a conception of dignity. As we have seen, dignity is rooted in the idea of distinctiveness and therefore presumes some division, crudely put, between haves and have-nots. It was the great promise of the concept of dignity arising from the convergence of Stoicism, Judaism, and Christianity to propose that the domain of the haves is comprised by all members of the human species, even if this insight was poorly understood, unevenly implemented, incompletely articulated, and disfigured by unjust practices. On the other hand, to extend the domain of value significance to everything, or even to all living things, or even minimally to all animals, stretches it so thin that its moral weight is dissipated.[31] Alternatively, appeals to respect as a moral criterion effectively surrender the categorical inviolability characteristic of dignity (Leopold 1989; P. W. Taylor 1986). I suspect this is why most holists simply abandon the concept of dignity rather than attempting to conceptually ground its extension.[32] Consequently, ethical holism, no matter its fruitfulness as an ethical outlook, provides a poor foundation for reconceiving dignity.

A response to biological nihilism appealing to many contemporary defenders of philosophical Darwinism is a variation of reconception that we can describe as Darwinian existentialism. On one level, existentialism and Darwinism would seem unlikely and mismatched partners. Where existentialism emphasizes subjectivity, personality, and passion, Darwinism, as a scientific outlook, stresses objectivity, the publicly accessible, and reasoned observation and description. Darwinism is committed to the attempt to know and describe the world more or less as it is, while existentialism, in whatever form, is an exploration and celebration of the self, revealed in the immediate experience of self-consciousness (Sartre 1985: 36). Understood in these terms, existentialism and Darwinism are almost conceptual opposites.

On another level, though, existentialism is a natural partner for neo-Darwinism, as several key affinities illustrate. First, Darwinism, like existentialism, proclaims the collapse of traditional notions of human exceptionalism and autonomy, and thereby both describes the predicament of nihilism and prompts the quest for a response to nihilism. More specifically, Nietzsche, the prophet of nihilism, frames his post-Enlightenment critique of ethics in patently Darwinian terms: genealogy and adaptationism (Dennett 1995: 461–67).[33] At an even deeper level, Nietzsche and Darwin shared the view that application of evolutionary principles to human beings posed an inescapably destabilizing threat to rationality and objectivity. Darwin at one point bared his "horrid doubt" that human rationality, if produced by mechanisms of selection aimed at adaptation, would be unreliable; Nietzsche imagined in a parable the corrosively skeptical implications of an adaptive view of rationality.[34] If human beings are the product of adaptive forces that have as their aim survival, not truth, disastrous implications follow: given our physical weakness and frailty, deception is more adaptive than truth, given our precarious circumstances and need for social life, self-deception is more adaptive than self-knowledge. The outcome of this is an unquenchable desire to seek truth and to locate the truly real, both of which are nothing more than fictive inventions.

Darwinist accounts of ethics, value, and religion are deflationary explanations that supplant the internal logic of these practices with externalist genealogies. But, of course, the inexorable self-enclosure of external genealogies applies to the Darwinist account of human beings as well: the snake swallows its own tail, and there is no longer any narrative outside of the globalizing skeptical story. This "naturalizing" trend covers not only human biology, but behavior, culture, philosophy, religion, eventually everything, including science.[35]

The Nietzschean collapse of value and death of god constitute no more and no less than the trajectory of philosophical Darwinism: the same materialism, atheism, and determinism, but with science now exposed as no less a myth than the now-debunked traditions of realism, theism, free will, and therefore human dignity. In this state, the only viable prescription is the interrogative imperative of the Madman who proclaims the death of God: "Must not we ourselves become gods simply to seem worthy of it?" (Nietzsche 1954: 96). And that in turn means that with values in collapse, with nihilism the outcome of applying the logic of Darwinian biological mechanisms to the scientific knower himself, values must be created. Simply choose (Sartre 1985: 28).[36]

Strikingly, "just choose" is essentially the message of many neo-Darwinists, who offer globalizing mechanical accounts of human behavior—governed by genes and environment—and then suggest that we can and should declare our independence from those masters. The most straightforward Darwinian existentialists paradoxically deny human freedom and yet proclaim the human capacity to rebel against genetic determinism (see, e.g., Dawkins 1989: 3, 201; Wright 1994: 10, 31, 102, 358; cf. de Waal 2006: 7–10; Lewis 1947a: 83–84); others take the more indirect approach of allowing for genuine agency but assert that values are by necessity arbitrary constructs (Wilson 1978: 71–78; Dennett 1995: 370–83; Wilson 1998: 118–20; Thompson 1999: 481–83; see also Rolston 1999: 140–41; Ruse 2001: 211–15). In either case, however, any value chosen is by definition a subjective construction, which can, of course, be deconstructed. This problem is not merely theoretical, as the mass-scale violations of human dignity in the

20th century exhibit. If human dignity is not a transcendent value, it cannot bear the moral weight necessary to serve as a check on the choices and behavior of those who would discount or ignore the standing of those who happen to impede their wishes (see Wilson 1998: 238–64). The dilemma we face is that, according to Darwinism, we "evolved genetically to accept one truth and discovered another" (Wilson 1998: 264). We were made for transcendent, abiding truths and yet are by Darwinism informed of the poignant emptiness of that proclivity. If that is true, of course, human dignity is indeed dead. We can construct all manner of moral principles and ethical values, but they will be constructs of convenience and social necessity rather than the sort of inviolable principle dignity must be.

A final possible Darwinian reconception strategy, Darwinian progressivism, emerges from a debate among contemporary neo-Darwinians concerning the question of evolutionary progress. While Darwin and other Darwinists of the 19th century were ambivalent about this question, they leaned toward progressivism.[37] The current changed radically, however, as the understanding of evolutionary processes was mechanized in the Evolutionary Synthesis and neo-Darwinists largely abandoned progressivism (Provine 1988). Stephen Jay Gould devised a defining metaphor for antiprogressivism: the evolutionary tape player that when rewound cannot be guaranteed to play the same thing as before. The implication of Gould's account is that evolution is a radically contingent process in which not only humans are improbable, any organism with highly developed intelligence is also highly improbable (Gould 1989: 45–52, 319).

By contrast, a significant number of contemporary Darwinians take issue with Gould on this. While granting that there is no guiding force behind evolution, contemporary Darwinian progressives assert that there are indeed "obvious moves" that evolutionary change will take. A standard example is the independent evolution of eyes or flight: vision or movement through an air environment is so positively adaptive that it counts as what Dennett calls a Good Move in Design Space (Dennett 1995: 306). Dennett's idea is a development of what Dawkins calls a biological arms race: a process in which competition between different organisms, most obviously predator and prey, leads to progressive development of their functional capacities, sometimes very rapidly (Dawkins 1986: 178–92; see Darwin 1987: 422–23 [Notebook E 95]; Wright 1990: 32).[38]

How is this relevant for human dignity? Darwin himself posed the relevant question in his notebooks: "The believing that monkey would breed (if mankind destroyed) some intellectual being though not MAN.—is as difficult to understand as Lyells doctrine of slow movements &c. &c." (Darwin 1987: 263 [Notebook C 74]). Even if human beings are not specially intended by God as the pinnacle of nature, beings very much like us might be the near-inevitable outcome of the evolutionary process. If intelligence is highly adaptive, and high levels of intelligence especially so, then perhaps we can link dignity to progress and designate as exceptional the organism that moves toward optimality at the top of the biological scale (see R. J. Richards 2009).

While this approach at least provides a way of reconceiving human exceptionalism, it suffers its own fatal shortcomings. Even if it is the case that natural processes will bring about some form of complex intelligence almost inevitably, humans are not the particular outcome of the process. In an elimination tournament, someone has to win, and that someone will win the prize, but all but the most obsessive sports fanatic understands that the outcome is shaped by all sorts of contingencies. Similarly, there is no clear reason why humans would gain distinctiveness and therefore dignity as a result of such a process; it is more like winning the lottery than being in a place of value in the cosmic hierarchy. A related problem is that any such win is by definition temporary. As long as the evolutionary process continues, other species will arise that will continue the move toward greater complexity and superiority. Dignity would thus be like winning this year's championship: good when it happens, but not permanent. A final problem is more menacing. Any conception of dignity linked contingently to the current occupant of the highest rung on the biological scale is an invitation to despotism. Especially if the current occupants realize that their position is both contingent and temporary, they will have motive to ensure that their position is protected. The easiest way to accomplish this would be to tyrannically enforce their superiority over others. By this means, dignity would license "species-cide," a result completely incompatible with the character of dignity. Darwinian progressivism is yet another dead end for reconceiving dignity.

RESTORING HUMAN DIGNITY AFTER DARWIN

C. S. Lewis noted that when a path of inquiry seems to have reached a dead end, retracing one's steps and reviewing the conditions of the catastrophe often provide the best solution strategy (Lewis 2001: 28–29; see also MacIntyre 2007: 1–5). If the Darwinian challenge to traditional human dignity can be neither evaded nor resolved by reconception, perhaps the solution lies in a restoration of the traditional concept that acknowledges and responds to the Darwinian revolution. Such a restoration is possible for both strands of the traditional concept of human dignity.

Restoring Human Exceptionalism: Human Rational Function

Descartes framed the modern debate on human exceptionalism by claiming that humans possess functional capacities different in kind from animals, as evidenced by the "machine tests" of communication of ideas and creative interaction with the environment (Descartes 1993: 32–33). Darwin directly challenged this claim in *The Descent of Man*, arguing both that rudimentary forms of human mental activity could be found in animals and that vestiges of animal instinct could be found in humans. Darwin's alternative to Descartes's clean-line distinction between humans and brutes was a gradualist account of incremental differences in function across the biological spectrum (Darwin 1882: chaps. 3–4). Darwinist critics of the traditional conception of human dignity often express pessimism that it can be revived post Darwin (Arnhart 1988: 173–74; Rachels 1990: 171–72).

Since the time of Darwin, a number of candidates for human functional exceptionalism have been proposed: not only the classic criterion, rationality, and the Cartesian markers of conceptual language and creative behavior (Descartes 1993: 32–33), but others, including conceptual abstraction, self-awareness, symbol use, tool use, aesthetic sense, culture, religiousness, complex sociality, deception, and humor (see, e.g., Rolston 1988: 65). Since the time of Darwin but especially in the last 50 years, extensive empirical research has contributed a great deal of data to the debate, demonstrating that many supposedly unique human functions also can be found in animals. For example, birds have been observed using tools, bees have been found to communicate about food sources using elaborate "waggle" dances, and many animals have been taught various complex tasks. Deceptive strategies have been studied in a number of species, at least some of which have a cognitive component. Other complex cognitive functions, such as self-recognition, have been extensively observed and documented in a number of species (Gould and Gould 1994; Griffin 2001; Schilhab 2004). While there are recurring design problems with research on animal intelligence, most notably a tendency on the part of humans to anthropomorphize animal subjects (Rachels 1990: 167–71; Wynne 2007; Wright 2006; but see de Waal 2006: 59–67), important tentative conclusions are possible, especially concerning higher animals. Higher animals clearly are conscious, meaning that they have an inner subjective life. They have desires that are shaped and modified by their perceptual experience of the world. They have complex social organizations and communicate in sophisticated ways that manage and negotiate social hierarchies.

Despite ongoing empirical evidence supporting human-animal continuity, a significant argument based on the structure of human cognition, and consistent with current empirical research, provides sufficient basis to support human functional distinctiveness. The notion of significant functional differences between humans and animals, even between humans and our closest genetic relatives, the great apes, is standard in contemporary biology. As E. O. Wilson puts it, regarding the vast increase in brain size from primates and extinct species of *Homo* to human beings,

> The growth in intelligence that accompanied this enlargement was so great that it cannot yet be measured in any meaningful way. Human beings can be compared among themselves in terms of a few of the basic components of intelligence and creativity. But no scale has been

invented that can objectively compare man with chimpanzees and other living primates. (Wilson 1975: 548)

The standard explanation for the great difference between humans and higher animals is the foundation provided by the larger human brain and the variations of language and culture that are underwritten by more expansive human brain power.

Research over the last half-century has proceeded with the assumption of functional continuity set out by Darwin in the *Descent*, and a great deal of such continuity has been established. As Darwin argued in the *Descent*, humans and higher animals share basic conscious experience at the levels of desire, emotions, perception, social relations, and agency (Darwin 1882: chap. 3; Griffin 2001). Despite these similarities, however, there is strong evidence for distinct structure to human cognition.

> Although humans and nonhuman animals share many similar cognitive mechanisms, our *relational reinterpretation hypothesis* (RR) is that only human animals possess the representational processes necessary for systematically reinterpreting first-order perceptual relations in terms of higher-order, rule-governed relational structures akin to those found in a *physical symbol system* (PSS). (Penn et al. 2008: 111)

Penn et al. find that animals generally share with humans cognitive capacities for making basic perceptual judgments, associating objects, perceiving spatial relationships, and understanding relationships between events and objects in their experience. At the same time, on the sorts of activities that require an ability to prescind from the concrete particulars of immediate perceptual experience and reflect using rules, principles, and logical connections, the performance of nonhuman animals is demonstrably inferior. As an example, chimpanzees were highly successful at associating a miniature replica of an object and the object itself but were much less successful when the comparison involved an abstract scale model of a physical location (Penn et al. 2008: 115). Research on theory of mind, the ability of subjects to understand and act on the state of another agent's mental state, suggests that primates are capable of formulating basic behavioral strategies in their interactions with other animals but show no indisputable evidence of comprehending the beliefs or intentions of others (Penn et al. 2008: 120). The reasonable conclusion is that human beings alone are able to reflect on perception and experience with "classical, inferential, role-governed, domain-independent systematicity" (Penn et al. 2008: 127).

Important relationships can be drawn between this research and other accounts of human and animal cognition. Korsgaard, for example, employs a cognitive model drawn from modern moral psychology and Kantian principles to distinguish the cognitive experience of lower animals, higher animals, and humans (Korsgaard 2006). Lower animals act purposively in the light of perception of the environment and the desires or functional objectives of the organism. A spider, for example, perceives a moth trapped in its web and moves toward it. We attribute a kind of agency to the spider even if we understand that its actions don't involve any complex cognition or mental states. By contrast, a higher animal not only acts purposively but also "can entertain his purposes before his mind, and perhaps even entertain thoughts about how to achieve those purposes" (Korsgaard 2006: 109). At this level, perception is bound up with genuine conscious awareness. While this account attributes genuine conscious awareness to higher animals, that consciousness is still distinct from and inferior to human awareness. While the purposes for animals, lower and higher, are given by their desires and emotions, and then activated by their perceptions of their surroundings, human experience allows cognitive distance from desires and emotions. In line with Penn et al., while animals are capable of quite complex perception of their environment and to some degree their own desires, humans are capable of considering how a desire serves as a ground for acting in relation to a particular object, and further capable of considering alternate grounds. Specifically, Korsgaard notes that human cognition facilitates posing questions about what one's purposes should be, and for judging competing purposes in the light of a principle. Thus, humans are not only aware of the world, we are aware of our stance toward the world and of the possibility of other stances (Korsgaard 2006: 110–13).

Recognition of the qualitatively distinct nature of human cognition helps us to see that the half-century attempt to interpret human cognition in terms of behavior (Ryle 1949; Turing 1981; Skinner 1971), all traceable back to Descartes's machine or zombie tests (Descartes 1993: 32–33), was misleading. Animals—and computers—are capable of very complex forms of interaction with their environments. But behavior is at best an indirect indicator of human cognition. The human difference lies less in particular patterns of behavior than in the qualitative character of cognition: that we organize our experience in terms of principles, and that we consciously employ such principles to guide our actions.

Penn et al. insist that their account, despite its assertion of human-animal discontinuity, is completely Darwinian (Penn et al. 2008: 164). This is because they affirm "that there are no unbridgeable gaps in evolution"; "All similarities and differences in biology are ultimately a matter of degree. Any apparent discontinuities between living species belie the underlying continuity of the evolutionary process" (Penn et al. 2008: 110, 129). This might seem paradoxical: their argument is, at the same time, for the claim that animals and humans are functionally discontinuous *and* that they are continuous.[39] It helps to keep Darwin's central principle in mind here. He argued in the *Descent* that "the difference in mind between man and the higher animals, great as it is, certainly is one of degree and not of kind" (Darwin 1882: 126). As he makes clear, his point is to assert a *developmental* or genealogical relationship between mental functions (Darwin 1882: 65–66).

For that reason, his objective was not so much to make a judgment about the relative qualitative capacities of humans and animals as to ensure the evolutionary interpretive principle that animal and human functions derive from a common source.

The nature of biological reductionist explanation can shed light on how human cognition can be both continuous and discontinuous within a Darwinian framework. The principle of reductionist explanation acknowledges different structures and functions at different explanatory levels (Dawkins 1986: 13; see also Wilson 1978: 7–13; Searle 1984: 20–22; Crick 1994: 7–12; Dennett 1995: 80–83; Haught 2003: 14–15). There is an order to the levels of reductionistic explanation such that certain explanatory projects require attention to particular levels. Even though explanation requires appropriate attention to the relevant level, each level is analyzable into its components at the lower level, and the structures and functions of the underlying level in each case account for the structures and functions at the next level up.

Appropriate differences between levels of explanation support the idea of emergence: that there are structures, functions, and properties at higher levels that are not present at the lower, because the lower-level structures are a condition for the structures of the higher (Searle 1984: 22; Percy 1991: 271). In the case of organisms, the notion of emergent qualities is a critical part of our understanding of their natures and functions. For example, cats and bats are mammals that are genetically close. They have similar morphology, organs, and bone structure, but the structural differences are significant, because they give rise to functionalities that are very different: cats prowl while bats fly. The flight of the bat is an emergent property of its structure. While it is true that a bat is composed of organic material, bone, and so forth, and that those materials are in turn analyzable in chemical terms, the bat's capacity for flight cannot be explained in terms of potassium or bone. It is a particular function of the structure of the bat as a biological organism.

By the same token, human and nonhuman cognitive functions are in one sense continuous, because they are composed of the same sorts of constituent elements and even rooted in similarly structured physical organs. On the other hand, they are discontinuous insofar as the functionalities they support are so distinct from one another. When Darwin says that animals and humans differ in degree and not in kind, the referent is genealogical and developmental continuity. As long as the structures admit of a line of developmental processes from one to the other, it is consistent with evolutionary principles. Developmental continuity does not imply functional equality.

The emergentist structural-functional account of organisms draws from Aristotelian biology. In *On the Soul*, Aristotle argues for a conception of the soul as the formal organizational principle of a living body: "the soul is the first actuality of a natural body which has life potentially. Whatever has organs will be a body of this kind" (Aristotle *De Anima* 2.1.412a26–28 [1968: 9]). He means by this that the soul is the structural principle of a living thing, which structures the body in such a way as to facilitate the life functions of that particular sort of organism. For example, a bat has particular structures—wings, lungs, eyes, brain, and so forth—that make possible the life functions of bats—flight, respiration, vision, perception. Only bodies ordered in certain ways can fly, or breathe, or see, or think, or—ultimately— be alive. In saying that soul is a principle, Aristotle means that it is not a thing: "the soul does not exist without a body and yet is not itself a kind of body. For it is not a body, but something which belongs to a body, and for this reason exists in a body" (Aristotle *De Anima* 2.2.414a20–22 [1968: 14]). The Aristotelian soul is not a thing that inhabits or is placed in a body, like the immaterial soul of Plato's *Phaedo*, it is a biological principle of organization. Insofar as it is not a dualistic principle of animation, Aristotle's conception of soul is compatible with Darwinian biology.[40] It is the informational principle of organization that facilitates emergent properties of functionality supported by the material components of the body it orders.[41]

A soul that is a principle of functional organization and that is the causal explanation of the functional powers of the thing is a sufficient basis for distinctiveness. On Aristotelian emergent terms, the human soul is not a thing added to the human body, nor is soul uniquely present in human beings. As a *biological* principle, soul is the particular functional principle of organization for all living things. It is not the presence of a soul that makes the human being distinctive on the Aristotelian view; it is the particular ensemble of functional capacities made possible by the particular structures and organs of the organism. In keeping with current empirical research, we can understand human beings as having qualitatively different emergent capacities, even though those capacities are completely rooted in material reality. Human organisms have brains of sufficient size and structural architecture to support rule-governed, second-order cognition (Gazzaniga 2009: 22). On this basis, human beings are loci of distinctiveness.[42]

We noted earlier that Darwinism is both a scientific theory, which asserts common ancestry of living things and graded relationships among species, and a philosophical outlook, which links the scientific view to a philosophical outlook involving materialism and atheism. The scientific theory has been enormously successful as a research program in modern biology and has proved extraordinarily fruitful in uniting a wide variety of disciplines conceptually and methodologically. While philosophical Darwinism has garnered the allegiance of many prominent evolutionists, it remains problematic on a number of levels, not least its implied value nihilism and the tensions between its materialist and mechanist postulates and its naïve epistemic realism. Since Descartes, the unhappy alternatives in philosophical anthropology have been Cartesian dualism, the materialism that descends from early modern mechanistic cosmology that has been embraced by philosophical Darwinism, and idealisms and phenomenalisms that descend from the Cartesian affirmation of subjective self-aware consciousness. Among these competitors, emergent functionalism is at least a plausible alternative.[43] First, it conforms to the explanatory framework of hierarchical reductionism common in contemporary biology. Second, it is broadly consistent with current research and philosophical articulation of functional human distinctiveness. Third, it elucidates the notion of hierarchical emergentism by clarifying the basis of functional properties in biological structure. Fourth, it takes a naturalistic view of the world that both affirms empirical reality and empirical methods of knowing and avoids the greedy or preposterous forms of materialism that find it necessary to debunk, dethrone, and disabuse. Consequent to that, it allows for a robust conception of human exceptionalism and, therefore, is consistent with a restored notion of human dignity.

Of course, there are problems integrating Aristotelian and Darwinist outlooks. Perhaps most importantly, Darwin's theory of transmutation of species requires, if translated into an Aristotelian framework, the view that forms or natures are malleable. Yet for metaphysical and epistemological reasons (largely following Plato), Aristotle takes the soul to be immutable (Frede 1992: 106). But there are possible solutions to this and other problems. Perhaps, for example, forms of living things can be understood as something like mathematical objects, which are instantiated in different ways in the changing empirical world without themselves being changeable. In this case, populations of organisms do indeed change or speciate, but the patterns of functionality they manifest exist independently, like ways of biological life waiting to be discovered.[44] At the very least, species post Darwin can be understood in terms of the "bare-bones essentialism" of natural kinds (Sulmasy 2006: 77).

The question of the relationship between the potential for certain functional capacities and the successful development of capacities is another such problem. If human dignity depends on the actual development of particular capacities of reflective rationality, then many members of the human species will fail to achieve the requisite mark and will therefore be denied the status of dignity (see Sulmasy 2006: 76–79; Meilaender 2009: 3–8, 87–104). The traditional Aristotelian response is that membership in a species or natural kind has to do with the natural capacities that are inherent in that kind. Even if development of the capacities is inhibited due to disease or accident, that does not diminish the individual's place within the natural kind. If the concept of natural kinds still has meaning in a post-Darwin context, this solution is still viable. Finally, one might object that an emergentist or Aristotelian conception of the human person is essentially materialistic, and therefore suffers all of the limitations and problems, including those for human dignity, of a materialist view.[45] While emergentism can be materialistic, however, it need not be; robustly Aristotelian emergentism is certainly not so, insofar as it acknowledges the reality of immaterial form.

Restoring the *Imago Dei*: Human Imaging of the Continuous Divine Cause of Nature

What are the post-Darwin prospects for restoring a concept of dignity rooted in the *imago Dei*? Alternate conceptions of God in the light of the largely autonomous natural sphere implied by Darwinian evolution underscore the problem of dignity: any divine agency sufficiently robust for bringing about organisms reflecting the *imago Dei* seems incompatible with Darwinian evolution. Is there a God for Darwin adequate to intend and bring into being rational organisms in his own image?[46]

Recall that the immediate object of the theological attack of the *Origin* was not theism *simpliciter* but special creationism or interventionist divine agency of the sort exemplified by Paley's teleological argument. As Darwin and other early evolutionists insisted, rejection of an interventionist God whose agency is special creation does not yield atheism by implication, for it is still possible for the laws that govern natural processes to have a supernatural origin. This first-cause theism (or deism) had some appeal in the latter half of the 19th century, perhaps because it promised to incorporate the scientific developments of the century within a broadly traditional framework. However, many in Darwin's time regarded the attack on special creation as an implicit attack on theism and found suggestions of a law-designing God but a step toward atheism. According to these critics, insofar as the ordered intelligibility of the natural world can be explained by processes in nature such as evolution by natural selection, divine agency has been supplanted, not simply reimagined. In any case, the wild profusion of evolutionary theories in the late 19th century gradually narrowed as the linkage of Darwinian natural selection to Mendelian genetics accentuated mechanisms, and perhaps as theists grew wary of the philosophical connections drawn by many fervent Darwinists to atheism, materialism, and other worrisome doctrines (Provine 1988: 53–62; Numbers 1988). With the decline of varieties of theistic evolution, the alternatives devolved to Darwinistic nontheism and non-Darwinian interventionist theism. Thus began the orthodoxy of philosophical Darwinism. In any case, a deistic first cause would seem insufficient to ground human dignity, for it is hard to call "created in God's image" organisms that emerge as the unintended outcome of mechanical natural processes, even if those processes were supernaturally ordained (see, e.g., Miller 1999: 196–97).

Strikingly, the interventionist God versus atheism dichotomy is largely replicated in contemporary discussions about God and evolution. While atheist and agnostic Darwinians argue that natural processes are sufficient in principle to explain all natural phenomena, those defending the compatibility between science and theism seek to find traces of divine action in a largely autonomous nature, whether at particular crux points such as the emergence of life or the formation of complex structures such as DNA, or in the interstices of quantum indeterminacy (Behe 1996; Meyer 2000; Miller 1999).[47] William Dembski's incredible talking pulsar offers a scenario in which supernatural intelligent agency is a reasonable inference to the best explanation (Dembski 1994); nevertheless, less flamboyant scenarios are subject to explanation by natural processes, and therefore are open to "God of the gaps" criticisms.[48]

There is a third way, however, via a first-cause creator different from the deistic law designer that was the default in Darwin's time. A theologically richer concept is provided by the Aristotelian-Thomistic first cause, the continuously creating primary cause of all that exists, the radical cause of all things in the universe at every moment, the source of intelligibility, and the continuous compresent origin of the causal powers of natural things.[49] God understood in this way is the first cause, first not in temporal sequence but in the order of being, underlying and providing the intelligibility for the entire system of natural causes. This God is therefore not the first-cause of deism, the cause that sets the entire universe into motion with a flick and then has no further interaction with it. Rather, the Aristotelian-Thomistic first cause is the radical ground for the being of the natural world and the causal source of the causal powers that natural entities themselves have (Thomas 1947). Because it is not reducible to nature, it is not pantheistic.

Why invoke such a conception of God? Perhaps most importantly, it provides a reasonable explanation for the otherwise surprising intelligibility that is assumed as a condition for natural science. E. O. Wilson, for example, marvels at "the fortunate comprehensibility of the universe . . . a world that proved surprisingly well ordered" (Wilson 1998: 47). The intelligibility of the world is a fact in need of explanation, however. As Aristotle and Thomas both fundamentally argue, that order and intelligibility derive from the agency of a rational divine cause. Second, this concept of God allows quasi-autonomous status to nature as an ordered and intelligible system of secondary causes, and therefore is consistent with Darwinism understood as a theory of biological development. Third, it emphasizes the notion of God as the sustaining cause of all that exists (Colossians 1:18–20), and therefore is broadly consistent with biblical and traditional orthodoxy. While the interventionist conception of divine agency is a more conventional reading of the biblical texts, it is not any more necessary than the anthropomorphic imagery frequently found in biblical narratives. Despite the common tendency of theists to think and speak concretely about God, he is no more likely to have a hand than he is to employ his hand to hold up the waters of the Red Sea. In any case, an interventionist account of divine agency suggests divine mutability, as God would differ prior to and after alleged interventions into nature. By contrast, conceiving of God as the immanently present continuous causal source of all natural things and events preserves divine omnipresence and natural autonomy. The supernatural and the natural are not two separate spheres of reality but are different ways of understanding the single sphere of being.[50]

This first-cause God is not a deist God responsible for setting up laws of nature and then allowing natural events to unfold independently in a way he neither foresees nor directs. He is not the cause of otherwise unexplainable particular aspects of nature but of all of nature at every moment. He is not merely a powerless self-emptying servant who expectantly waits for the universe to follow a trajectory of its own (see Haught 2008: esp. 51–60; but cf. Haught 2010) but is the omnipotent creator ex nihilo who gives natural objects the powers to pursue their natural ends.

Insofar as God is present as sustaining cause in all of physical reality, all things in some way image or reflect God. In their agency, they image God's power; in their functions, they image God's vitality; in their very existence, they image God's being. However, insofar as humans particularly reflect the rationality and agency of God, humans are a special reflection of God deserving of the title *imago Dei*, and therefore also subjects of inherent moral dignity (Meilaender 2009: 10–18). Understood this way, the first-cause continuous creator is compatible with Darwinian evolution, and consequently plausible as the causal source of the dignity of the *imago Dei* post Darwin.

DIGNITY BEFORE AND AFTER DARWIN

The consensus view of human dignity operated from antiquity through the Christian Middle Ages and the Renaissance and survived, albeit in attenuated form, into the secularizing and pluralizing milieu of modernity and the Enlightenment. In many times and cultural contexts, it functioned more as an ideal or aspiration rather than concrete effective policy (Starck 2002: 181). However, it cannot be denied that Christianity—or more accurately, the cultural synthesis of Judaism, classical paganism, and Christianity that was Christendom—led to improvements in political rights, laid the conceptual and social foundations for practices of tolerance and social welfare, and provided an intellectual framework for the abolition of slavery in the West and eventually in almost all the world (Desmond and Moore 2009; Stark 2003: 291–365). In a particular irony, it was instrumental in the emergence of the very science that laid the groundwork for Darwin himself, by developing a view of the physical world as ordered and intelligible (Lindberg 1986; Lindberg 2007; Stark 2003: 121–99).

The conditions of the eclipse of human dignity in the late 1800s and early 1900s are complex. The most obvious broad causal factor is the phenomenon of secularization, with its shift from cultural practices centered on religion and religious institutions to the state and to private affiliations, an overall decline in religious practice, and a related shift in the conditions of religious belief (C. Taylor 2007). While simplistic causal claims in history are rarely helpful or compelling, it is quite plausible to see in the decline of human dignity the background of Darwinist ideas. There are but a series of steps from the central claims of the *Descent*—infinite gradations throughout nature, fundamental continuity between humans and animals, the erasure of firm species lines, and consequent gradations *within* species; in short, the denial of human distinctiveness and the consequent collapse of the idea of human dignity—and the awful claims of the 20th century: distinctions between fit and unfit, between mentally fit and feeble minded, between

Aryans and parasites, between genuine humans making contributions to the social order and "insects" and "parasites" subject to "purging" (Gould 1981; Weikart 2004; Solzhenitsyn 1974: 27–28).[51] It is surely wrong to blame Darwin for social Darwinism, biological determinism in intelligence research, the eugenics movement, rapacious capitalism, Nazism, or Soviet Communism,[52] but it is obtuse to deny that Darwinian ideas were borrowed, developed, and altered to legitimize these and other horrible episodes of recent history (Ruse 2001: 173).

Darwinism as a matter of fact undermined the traditional concept of human dignity. Philosophical Darwinism amounted to an assault on the traditional concept of dignity. Other factors in the 19th and early 20th centuries were at work as well: reactions against Enlightenment rationality such as romanticism and existentialism chipped away at the Kantian articulation of dignity as autonomy, and religious skepticism and toleration for atheism pushed the notion of the *imago Dei* out of the secular consensus. It is impossible to say that the mass dehumanization movements of the 20th century simply would not have come about without Darwin. On the other hand, it is undeniable that Darwinian ideas were an important part of the justification of those movements.

The argument of this chapter is that human dignity can be asserted, post Darwin, if only there is a sufficient argument in favor of human distinctiveness *in some form*. Dualism in its various manifestations offers one such ground, but one that is increasingly unsatisfactory to the broadly empiricist bent of modernity. Darwinian existentialism offers another, but its subjective assertion of dignity is, even if not inconsistent with its own assumptions, insufficiently robust to command abiding assent. The best alternative is some form of emergentism: an acknowledgment that human beings are indeed animals, but animals of a very special sort. Principle-apprehending animals, philosophizing animals, science-doing animals—that is to say, *human* animals—can reasonably be understood as distinctive, even if they are, as Darwinism claims, genealogically continuous with all other living things. Such beings further can be understood as images of the continuously creative radical first cause of all nature. That is a sufficient basis for human dignity, even after Darwin.

Leon Kass offers a review of the recent fate of human dignity that is superficially similar to Freud's famous assertion of the outrages of science against human dignity:

> The notion of the distinctively human has been seriously challenged by modern scientists. Darwinists hold that man is, at least in origin, tied to the subhuman; his seeming distinctiveness is an illusion, or at most not very important. Biochemists and molecular biologists extend the challenge by blurring the distinction between the living and the nonliving. The laws of physics and chemistry are found to be valid and are held to be sufficient for explaining biological systems. Man is a collection of molecules, an accident on the stage of evolution, endowed by chance with the power to change himself, but only along determined lines. (Kass 1985: 37)

Where Freud's account is triumphalist declaration, however, Kass's is a cautionary warning. Darwinism can be articulated in the terms of reductionistic deterministic materialism; common descent can be the demise of any form of human exceptionalism. Scientists and philosophers can be seduced into embrace of scientism by the powerfully successful methodology of natural science.

The thesis of human exceptionalism has become something of a shibboleth in post-Darwin modernity.[53] Human beings know that we are functionally different from other organisms. We are the only organisms for whom the question, "Why and how are we different from other organisms?" is possible. And yet since Darwin overthrew the radical animal-brute dichotomy, the pendulum has swung far to the other side, with the claim that all human capacities are fundamentally continuous with animals. Strictly speaking, Darwinism does not require this claim. It is possible to affirm human exceptionalism even if human beings are organically related to nonhuman animals by common ancestry, through the concept of emergentism.

Further, while Darwinism conceptually destroyed the special creationist natural theology of Paley, evolution is not *essentially* atheistic, despite insistent rhetoric of the New Atheists. Darwin himself reflects the ambiguity: with the God-of-the-design-gaps crowded out by selection mechanisms and psychologically challenged by personal suffering and religious doubts, he abandoned in turn Christianity and first-cause theism but refused to identify himself as an atheist. Darwin's materialism wavered between a methodological commitment and a philosophical assumption, but there is no necessity for an evolutionist to be a metaphysical materialist. On philosophical grounds, to explain the intelligibility of the cosmos, proceeding from experience by inference to a first cause is still reasonable.

In the end, therefore, Darwinism struck a terrible set of blows against human dignity, with reverberations that echoed well into the 20th century. But Darwinism is completely compatible with human exceptionalism, and is *in principle* consistent with first-cause theism of the sort that could be a basis for the *imago Dei*. In sum, while Darwinism has been devastating for human exceptionalism and the *imago Dei*, it need not continue to do so. Darwinism can be consistent with dignity, even if within itself it lacks the resources to ground a robust conception of dignity.

NOTES

I would like to thank my Gonzaga colleagues Brian Clayton, Richard McClelland, Douglas Kries, and Michael Tkacz for extensive discussions on many of the topics of this chapter; additional suggestions were offered by Steve Dilley, Erik Schmidt, Brian Henning, Michael Collender, and David Kovacs.

1. For a sketch of the historical development of the concept of dignity and the modifications of the concept in the modern period, see my chapter "Human Exceptionalism and the *Imago Dei*: The Tradition of Human Dignity" earlier in this volume.

2. I grant the point of Secord that the "Darwinian revolution" must be placed in a wider social and historical context to be fully understood (Secord 2000: 4). At the same time, Darwin articulates and stands as the representative of an outlook that emerges in mid- to late-19th-century science and natural philosophy. Acknowledgement of the significance of such figures as Robert Chambers, Charles Lyell, and Thomas Huxley in the development of what will come to be labeled Darwinism does not require a diminished assessment of Darwin's crucial role (see, e.g., Chambers 1844; Lyell 1863; T. H. Huxley 1887; F. Darwin 1887; Secord 2000; Himmelfarb 1968, Moore 1979; Desmond and Moore 1991). I further grant that the Darwinian revolution was not a simple replacement of a supernaturalist design worldview with a "secular scientific paradigm" (Brooke 2009b: 259). The story is much more complex and much more interesting than a simple dichotomous narrative.

3. As modern science increasingly described the physical universe in mechanical terms, it became appealing to conceive of human beings as qualitatively distinct from animals, and more broadly from physical nature, in what became a standard dichotomous contrast of humans to mechanical "brutes." On modern scientific mechanism and the modern shift to explanation in terms of efficient and material causes, see Bacon 1964; Descartes 1993: 24–34; Ashworth 2003; Lindberg 2007: 365; and Provine 1988. The classic treatment of the dualist conception of reality that follows by implication is the philosophy of Descartes (see Descartes 1993: esp. 26–33; Descartes 1970: 36–37, 53–54, 63–64, 206–8, 244–45); while not all modern thinkers were metaphysical dualists, the competing impulses of mechanistic science and a general sense of human uniqueness led to a broadly dualist conception of human-animal differences (Keynes 2002: 37–40). While dualist conceptions of philosophical anthropology pre-dated the modern era (see Cooper 2000; Green 2008), earlier dualisms were motivated by the question of personal survival after death, while modern dualistic thinking was motivated much more clearly by an attempt to understand human functional powers, especially reason, as exceptions to the mechanical cosmology of modern science.

4. On evolution as an explanatory replacement for special creation, see Darwin 1859: 55, 115, 243–44, 275–76, 469–70, 472, 480–82, 488–89; T. H. Huxley 1887; Ospovat 1981; Dawkins 1986: 3–6; Brooke 2009a: 203; Brooke 2009b: 262–63; Haught 2010: esp. 11–27; and Worster 1985: 117–19. Darwin's argument was that it was not necessary to invoke God where biological processes would suffice to produce biological complexity (Dawkins 1986: 5; Brooke 2009b: 263). While there were philosophical and theological challenges to theistic design and creation in the 19th century (Hume 1988; Chambers 1844; Wilson 1998: 32–34; Secord 2000; Desmond and Moore 1991), the notion of theistic special creation was a natural foil for Darwin's articulation of biological evolution because of a persistent popular sense that the beauty and complexity of nature provides direct evidence for the intelligent agency of God and that study of nature is an appropriate form of piety (Browne 1995: 130; Brooke 2009a; Livingstone 2003; Secord 2000; R. A. Richards 2009: 176–77; Roberts 2009).

5. The passage is worth citing in full to capture the immediacy and fervor of Darwin's insight, as evidenced by the punctuation and spelling:

> Once grant that species one genus may pass into each other.—grant that one instinct to be acquired (if the medullary point in ovum. has such organization as to force in one man the developement of a brain capable of producing more glowing imagining or more profound reasoning than other—if this be granted!!) & whole fabric totters & falls.—look abroad

study gradation study unity of type study geographical distribution study relation of fossil with recent. the fabric falls! But Man——wonderful Man. 'divino ore versus cœlum attentus' is an exception.—He is Mammalian.— his origin has not been indefinite—he is not a deity, his end under present form will come, (or how dredfully we are deceived) then he is no exception. (Darwin 1987: 263 [Notebook C 76–77])

The Latin quotation (transcribed slightly differently in Gruber 1974: 449) is commonly translated "with divine countenance turned toward heaven"; while the source is unknown, it possibly derives from a hymn or poem with which Darwin became acquainted during his studies at Cambridge from 1828 to 1831, a speculation perhaps justified by the fact that Darwin was "ecstatic about the anthems" at King's College chapel (Desmond and Moore 1991: 57).

6. The notion of a supernatural law giver responsible for the regularities in nature that in turn bring about complex life forms was a theme of the wildly popular book *Vestiges of Creation*, published anonymously by Robert Chambers in 1844 and reissued in multiple editions for decades thereafter ([Chambers] 1844: see esp. 154, 161; Secord 2000). Darwin himself floated a similar idea in his research journals in the late 1830s, arguing that divine dignity was more consistent with first-cause God rather than a craftsman God meddling in creation and "warring against" the very laws he had ordained (Darwin 1987: 432–43 [Notebook D 36–37]; see Desmond and Moore 1991: 261). This "first-cause" theism—essentially the deistic idea than largely autonomous nature was the creative product of supernatural intelligent creative agency—was appealing because it retained the possibility of a divine creator in combination with the largely autonomous nature thought by many as essential to 19th-century science. As a compromise, however, it dissatisfied both traditionalists because it seemed to supplant divine agency in the world and scientific and religious radicals because it stopped short of declaring theology irrelevant to the scientific project (Himmelfarb 1968: 216–22; Desmond and Moore 1991: 320–23; Hull 1973: 10–11).

It is true that Darwin sometimes criticized the notion of theistically guided evolution. In a letter to Joseph Hooker regarding the origin, Darwin expressed regret for having used the "Pentateuchal term of creation, by which I really meant 'appeared' by some wholly unknown process" (F. Darwin 1887: vol. 3, 18). An extended analogy in *Variation of Animals and Plants under Domestication* argued that adaptiveness does not entail divine foresight, foreknowledge, or design (Darwin 1875: 425–28). Many other comments hostile or unsympathetic to theism can be found in Darwin's correspondence (see Brooke 2009b: 256–57). However, Darwin's statements critical of theism cannot be taken as any more authoritative as the text of the *Origin* itself or other comments sympathetic to theism by Darwin in other contexts (see Brooke 2009a: 206; Brooke 2009b: 259).

Darwin's disagreements with his American champion, Harvard botanist Asa Gray, provide an instructive case. Gray argued that evolution and theism were perfectly compatible, and attributed divine purposiveness to biological change (Gray 1861). Darwin made clear in correspondence that he did not agree with Gray's account of divinely guided evolution (F. Darwin 1887: vol. 1, 313–15, vol. 3, 62, 84–85; cf. F. Darwin 1887: vol. 3, 189; Gray 1861: esp. 34–43). Darwin's objection to Gray's view was not to the bare claim that natural processes are ordered by divine intelligence, but to Gray's interventionist conception of God's governance. Darwin was open to divine governance of natural processes but also wanted to maintain the relative autonomy of nature with respect to any such creator. In any case, Darwin did not find Gray's theistic evolution impossible or incoherent.

7. The question of Darwin's theology has increasingly drawn the attention of scholars, in part because the available data suggest a fascinating, if convoluted, story (see Livingstone 2003; Brooke 2009a; Brooke 2009b; R. J. Richards 2009: esp. 54–66; Dilley 2011). There were pronounced Unitarian and free-thinking traditions in his family and his early student associations (see Himmelfarb 1968: 32; Desmond and Moore 1991: 12, chap. 3; Brooke 2009a: esp. 197–98). Despite this family history, Darwin describes in his autobiography a pilgrimage from "orthodox" theism appropriate for a man training for a career as an Anglican minister to "first-cause" theism to agnosticism (Darwin 1958: 85, 92–94; see Brooke 2009a: 204; Brooke 2009b: 263). While Darwin forcefully rejected particular aspects of Christian theology, he maintained friendships and professional alliances with theists, avoided needless conflict with either theists or institutional Christendom, and insisted on the conceptual compatibility of evolution and theism (Darwin 1958: 86–87; F. Darwin 1887: vol. 1, 304, 306–7, 313, 317; F. Darwin 1887: vol. 2, 288–89; Darwin 1876: 421–22; T. H. Huxley 1887: 202–3; Himmelfarb 1968: 232; Secord 2000: 10; Keynes 2002: 310–13). Confessing that his theological thinking was "a muddle," he repeatedly declined appeals to lend his name and reputation to the cause of public atheism (Darwin 1958: 92–93; F. Darwin 1887: vol. 1, 304, vol. 2, 353–54; Brooke 2009a: 204; Brooke 2009b: 256–57; Aveling 1883; F. Darwin 1887: vol. 1, 317 fn; Desmond and Moore 1991: 656–58).

8. The editors of the *Cambridge Companion to Darwin* note perceptively: "Some scientific thinkers, while not themselves philosophers, make philosophers necessary. Charles Darwin is an obvious case" (Hodge and Radick 2003: 1; see also 11). Of course, one could argue that while Darwin was not a philosopher, he was not a "professional scientist" either, as that role was emerging in the time he lived. At the same time, it is clear that Darwin read philosophers and philosophical theologians such as Locke and Paley in the course of his college education and discussed and cited philosophers such as Hume in his research notes and publications (Desmond and Moore 1991: 78, 87–88; Darwin 1987: 591–92 [Notebook N 101]; Darwin 1882: 101 fn 23; Huntley 1972).

9. Examples can be multiplied, but here are a few representative ones. On materialism, Darwin on the one hand insists that his work addresses only scientific questions, leaving theological issues minimally affected (F. Darwin 1887: vol. 1, 307), but he frequently alludes to or endorses materialism, if rarely in his published work (Darwin 1987: 291 [Notebook C 166]; see also Darwin 1987: 614 [Old & Useless Notes 37]; 616 [Old & Useless Notes 39]; see also Gould 1981: 324). Regarding atheism, we have addressed the fundamental ambivalence that grounded Darwin's agnosticism, but it is worth noting that it occurred to Darwin as well as to others that explaining phenomena heretofore explained by divine agency in the terms of natural mechanisms at the very least crowded God out of explanatory "gaps" (Darwin 1958: 85, 92–94; Darwin 1987: 300 [Notebook C 196–97]; see also Drummond 1895: 333; Plantinga 1997b; Hahn 1986: 256). On determinism, the machine metaphor for natural events characteristic of modern science (LaPlace 1902: 4; Ashworth 2003) is one Darwin found congenial, particularly in his discussions of the mechanisms of human "motives," instinctive tendencies or behavioral leanings that are products of natural selection just as much as morphological features are (Darwin 1987: 606–8 [Old & Useless Notes 25–27]; Darwin 1987: 536 [Notebook M 72–74]), though he did allow the possible role of some form of free agency (Darwin 1882: 67–68).

10. See Aveling 1883: 3; Desmond and Moore 1991: 217, 486; Gould 1981; Stark 2003: 186. One striking example of a perceptive analysis of the probable philosophical trajectory of Darwinian biological ideas, particularly as those

ideas might impact the grounds for human dignity, is the review of the *Origin* authored by Bishop Samuel Wilberforce ([Wilberforce] 1860: see esp. 258).

Perhaps the oddest dissenter from philosophical Darwinism is Huxley, famed as "Darwin's Bulldog." He argued that evolution strictly speaking was opposed only to biblical literalism, not theism broadly understood: "the doctrine of Evolution is neither Anti-theistic nor Theistic" (T. Huxley 1887: 202–3). He rejected dogmatic atheism on epistemic grounds, coining the term "agnostic" for his position, and rejected materialism and physical determinism as well (L. Huxley 1901: vol. 1, 443–44).

11. The tendency of neo-Darwinists to move from scientific principles of methodology to metaphysical philosophical commitments is so common that citations might seem superfluous, but a handful of obvious representative texts would include Dennett 1995: 59; Wilson 1975: 3; Crick 1994: 3; Dawkins 1986: 6; see also Hull 1991.

12. Contemporary Thomists such as William Carroll often point out that perceived conflicts between science and religion are complicated by a failure to properly distinguish appropriate areas of disciplinary competence or orders of knowing (Carroll 2000: 322–23). The distinction I make here between Darwinism understood as biological theory and Darwinism construed as a philosophical outlook is a similar point.

13. There is a minority view that Darwin's theory supported and enhanced the development of human dignity (Earls 2010). Darwin's thesis of common descent offered biological evidence for commonality between human beings and evidence to complicate theories of racial difference, particularly theories based on scientific claims of superiority and inferiority, and also the notion that races were products of separate acts of special creation. It is further true that Darwin personally regarded slavery and other forms of inhumanity as abominable and was willing to express such views publicly (see, e.g., Darwin 1839: 22, 26–28, 121, 592; Desmond and Moore 2009; Desmond and Moore 1991: 120, 122, 238, 241, 328–29). But the *mode* of equalization that Darwinism brought about was to lower humans to the level of animals rather than raising supposedly subhuman beings to a fully human level (Desmond and Moore 1991: 238, 442; Desmond and Moore 2009: xviii). Perhaps most tellingly, Darwin and Darwinists retained the idea of gradations *among* human beings. Even though, in an oft-remembered comment, Darwin instructed himself never to speak of higher or lower (Darwin and Seward 1903: 114 fn 2; see also Darwin 1987: 189 [Notebook B 74]; Grene 1961: 42; Gould 1977: 36–37; compare Himmelfarb 1968: 220; R. J. Richards 2009), we find such hierarchical language throughout Darwin's private notes and published writings, in many cases specifically applied to intrahuman differences (see, e.g., F. Darwin 1887: vol. 1, 316; Darwin 1882: chap. 3; Lyell 1863: 492). Some Darwinists took intrahuman racial hierarchy in disturbing directions, such as the comment by Haeckel, Darwin's German champion, that "the differences between the highest and the lowest humans is greater than that between the lowest human and the highest animal" (Weikart 2004: 105–6). For these reasons, it is, on balance, hard to see Darwinism as offering a new ground for human dignity. Strictly speaking, evolutionary doctrine neither supports nor refutes the moral brotherhood of all races (Desmond and Moore 1991: 542).

14. Human dignity has played an increasingly significant role in medical ethics despite criticisms that the concept is underdeveloped, unworkably vague, or even incoherent (Macklin 2003; see also McCrudden 2008; Sulmasy 2006; Pinker 2008; Meilaender 2009; Somerville 2010).

15. My thesis indicates by omission what I cannot address in this chapter. For example, I will not here review or evaluate modern scientific, religious, or philosophical

critiques of the neo-Darwinist project. In the interest of full disclosure, however, I will say that I find Darwinian evolution powerful as a scientific account of the "consilience of evidence" of modern biology (Wilson 1998), while also finding quite plausible a number of philosophical critiques that have been leveled against the globalizing and reductive tendencies of neo-Darwinist thinkers. Where possible and relevant, I discuss these objections in what follows.

16. Many critics have noted that the concept of human dignity was reintroduced into discussions of international law in the post–World War II context, especially in connection with the United Nations, and into the field of medical ethics in the 1970s, and that the philosophical grounds of the reintroduced concept have been very unclear (see Morsink 1993; Dicke 2002: 112–14; Arieli 2002; Chaskalson 2002; Kretzmer and Klein 2002; Macklin 2003; President's Council on Bioethics 2008). It is my view that Darwinian evolutionary theory is one of the most important factors in the gap that occurs from the late 19th to the mid-20th centuries in the long tradition—over 2,000 years—of human dignity. Jacques Maritain, who participated in the drafting of the Universal Declaration of Human Rights and other United Nations documents, noted that the postwar practical agreement on the enumeration of rights was only possible by setting aside as "futile" the project of finding a "common *rational justification* of these practical conclusions and these rights" (Maritain 1951: 76; see also United Nations General Assembly 1948). I believe that Darwinism in some of its manifestations bears significant responsibility for the difficulty of providing a philosophical account of the appeals to dignity in those documents. So far as possible, I will sketch out arguments for these views in this chapter.

17. Even professed nihilists, such as Loyal Rue, admit that myths must sustain human individual and social life (Rue 1994: 274–79). Rue suggests that it is possible for individuals to live without values and for cultures to have short periods of nihilism. I doubt even that is possible, because a complete lack of value would not provide sufficient justification for any course of human action or its opposite.

Some nihilists are unaware that their position is nihilist or are blind to the implications of the nihilism of their view. B. F. Skinner famously counseled that in the wake of Darwin we should not only recognize the effect of environment on biology and morphology, we should understand that behavior is the mechanical product of environmental selection as well (Skinner 1971: 14–17). Acknowledgment of this reality will lead us to abandon the imaginary "I" or "self" traditionally thought to be the core of behavior. Dennett notes that this is a "greedy" form of reductionism that completely mechanizes human experience and agency without accounting for the complexity of mental function (Dennett 1995: 395, 468–69). As such, it cannot account for the agency of the conditioner who makes choices about desirable and undesirable behavior for selection, who is himself a mechanism whose behavior is the product of environmental selection forces. To treat human beings in this way not only abolishes freedom and dignity, as Skinner admits, it abolishes human nature as well, and renders incomprehensible the project of understanding human behavior or anything else (Lewis 1947a: chap. 3).

18. Arnhart's stance regarding dignity is rather complex. Since he takes human dignity as primarily a religious value, acknowledging only the *imago Dei* thread of the concept, he links the fate of dignity in Darwinian ethics to the question of evolution-religion compatibility. Since he thinks that evolution is not necessarily atheistic, he argues that religion can enhance the ethical naturalism that he advances, providing a supporting role for the concept of human dignity (see Arnhart 1998: 249).

19. As I will clarify shortly, this method of approaching the question does not mean that I am abandoning the *imago Dei* as a possible post-Darwinian basis for dignity.

20. In this connection, critics of evolutionary theory especially should avoid fostering the perception that their opposition to evolutionary theory is opposition to science, particularly from a religious point of view. While the "war narrative" has enjoyed great popularity over the last century, in significant part in the aftermath of the Darwinian revolution, and promoted by Darwinian polemicists and their creationist counterparts (for example, on the one hand Draper 1874; White 1896; Dawkins 1986; on the other [Wilberforce] 1860; Johnson 1991; see also Pennock 2000), facts of the matter show that religion-science conflict claims are historically overstated (Lindberg and Numbers 1986; Stark 2003; Lindberg 2007; Numbers 2009). If it is possible to defend a robust account of human dignity, and to combat philosophical Darwinism, without promoting conflict between science and religion, then that is the best course of action. Augustine wisely counseled early Christians of the value of knowing natural philosophy or science so that when they interpreted Christian scriptures or doctrines they would not "talk nonsense" in such a way that would bring scorn on the faith (Lindberg 1986: 31). A similar principle governed Galileo's understanding of the interpretation of the scriptures in the light of science. His view was that any apparent conflict between scripture and empirical scientific observation should indicate a problematic interpretation of the scripture, since truth is unitary and both scripture and our experience—when carefully conducted—must be true (see Galileo 2008: 160–61). The modern Catholic Church, perhaps chastened by the Galileo experience, has cautiously avoided antiscientific rejectionism, even while affirming traditional Christian theological positions (see Artigas et al. 2006; Pius XII 1950; John Paul II 2003; Benedict XVI 2008; de Duve 2008; Benedict XVI 2005).

21. Lyell's insistence on protecting the human mind from the implications of evolutionary biology caused Darwin considerable frustration (Lyell 1830: 155–56; Lyell 1863: 494, 506; see also Himmelfarb 1968: 258–60; Desmond and Moore 1991: 475, 442, 515; Numbers 1988: 628). The co-discoverer of natural selection, Alfred Russel Wallace, held a version of dualism, though on scientific rather than theological grounds. According to Wallace, the massive functional gap between human and nonhuman cognition could not be explained by selection alone, particularly because savage humans that did not need the advanced intelligence required for complex social life nevertheless possessed such intelligence (Wallace 1871: 183; Wallace 1895: 424; Himmelfarb 1968: 375–76; Desmond and Moore 1991: 569–70). Wallace's argument might sound like an appeal back to traditional theism, but Wallace abandoned theism and thought of the source of human intelligence, as well as the origin of life, as a spiritual force of some sort. In the sense that I am arguing here, Wallace's dualism is not an evasion strategy, because he fully embraces human evolution, but finds human intelligence a problematic datum.

22. Dualism might be understood as the position required by Pius XII's comment about the openness of the Roman Catholic Church to research on evolution: "the Teaching Authority of the Church does not forbid that, in conformity with the present state of human sciences and sacred theology, research and discussions, on the part of men experienced in both fields, take place with regard to the doctrine of evolution, in as far as it inquires into the origin of the human body as coming from pre-existent and living matter—for the Catholic faith obliges us to hold that souls are immediately created by God" (Pius XII 1950: §36). However, this depends on how we understand "soul,"

whether in a Platonic-Augustinian sense or an Aristotelian-Thomist sense, and further depends on whether Saint Thomas's position is interpreted as a form of dualism. As I indicate below, I think that a broadly Thomistic view does not require such a dualistic reading. While it is true that Thomas's view of the soul as a "substantial form" involves its incorporeal reality (*Summa Theologica* 1.75.2; see also Cooper 2000: 11–13), this need not require real existence apart from the body as an inherent feature of soul.

23. As my argument below will make clear, I do not mean to rule out all forms of dualism as logical possibilities. Substance dualists with robust accounts of mind-body interaction, particularly those working in the Aristotelian, Thomist, or emergentist traditions, seek not to evade the biological dimension of human nature but to account for it in the light of human mental experience (see, e.g., Braine 1992; Stump 1995; Swinburne 1997; Moreland and Rae 2000). My critical comments here are directed more toward prophylactic dualisms that seek to shield human nature from the Darwinian revolution.

24. Darwin himself articulated a version of this view, suggesting to a correspondent that "the theory of Evolution is quite compatible with the belief in a God," and "Science has nothing to do with Christ" (Darwin 1887: vol. 1, 307).

25. The NOMA proposal is sometimes defended as a way to avoid the so-called naturalistic fallacy, which asserts that facts do not imply values. This idea, which derives from Hume and G. E. Moore, and which is something of a shibboleth in contemporary philosophy, is obvious and trivial in one sense but very implausible in another (Rachels 1990: 92–97; Dennett 1995: 467–69; Arnhart 1998: 69–83; Wilson 1998: 249–51; Thompson 1999: 477–78, 480–81). It is certainly true that a fact or set of facts does not yield as an outcome a value or ethical principle by direct implication. At the same time, values do not just float abstractly free from the context of the world, but rest within a concrete set of meanings and contexts. For example, the facts of an organism's structure and functionality give us insight into its health and vitality, what the ancients called its "excellence" or "virtue" (see, e.g., Plato *Republic* 1.352d–353e). Examples such as this show that while facts and values are not interchangeable, they are not nearly as isolated from one another as proponents of the naturalistic fallacy contend.

26. See "Human Exceptionalism and the *Imago Dei*: The Tradition of Human Dignity" earlier in this volume.

27. There is some evidence that Gould's irenic overtures to religious people were less than fully honest. While Gould laced his works with references to the Bible, for example (much to the annoyance of many of his religiously skeptical scientific colleagues), he surely did not regard Judaism or Christianity as authoritative sources of morality, particularly as moral constraints on the practice of science. Even more telling, Gould admitted in an interview that he regarded religious beliefs as comforting delusions (Miller 1999: 170).

28. "Biocentrism" has come to be used to describe nonholistic biological egalitarianism (P. W. Taylor 1986; Rolston 1988: 64). Even that narrowly defined view shares with what I call here "Darwinian holism" an abandonment of human exceptionalism and a robust affirmation of the value of nature as a systematic whole (see especially P. W. Taylor 1986: 99–119, 129–55). In practice, however, biocentrism more broadly embraces a biologically holistic outlook, particularly as it is shaped by Darwinism (Worster 1985).

29. The varieties of holism would include treatments such as Leopold 1989; Worster 1985: esp. 114–87; P. W. Taylor 1986; Rolston 1988; Rachels 1990; Rue 1994; Goodenough 1998; Raymo 1999; Thompson 1999; and Haught 2003. Not all of the proponents of these views hold them as *evolutionary* positions. For example, Singer's animal ethics is not Darwinian in its basis,

but it appeals to utilitarianism and the capacity of animals to suffer (Singer 1986). Of course, most of these authors would affirm that their views are, at a minimum, *consistent* with a Darwinian outlook.

30. Some Darwinian holists see the objective as the extension of rights to animals, and perhaps even beyond to plants, natural objects, and ecosystems, while others make the argument in terms of appropriate ethical consideration of the interests of living things.

31. Related to the extension problem is what we might call the contiguity problem. A number of holists have embraced the idea of the "expanding circle" of ethics, in which humans acknowledge the narrowness of our past and current outlook and extend value beyond these limits (Leopold 1989: 202–5; Singer 1981: esp. 111–24; Rolston 1988: 50–51, 120–21; Wright 2009). But this expansion of value can be crude or fine-grained. So far as it is crude, and denies any form of human exceptionalism or superiority, it will be deeply problematic. Darwin noted that living things vary by gradations, with incremental differences from one organism type to another (Darwin 1882: 65–66). As a matter of fact and of human biology, our valuations of things reflect the relationships that things have to us in this graded hierarchy: we naturally value things close to ourselves and disvalue things different from us in rough degree to their distance from us on the gradation scale (Rolston 1988: 120–21; Ruse 2001: 200–1; Lee and George 2008: 417). Primatologist Frans de Waal notes that the expanding circle of human morality faces inherent limits, because the "own group orientation" that has been the ground of human social behavior for millions of years cannot be easily extended to all life on earth (de Waal 2006: 163). Even if a Darwinian holist attempts to correlate dignity to whatever provides the basis for intrinsic value—sentience, biological flourishing, being a striving, living thing—it is hard to see how the same concept can reflect kinship preferences and yet have a universal or near-universal scope.

32. Rachels, for example, seems to think of the concept of dignity as *necessarily* construed in human terms, such that abandonment of human dignity leaves no meaningful path for employing the concept of dignity at all (Rachels 1990: 171–72). Similarly, Taylor seems to think that dignity as an ethical concept is inseparable from claims of human superiority (P. W. Taylor 1986: 129–52).

33. As Dennett says, if Nietzsche is the father of existentialism, Darwin is its grandfather (Dennett 1995: 62).

34. Consider a now-notorious comment Darwin made late in life:

> with me the horrid doubt always arises whether the convictions of a man's mind, which has been developed from the mind of the lower animals, are of any value or at all trustworthy. Would any one trust in the convictions of a monkey's mind, if there are any convictions in such a mind? (F. Darwin 1887: vol. 1, 316)

Darwin's point seems clear: to the extent that an evolutionary account of human beings and human rationality is taken seriously, it undermines our confidence in our ability to grasp the truth. Similarly, Nietzsche, in a fragment written in 1873, tells a Darwinian fable in which "clever animals invented knowledge" to compensate for the weakness and ephemerality of their existence. The moral of the parable is that the human pride in our ability to seek and know truth is self-deceptive, because the purpose of our intellect, like all organs and organic functions, is survival. The capacity to know is therefore instrumental, and even it is overshadowed by the adaptive significance of our ability to deceive (Nietzsche 1954: 42–43).

Darwin's "horrid doubt" has earned notoriety thanks primarily to an argumentative development of it formulated by Alvin Plantinga (1993: 216–37). Briefly put, Plantinga argues that belief in the evolutionary emergence of functional capacities, if understood in a purely naturalistic way, corrosively undermines confidence in human cognition. Other versions of the argument have been offered by a range of thinkers before and after Plantinga (see Lewis 1947b; Percy 1991: 271–91; Nagel, 1997: 127–43; Hasker 1999: 5–20; Koons 2000; Reppert 2003; Kass 2007: 41), and one often finds scattered references that allude to or mention but do not develop the argument (Descartes 1984: 288; Wright 1994: 365; Livingstone 2003: 188; Haught 2010: xiii). In a striking irony, Freud was aware of the problem but projected it away from his own materialism to unnamed critics of psychoanalysis advocating what Freud called "intellectual nihilism" (Freud 1964: 175–76). Plantinga's version of the argument has drawn significant response (Ruse 2001: 106–10; Beilby 2002). My own view is that the argument does not refute Darwinian evolutionary biology per se, but rather shows that reductionistic forms of metaphysical materialism, including reductionistic and materialistic forms of Darwinism, undermine the ordinary presumption of epistemic reliability.

35. I'm indebted here to the account of objectivity in Nagel 1997, who points out that to step outside of the framework of objectivity—for example, by proffering a ruthlessly naturalized account of reason—is to abandon the necessary conditions for logic, language, and science. Nagel is a reluctant objectivist, for he admits that an objectivist picture of the reality tacks uncomfortably close to theism (Nagel 1997: 130; see also Wilson 1998: 238–64).

36. On Haught's reading of Gould's NOMA, the sphere of religion essentially paints meaning on an otherwise meaningless world of facts, but such values are not objective to reality, they are human subjective creations (Haught 2003: 6–7; see Gould 1977: 12–13). If Haught is right, and I think he is, NOMA collapses into Darwinian existentialism. From a very different angle, Rolston recognizes the problematic status of values as human generated ("anthropogenic"), but argues that they are objectively grounded in objects even if humans provide the "value ignition" of recognizing those objective values (Rolston 1988: 112–17). Thus, a theory of value sympathetic to the Darwinian outlook need not be subjectivist or "existentialist" in the sense I criticize.

37. On the progressive side, see, for example, Darwin 1882: chap. 3; F. Darwin 1887: vol. 2, 177; on the nonprogressive side, see Darwin 1987: 189 [Notebook B 74]; Darwin 1987: 576 [Notebook N 47].

38. While theistic guidance is not necessary for the progressive development of biological arms races, of course it is conceptually compatible with it (Ruse 2001: 85–86; Haught 2008: 134–40).

39. For the record, Korsgaard takes a similar position, affirming both that animals are more similar to humans than most people think and that there are deep discontinuities between humans and animals (Korsgaard 2006: 103).

40. Speaking of Steven Pinker, Kass notes that he

appears ignorant of the fact that "soul" need not be conceived as a "ghost in the machine" or as a separate "thing" that survives the body, but can be understood (à la Aristotle) to be the integrated powers of the naturally organic body. He has evidently not pondered the relationship between "the brain" and the whole organism, or puzzled over the difference between "the brain" of the living and "the brain" of the dead. He seems unaware of the fact of emergent properties, powers and activities that do not reside in the materials of the organism but emerge only when the materials are formed and organized in a particular way; he does not understand that the empowering organization of materials—the vital form—is not itself material. (Kass 2007: 42)

41. Statistical study of the proportional relationships of distinct parts of the brain across species shows a clustering correlation between brain parts and functionalities. In particular, such research shows a statistical discontinuity between human brains and those of other animals, including primates. An important implication of this is that the DNA commonality between humans and chimpanzees, 98.6 percent, may very well misleadingly imply a greater-than-accurate human-chimpanzee morphological and functional similarity (Oxnard 2004: 1150–56).

42. Rolston argues that human functional distinctiveness merits describing humans in terms of the Latin *genius* and the German *Geist*, with all the connotations of power, spirit, and creativity (Rolston 1999: 140). This argument is an extension of the position of his earlier book, *Environmental Ethics*, which stresses that the human capacity to "oversee" is biologically exceptional, even as it allows us to understand our place in nature (Rolston 1988: 71–74, 338). Both of these views are broadly consistent with the outlook I am articulating here that affirms human exceptionalism, and consequently dignity, within a Darwinian framework.

43. Recent proponents of an Aristotelian/emergentist view include Arnhart 1988, 1998; Nussbaum and Putnam 1992; Hasker 1999; MacIntyre 1999; Frede 1992; and Kass 2007.

44. The idea of a library of life plans or patterns of functional organization waiting to be instantiated in nature is one that was raised in Darwin's time. For example, Edward Forbes, Richard Owen, and Louis Agassiz held similar views (Desmond and Moore 1991: 331, 368; Brooke 2009b: 258). Darwin mentions this sort of view when discussing biological classification in the *Origin* but argues that the reality behind biological classification is not the Mind of the Creator but relationships of evolutionary descent (Darwin 1859: 413–14). Darwin's alternatives are not exclusive. Broadly speaking, Darwin and his evolutionary allies—in particular the combative Spencer and Huxley—regarded the idea of a system of eternal biological forms as a dodge to preserve human exceptionalism (Desmond and Moore 1991: 472–73). Jumping ahead in time, the idea of a system of eternal biological forms is quite similar to Dennett's Library of Mendel (Dennett 1995: 107–13), though Dennett would reject an essentialist Aristotelian interpretation of the idea (cf. Haught 2008: 59).

45. For example, Dilley 2008 objects to Arnhart's Aristotelian emergentism on the ground that it falls into determinism and the denial of meaningful moral responsibility.

46. Miller frames this question as the problem of how a divine cause can affect random, unguided natural processes in such a way as to produce an intentionally desired outcome: "Surely we could not be *both* the products of evolution and the apple of God's eye?" (Miller 1999: 233). I will suggest that this formulation can be misleading.

47. I do not mean to suggest that intelligent design (ID) theorists or others seeking to identify divine agency in natural events deny God's continuous creative activity in nature. I mean only that they seek to *detect* divine agency by identifying particular natural events for which mechanical (chance or law-governed, to use the language of the Explanatory Filter) explanations fail, giving us reason to infer intelligent divine agency. Thus, on the level of natural explanation, they agree with atheist critics that many, if not most, natural

events can be understood in terms of natural causes but argue that at least some natural events cannot be so understood. Nature conceived this way is largely autonomous.

48. Some ID theorists suggest that divine agency operates through fine-tuning of the conditions for physical reality and emergence of life (see Behe 2007: 207–40). However, it is beyond doubt that the general strategy of ID theory is to isolate particular elements of nature for which natural explanations are insufficient. This strategy makes sense only if God—or whatever intelligent designer is otherwise at work—is proffered as the alternate cause of the phenomena in question. If this is the strategy, then the phenomena in question are gaps, and alternate natural explanations that close the explanatory gap replace God. Thus, ID theorists hold the same sort of vulnerable position as Paley-style natural theologians held with respect to Darwin.

49. I have in mind the concept of God deriving from Aristotle (especially *Physics* 8.6; *Metaphysics* 12.6–9) and critically developed by Thomas Aquinas (*Summa Theologica* 1.22; Thomas Aquinas 1997; Carroll 2000; Tkacz 2008). There are interpretive and historical complexities with the continuity between Aristotle's and Thomas's conception of the divine first cause. For example, Aristotle speaks of the Unmoved Mover not as the creator but as the final cause, the end of natural motion. While I believe that these apparent conflicts can be worked out, that is not necessary for the argument I offer here. I need appeal only to the plausibility of the concept of God Thomas offers. I am further aware that there are conceptual problems regarding the compatibility of Thomas with Darwinian evolution, on issues such as immutable universals and divine agency (see, e.g., Gage 2010). I have tried to address those objections in the course of my discussion.

50. A standard objection to the orthodoxy of this conception of God is that it rules out the possibility of miracles. However, it does not preclude miraculous events, events that particularly reveal God's nature and purposes; it rejects the interpretive claim that such events are interruptions of or interventions into the natural order. They are revelations of God via the natural order, but, of course, in their own way, all natural events are revelatory of God (Psalm 19; Romans 1:20). Even the miracle of the Incarnation, of Emmanuel or "God with Us" does not imply that God was before the Incarnation not with us. Its significance lies in the fact that it is a fuller self-disclosure of God in perceivable reality.

51. Rue stresses the role of deception in "Nazi medicine," but the willingness to deceive and to engage in self-deceptive practices had as an antecedent the view that those subjected to the "therapeutic imperative," the Jews and other "parasites," were nonhumans, living beings in gradations below the designation of value, and thereby bereft of any moral protection (Rue 1994: 79–81).

52. Individuals do not control the social movements that spring from their ideas (Himmelfarb 1968: 380–81). It also should be noted that Darwin himself was, by all accounts, a gentle and humane man, deeply affected by the sunken condition of "savage" humans, the terrible treatment of humans under conditions of slavery, and unnecessary pain inflicted on animals (Desmond and Moore 1991: 615–18; Desmond and Moore 2009; Darwin 1987: 286 [Notebook C 154–55], 308 [Notebook C 217]; Darwin 1881; F. Darwin 1887: vol. 2, 374, 377; Arnhart 1998: 189–93).

53. Responding to the argument of Penn et al. 2008 asserting a discontinuity between human and animal cognition, Bickerton notes that such a thesis is "politically incorrect (somewhere between Holocaust denial and rejection of global warming)" (Bickerton 2008: 132).

Part III

The Rhetoric of Human Dignity in Bioethics

10 Human Dignity and the New Reproductive Technologies

Audrey R. Chapman

Sexual intimacy and procreation are central to human identity and functioning, and presumably have potential implications for human dignity as well. Not being able to conceive sets infertile individuals and couples apart from the fertile majority. It is estimated that some 7 million individuals, about 12 percent of those of reproductive age, have fertility problems. Research has shown that involuntary childlessness can have a significant impact on peoples' lives (Vickers 2010). As Thomas Murray comments,

> There are times when adults hunger for children. . . . When we speak of the suffering of people who want to have children but cannot, "suffering" is neither metaphor nor hyperbole: people who crave children to raise and love but cannot have them suffer because, for many of us, our children are a vital part of our own flourishing (Murray 1996: 14).

Efforts to ease the suffering associated with infertility have given rise to the development of a wide range of reproductive technologies.

For most of human history, individuals could do little to intervene in the process of reproduction, except some limited and initially often unsuccessful initiatives to prevent or terminate pregnancy. This situation changed dramatically in 1962 with the introduction of oral contraception and then in 1978 with the birth of Louise Brown, the first child conceived through in *vitro* fertilization. Subsequently, there has been a veritable revolution in technologies that provide control over the reproductive process, including technologies that avoid reproduction, assisted reproductive techniques for attempting to overcome infertility, and technologies for the selection and, potentially, the alteration of traits in offspring.

The wide range of different types of reproductive technologies introduced or contemplated has been the subject of great controversy and considerable unease. John Robertson notes the ambivalence they engender: "there is something profoundly frightening about technological control over the beginning of life. Anxiety over these techniques abounds, even as a growing number of persons seek them out. We are both fascinated and repelled" (Robertson 1994: 3). On the one hand, many in society, particularly persons who suffer from infertility or who risk having children with a genetic disease, have welcomed the development of these technologies. Others, however, have decried human intervention into reproduction for a variety of ethical and theological reason. Questions have also been raised about the safety of some of these technologies both for the women using them and for the prospective children.

This chapter will consider how the concept of human dignity has been used in the debates about reproductive technologies. It should be noted that although concerns associated with reproductive technologies have given rise to a very extensive ethical literature, these writings rarely deal explicitly with the topic of human dignity. Because ethical evaluations of particular reproductive technologies frequently intersect contentious foundational issues, such as the status of the embryo, the importance of naturalness in procreation, the autonomy of the individual, and the right to make private reproductive choices, most of these writings focus more on these topics than on human dignity. References to human dignity, when they occur, tend to be in passing, often offered primarily as rhetorical reinforcement of the viewpoint put forward. Many authors even neglect to clarify their conceptualization of dignity. As an example, a 2002 report by the President's Council on Bioethics with the title *Human Cloning and Human Dignity: An Ethical Inquiry* (President's Council on Bioethics, 2002) fails to conceptualize human dignity or address the specific ways in which human cloning may impinge on human dignity.

Yet another complication is that a specific reproductive technology could potentially affect the dignity of a variety of subjects—the woman or couples opting to use them, the child they bring or do not bring into being, and the wider society. That a particular technology has implications for the dignity of one of these categories of subjects does not mean it potentially affects the others. Also, when it does, the consequences can be quite different. Nevertheless, most of the writings on reproductive technologies that do discuss human dignity fail to identify the subject(s) or make these distinctions. Instead human dignity is often discussed in a kind of subjectless vacuum.

The failure of discussions of human dignity and reproductive technologies to offer a robust conception of human dignity with specific criteria to use in evaluations parallels applications of human dignity to other scientific developments. An 18-month international seminar on this topic which I co-organized found that policy documents and legal instruments rarely provide an explicit definition for dignity or show how human worth might be degraded or supported by a given technology or scientific activity. Dignity's intrinsic meaning in such documents is often left to an intuitive understanding (Caulfield and Chapman 2005: 736–38). In general, discussions about the impact of scientific discoveries and new technologies on human dignity take one of two approaches. The concept of human dignity is often used in the conventional legal and ethical manner to emphasize the right of individuals to make autonomous choices. This conception treats human dignity as a means of empowerment. Some scholars have gone so far as to suggest that this is the only appropriate normative use of the idea of dignity (Macklin 2003). Alternatively, dignity may reflect a broad social or moral position that a particular type of activity is contrary to public morality or the collective good. Statements that a particular technology infringes human dignity convey a sense of general social unease, perhaps even a politically palatable articulation of the "yuk factor" (Caulfield and Chapman 2005: 0737). It may also register concerns about activities that seem to threaten "those parts of the human condition that are familiar and reassuringly human," without detailed explanation of why and how the activities are troubling (Brownsword 2003).

The chapter has three sections. The first is organized around three of the foundational ethical issues referred to above. The second section focuses on selection technologies and reproductive cloning, both of which—particularly the possibility of proceeding with human reproductive cloning—have elicited some explicit commentary on its human dignity implications. For reasons of both space limitations and maintaining focus, the analysis will not include a related topic that has also raised human dignity issues: the use of reproductive technologies for human enhancement. The third section offers reflections on the requirements for a more robust discussion of the implications of reproductive technologies for human dignity.

HUMAN DIGNITY IN THE CONTEXT OF ETHICAL DEBATES OVER REPRODUCTIVE TECHNOLOGIES

The Status of the Embryo

Conceptions of the embryo frame much of the debate about the implications of reproductive technologies for human dignity, as it does many other bioethical issues. With some simplification, the range of positions may be grouped into three principal views. The first of these approaches, which is often identified with the Roman Catholic Church, holds that, from the moment of conception, the embryo is a full human being and therefore entitled to be treated with human dignity. The second approach, more typical of liberal Protestant denominations and the Jewish faith as well as many secular bioethicists, considers the development of personhood to be an ongoing process and that the early embryo or fetus only gradually becomes a full human being deserving of human dignity. A third group, composed primarily of scientists, holds that the embryo is no more than a collection of cells, albeit cells with a potential to develop into a human being (Chapman et al. 1999: 12–16; House of Commons Science and Technology Committee 2005: 15–17).

The view that the embryo from the moment of conception has significant moral status and dignity usually has one or more of three groundings: their human genetic composition, their ensoulment, and their potentiality to become a full human being. All these have been debated and criticized (Steinbock 2001), but what is relevant here is that all three foundations lead to the same conclusion: it is wrong to harm or kill embryos because embryos are human and all human beings have a right to be protected. Some holding to this position equate the intentional destruction of the embryo with murder. It follows that those in this group reject any reproductive technologies that could potentially cause harm to the embryo or fetus. This includes most, possibly all, reproductive technologies because they are viewed as taking a cavalier attitude toward prenatal forms of human life and thereby diminishing respect for human life. In vitro fertilization (IVF), for example, externalizes fertilization and then freezes, tests, and discards some of the embryos. Prenatal screening is viewed as a search-and-destroy mission aimed at detecting flawed embryos and avoiding their implantation because they are imperfect. Also prenatal testing of embryos can lead to abortion, and contraceptive drugs such as RU486 make abortion easier (Robertson 1994: 12).

For those arguing that the embryo deserves to be treated with full dignity, the acceptability of prenatal testing depends on the risks it poses to the unborn child and the purposes for which it is done. The official Catholic position is that it would be considered to be licit if prenatal diagnosis respects the life and integrity of the embryo and the human fetus and is directed toward its safeguarding or healing (Vatican Congregation for the Doctrine of the Faith 1987: sec. 1, no. 2). Catholic teaching rejects any form of prenatal testing that endangers the fetus or may lead to rejection of the pregnancy. Embryo selection technologies such as preimplantation genetic diagnosis (PGD) that seek to avoid implantation of an embryo with a serious genetic problem are also considered wrong because the process involves IVF and will result in the destruction of those embryos that are considered unsuitable (United States Conference of Catholic Bishops 1996). Catholic moral tradition not only prohibits killing the embryo and the fetus, it also proscribes any procedure that is directed to altering its basic human nature. However, if genetic manipulation, the deliberate alteration of genes during intrauterine life, could be directed toward correcting genetic diseases and if it were to be conducted with proper safeguards, it might be deemed ethically acceptable. This even includes germline modification conducted for therapeutic purposes. However, it seems unlikely that any interventions will be able to meet this standard. It would require the procedure be done without requiring IVF or other procedures that the church considers to be contrary to the dignity of the resulting human being (Moraczewski 2003: 2009–10), something that at least currently seems technically unfeasible.

The gradualist or developmental approach of the dignity of the embryo acknowledges the special status, if not the full human dignity, of the human embryo. For example, according to Bonnie Steinbock, "We may choose to respect embryos and preconscious fetuses as powerful symbols of human life, but we cannot protect them for their own sake" (2001: 10). This view leaves its adherents in a position of trying to define what this special status means and what kinds of respect it entails. What is relevant to this discussion is that it does not ipso facto lead to the rejection of all or most reproductive technologies. Many of those with this perspective accord individuals considerable reproductive liberty. At the least, it requires that embryos not be used without first carefully evaluating the rationale and the consequences for doing so. For some this process of discernment about reproductive technologies may be associated with how life expresses personal integrity and social dignity (Bettenhausen 1998: 96).

The Importance of Naturalness in Procreation

By definition, reproductive technologies are unnatural interventions into the procreative process. For most people this does not raise problems, but for some it does. One source of opposition to the use of reproductive technologies is based on the belief that the inception of human life is not to be tampered with or subjected to unnatural procedures. To do so is considered to go beyond our role as human beings. Related to this concern, some ethicists worry that technical intervention potentially interferes with the fundamental bond between parent and child.

On the secular side, Leon Kass has been a strong advocate of the importance of naturalness in procreation or, as he puts it, the dignity of human embodiment, human procreation, and human finitude. Kass believes that artificial interventions in human procreation amount to baby manufacture. For him, to say yes to baby manufacture is to say no to all natural human relations and to reject the deepest meaning of human sexual coupling, namely human erotic longing (Kass 2002: 19). Although Kass purports to regard modern science as one of the great monuments to the human intellect and the field of modern biology as unrivaled in the wonderful discoveries it can and will increasingly offer us (Kass 2002: 25), he offers a critique of biomedical science and technology, including in *vitro* fertilization and genetic technology, as posing challenges to identity and individuality, bodily unity and integrity, and the dignity of the body. He believes that new biotechnologies threaten "what we might summarily call human dignity" (Kass 2002: 22), but even in a book that has "defense of dignity" in its title, he does not explicitly conceptualize his notion of human dignity. His most comprehensive discussion of the sources and meaning comes later in his contribution to a collection of essays commissioned by the President's Council on Bioethics, for which he was the founding chair. Kass grounds basic human dignity in our equal membership in the human species and the higher dignity of being human in active human vitality, creation in the divine image and the characteristics that entails, human aspiration sired by a divine spark, and intimations of transcendence. Kass also associates human dignity with the sanctity or respect owed to human life as such, including the early stages of human life (Kass, 2008: 297–332).

Gilbert Meilaender also argues that how we come into being is of central importance for the sense of what it means to respect or, alternatively, to undermine human dignity. For him the language of human dignity functions as a placeholder for a larger vision of what it means to be human (2009: 33). He conceptualizes the human being as neither beast nor God and links human dignity with the acceptance of this in-between state (Meilaender 2009: 5). He worries that reproductive technologies, which he associates with the ready acceptance of the abortion of "defective" fetuses, violate the fundamental human dignity we share. Moreover, he claims that not accepting naturalness in procreation will necessarily interfere with the bond between parent and child and the relationship between the generations:

> The character of human life is degraded or diminished if we envision the relation between generations in a way that makes some strong and others weak, in a way that makes some a "product" or an "artifact" of the will and choice of others (Meilaender 2009: 33).

He offers the dichotomy of choice on the one side and love and acceptance of children on the other (Meilaender 2009: 26), ignoring that most couples who seek the assistance of reproductive technologies do so because they are otherwise unable to conceive a child. Like others emphasizing naturalness, Meilaender, associates what he terms "distinctively human procreation"— that is, begetting a child through copulation—as necessary to uphold human dignity. This seems to be at least in part because Meilaender, like Kass, associates being human with being embodied in a completely natural manner.

The concern that biotechnology threatens the natural order of things and tampers inappropriately with God's creation has been a concern of several religious thinkers and some Christian denominations, particularly the Roman Catholic Church. Advocates of this position believe that natural forms of biological conception through the sexual union of a man and woman in marriage follow God's sacred plan and thereby have a special dignity. It follows from this position that use of contraception to prevent procreation or any of the technologies that intervene in human reproduction violates human dignity. Teachings of the Roman Catholic Church, particularly *Donum Vitae* (The Gift of Life), emphasize the inseparable connection between the conjugal act through which the spouses mutually express their personal love in the "language of the body" and its openness to procreation (Pontifical Academy for Life 1987). According to the Pontifical Academy for Life, among all the fundamental rights that every human being possesses from the moment of conception, the right to life is the primary right because it is the precondition for the existence of all other rights. Moreover, the inalienable dignity of the person, which belongs to every human being from the first moment of his or her existence, "requires that his origins should be the direct consequence of suitable personal human action; only the recipro-cal gift of the married love of a man and a women, expressed and realized in the conjugal act with respect for the inseparable unity of its unitive and procreative meanings, is a worthy context for the coming forth of a new life" (Pontifical Academy for Life 2004).

Given its views on procreation, the Catholic Church has consistently opposed any artificial reproductive techniques that fuse male and female gametes in a manner other than the sexual union of a man and woman unless the technical means is not a substitute for the conjugal act but serves to facilitate and help that act attain its natural purpose. This disapproval includes technologies, such as *in vitro* fertilization, that seek to address problems of infertility. There is also the concern that development of the practice of *in vitro* fertilization has required innumerable fertilizations and destructions of human embryos (United States Conference of Catholic Bishops 1996).

Robert Kraynack is another author with a similar position. For him human dignity is based on God's "mysterious election" rather than in any essential characteristics or attributes of the person. He considers sexual reproduction to be both a natural biological process and a divine mystery, because God could have made the human species to reproduce in other ways. He views biotechnology as a threat to the natural order of things, because it implies that everything can be reinvented by the human will and by science. Instead he advises that we should have a cautionary sense of awe before the mystery of life and procreation. Nevertheless, Kraynack is open to some technologies that modestly follow the course of nature and respect the mysterious unity of man (human or person) as body and soul. In contrast with the official Catholic position, he considers *in vitro* fertilization as defensible in terms of respecting human dignity, because it is essentially replicating the natural processes in couples who cannot conceive on their own (Kraynack 2008: 61, 74, 80, 81).

Autonomy of the Individual and the Right to Make Private Reproductive Choices

One of the most significant threads of the linkages between human dignity and reproductive technologies is the right of the individual to make private reproductive choices. In our "individualistic" society, the autonomy and freedom of the individual constitute one of the central values. The philosophical view that individuals should have the right to make private choices free from the interference of the state can be traced to John Stuart Mill's statement in *On Liberty* that "the only purpose for which power can be rightfully exercised over any member of a civilized community, against his will, is to prevent harm to others" (Mill 1869: 15). Of the four dominant principles of biomedical ethics, respect for the autonomous choices of other persons is dominant (Beauchamp and Childress, 1994). Some ethicists equate human dignity with the exercise of autonomy (Macklin 2003).

A variety of human rights documents whose foundation is "the inherent dignity and inalienable rights of all members of the human family" (United

Nations 1948: Preamble) lend support to the importance of a broad interpretation of individual freedom. To promote the dignity and rights of individuals, the Universal Declaration of Human Rights states that "No one
shall be subjected to arbitrary interference with his privacy, family, home
or correspondence" (Article 12). The International Covenant on Civil and
Political Rights, to which the United States is a state party thereby making its
provisions legally binding, enumerates the right of men and women of mar-
riageable age to marry and to found a family (1966: Article 23). The Convention on the Elimination of All Forms of Discrimination against Women
accords equal rights to both spouses to decide freely and responsibly on the
number and spacing of their children and to have access to the information
and means to exercise this right (1979: Article 16 (e)). The interpretation of
the right to health in the International Covenant on Economic, Social and
Cultural Rights includes reproductive rights, defined as "the right to control's one health and body, including sexual and reproductive freedom, and
the right to be free from interference such as the right to be free from torture, non-consensual medical treatment and experimentation" (Committee
on Economic, Social and Cultural Rights 2000: para. 1). It should be noted,
however, that none of these documents specify a right to use particular reproductive technologies. Nor is the United States a state party to either the Convention on the Elimination of All Forms of Discrimination against Women or
the International Covenant on Economic, Social and Cultural Rights.

Advocates of procreative liberty argue that the freedom to pursue to
decide whether to have offspring and to control the use of one's reproductive capacity is one of the most important expressions of individual liberty
on the grounds of its importance to individual meaning and dignity. Like
some of the writings discussed above, advocates of reproductive freedom
rarely offer a robust conception of human dignity or show a clear connection with procreative freedom. Moreover, these writings rarely address the
implications of untrammeled reproductive freedom for the child that will be
born or not born and the wider society.

According to one of these advocates, John Robertson, any restriction, regulation, or imposition of these technologies interferes with or limits procreative
freedom and therefore needs to be justified (Robertson 1994: 16). It is his view
that such strong justification "is seldom present" (1994: 4). Further, Robertson claims that reproductive freedom incorporates a right to have access to
just about any of these technologies, including those that select offspring characteristics and thereby serve as "quality control devices" to enable couples to
have healthy children, such as prenatal screening, preimplantation diagnosis, gene therapy, and germline interventions (1994: 33, 149–72). Robertson
does qualify his interpretation of procreative liberty: it does not extend to the
positive right to have the state or particular persons provide the means or
resources necessary to have or avoid having children (1994: 23).

A number of justifications have been put forward for limiting reproductive
freedom to use specific forms of reproductive therapies. Other critiques have

raised cautions about the moral, social, and health consequences. However, these writings rarely argue their positions from a human dignity perspective. As noted, some of the thinkers with ethical reservations based their views on the need for naturalness believe that the use of noncoital technologies is unethical and constitutes a violation of human dignity, because such technologies separate the unity of sex and reproduction. Meilaender also makes two claims about an emphasis on autonomy: it assumes that human nature has no telos, no way of life that constitutes its particular flourishing, and that there is a danger in allowing autonomy to preempt other aspects of human dignity (2009: 29). Others object to rights language as an inappropriate prism through which to view parenthood and family life. Tom Murray comments that individual rights are blunt and clumsy instruments with which to understand the moral intricacies of complex family relationships and suggests that rights more appropriately illuminate our moral relationships with the state and with strangers in the marketplace, and not family life. He warns that asserting a right to unlimited choice without regard for its meaning and impact could threaten what we value about families (Murray 1996: 18–28). Feminist appraisals of reproductive technologies raise another set of concerns, some of which derive from the effect of specific technologies on women's health and well-being and others about how their use will affect women's agency, roles, and power (Donchin 1996).

A report of the House of Commons Science and Technology Committee titled *Human Reproductive Technologies and the Law* listed eight rationales that have been given for limiting reproductive freedom, some of which, such as the protection of the embryo, were discussed above. Additional justifications identified are: the need to protect the welfare of children who are born using reproductive technologies, a concern that when embryos and gametes are donated a genetic link should be maintained, worry that the use of these technologies will expose patients to excessive levels of risks, concerns about the potential impact of individual reproductive decisions on society, and the need to supervise morally controversial aspects of assisted reproduction (2005: 18). The Committee concluded that in a society that is both multifaith and largely secular there will never be consensus on the level of protection accorded to the embryo or the role of the state in reproductive decision making. It recommended that reproductive technologies should proceed under a precautionary principle requiring that alleged harms to society or its patients need to be demonstrated before reproductive freedom is limited (2005: 22).

So where does this leave the discussion of the relationship between reproductive freedom and human dignity. Clearly there are many valid reasons for regulating reproductive technologies to prevent human harm, something which federal and state governments in the United States are, unfortunately, loathe to do. It is arguable, however, whether all such regulations would constitute limitations on reproductive freedom. Importantly for the discussion here, while many of the concerns discussed above are valid, they are not put forward within the framework of human dignity.

HUMAN DIGNITY IN THE DEBATE ABOUT SPECIFIC REPRODUCTIVE TECHNOLOGIES

Selection Technologies

A variety of reproductive technologies offer the possibility of screening embryos—both those implanted in a woman's womb and early-stage blastocysts created through in *vitro* fertilization—to test for birth defects, chromosome abnormalities, and genetic diseases. Two pregnancy tests are in widespread use, both with a small risk to the mother as well as to the fetus: amniocentesis, usually performed between the 12th and 16th week of pregnancy, and chorionic villus sampling, which can be done as early as the 10th week of pregnancy. In addition, noninvasive prenatal genetic diagnosis through analyzing a maternal blood sample in the first trimester of pregnancy is now on the horizon (Benn and Chapman 2010). Preimplantation genetic diagnosis (PGD) offers a technique to examine in *vitro*–fertilized eggs for some types of gene abnormalities as early as the fourth cell division. Overall prenatal diagnosis is now possible for some 1,000 genetic diseases and metabolic disorders, and others are likely to be available in the years ahead. Many of these tests also reveal gender and paternity, and likely in the future will be able to provide some information about specific traits.

Most couples or women choosing to use these selection technologies do so because of their desire to have a healthy child. However, the ability to diagnose developmental and genetic disorders is far more advanced than the capacity to treat them *in utero* or in a Petri dish. Therefore, parents who receive problematic test results currently have the painful choice of destroying an early-stage embryo (or a fetus) or continuing the pregnancy and bearing the child with the problem. The question as to whether a pregnancy termination is ever warranted intersects with the always-contentious debate over the morality of abortion, which in turn is bound up with the issue of the status of the embryo discussed above.

Ethicists have been concerned for some time that reproductive testing may change the way parents view children: that children will be seen less as a blessing to be embraced whatever their characteristics and more as a commodity to design to meet certain specifications (Sandel 2007). Some disability advocates contend that prenatal genetic testing, with its implicit aim of preventing the birth of disabled babies, undermines the worth of individuals living with disabilities (Chachkin 2007).

Preimplantation genetic diagnosis (PGD) offers somewhat different ethical issues. It usually enables the prospective parents to select among a variety of fertilized early-stage embryos or blastocysts to determine which should be implanted. The others can be frozen for future use or discarded. Some opponents of PGD characterize this selection process as a form of eugenics.

One critic of testing and selection technologies, which he terms "quality control," the German philosopher Jürgen Habermas, does place some of his objections within a human dignity framework. According to Habermas, human dignity is connected with a relationship symmetry that requires a community of moral beings who can place one another under moral obligations and expect one another to conform to norms in their behavior. He argues that knowledge of one's own hereditary factors may be restrictive for the choice of an individual's way of life and may undermine the symmetrical relations between free and equal human beings (2003: 33). Habermas also equates selection with negative eugenics. He puts forward two possible consequences of genetic selection technologies: genetically programmed persons might no longer regard themselves as the sole authors of their own life history, and they might no longer regard themselves an unconditionally equal-born person in relations to previous generations (2003: 9). While he states that the fertilized egg cell does not possess human dignity in the strict sense, he proposes that it be given a restrictive concept of dignity, which he terms the dignity of life, to protect against instrumentalization (2003: 37–38). He considers PGD to be one-sided and therefore instrumentalizing, because there can be no assumption of an anticipated consent (2003: 68).

New noninvasive approaches to prenatal diagnosis raise a number of issues with implications for human dignity. Early noninvasive testing may normalize prenatal testing and selective abortion and in the process trivialize the significance of terminating a pregnancy. Testing for sex determination is already available on a direct-to-consumer basis that can be carried out without guidance from a clinician. While there may be legitimate medical reasons for determining the sex of a fetus to avoid having a child with a sex-linked or sex-limited condition, the technology may encourage couples with a strong preference for a child of a particular sex to abort a pregnancy when the fetus is the "wrong" sex. The widespread use of reproductive technologies for purposes of sex selection has already produced serious sex ratio imbalances in regions of India and China, potentially further diminishing the status of girls and women (Benn and Chapman 2010). The more limited use of sex selection technologies for family balancing that presumably would be more evenhanded in choosing the sex of babies would still have consequences for human dignity, because it is likely to abrogate a child's right to an open future. In such cases, it is likely that the selection will be accompanied by gender stereotyping that will affect parental expectations and the manner in which the child will be raised. The combination of these factors will make it difficult for the child to develop in ways that are different than the parental expectations (Davis 2010: 123–49).

Reproductive Cloning[1]

The February 1997 announcement that scientists had been able to clone a lamb from the somatic cell of a mature sheep generated a furor over the prospect of human cloning. In early March, President Clinton requested that the newly formed National Bioethics Advisory Commission (NBAC), an independent body of experts, study the issue of human cloning and provide public policy recommendations. Within a few months of the initial announcement, a wide range of secular and religious ethicists had formulated at least preliminary thoughts about the ethical implications of human cloning. When the NBAC issued its report in June 1997, it acknowledged that much of the negative reaction to the potential application of cloning in humans could be attributed to fears about harms to the children who would be conceived through cloning, particularly the potential psychological harms associated with the children experiencing a diminished sense of individuality and personal autonomy. The NBAC recommended the enactment of federal legislation to prohibit anyone from attempting to clone a human child, because the creation of a child using somatic cell nuclear transfer would expose the child to unacceptable risks (National Bioethics Advisory Commission 1997: 107–8). The Council of Europe, UNESCO, and the UN General Assembly also adopted documents opposing human reproductive cloning as one of the "practices which are contrary to human dignity (and which) shall not be permitted" (UNESCO 1997: Article 11).

Secular ethicists presented a wide range of perspectives. On the negative side, Leon Kass, then a physician and ethicist teaching at the University of Chicago, urged the NBAC to recommend a sweeping ban on human cloning, warning that it "represents something radically new" that presents an unacceptable threat to human identity and individuality. According to Kass, cloning would represent a major step toward making human beings a man-made thing, a kind of technological product (Kass 1997: 108). Like some of the religious critics of cloning, Kass was particularly concerned about the effect of "making" rather than "begetting" children on family relationships. In contrast with those who disparaged the public's reaction to cloning, Kass defended "the wisdom of repugnance" (Kass 1997b).

Religious thinkers put forward a series of concerns about the ethical implications of cloning: the risks to human dignity for both the cloned person and the rest of society, the potential effects of cloning on identity and individuality, the disruptive impact of cloning on procreation and the family, and questions about whether cloning exceeds the limits of appropriate human intervention into the natural order. Many religious thinkers expressed views that cloning would violate human dignity. Despite these concerns, however, there was complete consensus that if any humans were to be cloned, the resulting person would have the inherent value, dignity, and moral status common to all humanity and should be vested with the same civil rights and protections. Nonetheless, there were also groups and thinkers who, despite their ethical concerns, had a qualified acceptance of cloning. Generally, thinkers in this group did not consider the act of human cloning to violate human dignity but instead were concerned about the societal and health consequences (Chapman 1999: 93–95).

The Roman Catholic Church's claims that human cloning is unnatural, contrary to the divine plan, and would violate the inherent dignity of human beings represented one of the most prominent sources of opposition within the religious community. Testifying to the NBAC on behalf of the National Conference of Catholic Bishops, the Rev. Dr. Albert Moraczewski emphasized that cloning would violate human dignity by exceeding the delegated dominion given to the human race, by jeopardizing the personal and unique identity of each individual, and by providing an opportunity for genetic manipulation of the nuclear genome, perhaps with eugenic intent. He also rejected this technology, because cloning implicitly involves the assertion of a right over another person (Moraczewski 1997: 185–97).

In general, concern about the implications of cloning for human dignity in the religious community appears to have two primary bases: the manner in which cloned persons are likely to be treated by other members of society and the possibilities that cloning will contribute to a general devaluing and commodification of human life. Whatever the motivations of the progenitor, there was apprehension that treating human life as a product to be manipulated and manufactured would diminish a sense of the sacredness of life. Members of the religious community also shared the NBAC's caution that efforts to produce a child through somatic cell nuclear transfer would amount to experimenting on human beings to perfect a new technology. The question was asked more than once, what will happen to the failures, to the bad results? There was worry that a cloned child would be treated as less than a fully equal and unique person; that is, even if a cloned human being were to be accorded all of the legal rights and protections afforded to others, it might be difficult for its creator or a society to accept its full humanity and dignity. The concerns with objectification and commodification derived from the apprehension that the clone would be treated as an object or a means to fulfill the will of another human being and not respected as an end in itself (Chapman 1999: 100).

Not all the contributions to the debate or later reflections on the issue agreed that reproductive cloning necessarily represents a threat to human dignity. An Expert Group on Human Rights and Biotechnology convened by the UN High Commissioner for Human Rights that met in 2002 was less inclined to rush to judgment. It called for caution in considering a legal response, noting that premature attempts at prohibition and comprehensive regulation would have disadvantages that would be difficult to correct once an international convention were to be adopted. The experts recommended that a serious and detailed analysis of the reasons why reproductive cloning gives cause for concern be conducted. They also suggested that there be exploration first as to how far individuals should be allowed to use biotechnology to determine the traits of their children (UNHCR's Expert Group on Human Rights and Biotechnology 2002: 156–58).

Ruth Macklin, who, as noted above, equates human dignity with autonomy, argued in her testimony to the NBAC that there was no reason for society to panic at the prospect of cloning or, for that matter, to ban it.

Disparaging the concerns of those who opposed cloning, Macklin claimed that much of the ethical opposition seemed to grow out of "an unthinking disgust—a sort of 'yuk factor'" (Macklin 1997a: 64). While Macklin acknowledged that many theologians contended that to clone a human would violate human dignity, she disagreed. According to Macklin, "It is not at all clear why the deliberate creation of an individual who is genetically identical to another living being but separated in time would violate anyone's rights" (Macklin 1997b: 98). In her testimony before the NBAC, Macklin underscored that the ethics of cloning must be judged not by the way in which parents bring a child into existence but by the manner in which they treat and nurture the resulting child (1997b: 100).

Deryck Beyleveld and Roger Brownsword's analysis of the human dignity implications of reproductive cloning takes issue with two frequent claims: first, that cloning involves treating the clone not as an end in itself and thus not as a possessor of human dignity but merely as a means to the ends of others; and, second, that cloning involves removal or reduction of the autonomy of the clone, thereby affecting its dignity. Their reasoning is that the clone is not an agent, because the clone does not exist until cloning is complete. They argue that there must be something about cloning as such that makes it impossible for the clone to be treated as an end when it comes to display the capacities of agency or which necessarily runs counter to the clone being treated as an end. They also question whether cloning necessarily eliminates or reduces the capacity for free action of the clone in which its dignity resides. Instead they claim that whether a person's genetic makeup is chosen by others or is determined randomly is entirely irrelevant to the freedom he or she has and requires as an agent and thus someone in possession of dignity. Moreover, they point out that modern genetic theory provides no warrant for assuming that any phenotypic characters or traits are determined entirely or even mainly by genetic factors. In view of the importance of genetic-environmental interactions in shaping phenotypic characteristics, they suggest that if there are any problems with preselection of genes in regard to human dignity, then there must be equal problems with the preselection of environmental influences. They conclude that the impact of cloning will depend primarily on the intentions of the parents and whether they will enable their child to develop its own life plans. Therefore, while human cloning might sometimes be contrary to human dignity, these cases do not provide warrant for holding that cloning as such is contrary to human dignity (Beyleveld and Brownsword 2001: 158–64).

CONCLUDING REFLECTIONS

Hitherto the discussion of human dignity and reproductive technologies has rarely met rigorous conceptual and analytical standards. In most cases, the notion of human dignity being applied is not well developed. Nor do authors specify criteria for assessing the impact. Analysis of the implications of the technologies is often subjectless in the sense that writings on this topic do not explain whose dignity is being violated or upheld. In theory, reproductive technologies could affect the dignity of the woman or couple using the technology, the child conceived, or the society to which they belong. Depending on the subject or subjects, the type and severity of dignity issues could be quite different as well as the potential remedies. While the authors writing on human dignity and reproductive technologies usually come to clear-cut conclusions, particularly if they are critical, the principles grounding the assessment are less clearly presented. Nor is there necessarily clarity as to whether the technology will affect the subjective and/or objective dimensions of human dignity. To be more specific, are the particular dignity issues located in the subjective perceptions of the couple, the child they conceive, or the wider society? Or, alternatively, do the human dignity implications relate to objective outcomes, such as the child or the parents being injured by the technology or the way any of these subjects are likely to be treated by one another or the wider society?

I would like to propose that a meaningful consideration of the implications of a specific reproductive technology requires four elements: (1) a well-defined conception of human dignity with specific criteria that can be used for evaluative purposes; (2) a clear specification as to whose dignity may potentially be affected—the woman or couple opting to use the technology, the child whose future will be affected by the technology, and/or the wider society; (3) a nuanced discussion of the specific reproductive technology being assessed with a clear distinction between the technology itself and the way it is likely to be used; and (4) a distinction between subjective dignity and objective dignity—that is, the difference between the resultant effect on the subject herself or himself and the manner in which the subject is viewed by others along with their perceptions of the wider implications for the species. Stephen Malby offers a multipart model of human dignity that has several of these elements (Malby 2002: 113–16).

Few conceptions of human dignity have specific criteria that can be applied to the evaluation of reproductive technologies. Martha Nussbaum's capabilities approach, however, holds promise (2000, 2006, 2008). Nussbaum conceptualizes human capabilities as "what people are actually able to do and to be" as a measure of the extent to which they can live a life that is worthy of the dignity of the human being (Nussbaum 2006: 5). In some of her writings, she identifies capabilities with the basic social minimum governments should provide for their citizens or, to put it another way, the minimum political entitlements due to every resident. Nevertheless, I think the capabilities approach can be adapted for other applications to human dignity as well.

It should be noted, though, that Nussbaum does not herself discuss capabilities or human dignity in relation to reproductive technologies. It is not that Nussbaum is uninterested in reproductive issues. As a feminist, she has a great deal of concern for the way that women are treated, and particularly for the vulnerability to violence and sexual abuse of poor women in the third world. One of her most important books in which she develops her capabilities approach, *Women and Human Development* (2000), takes the predicament of poor women in India as a point of departure.

Nussbaum is a "universalist"—that is, she holds that all persons possess full and equal human dignity by virtue of their common humanity, including a wide range of children and adults with severe mental disabilities, but she does not vest embryos or fetuses with human dignity. Because all persons possess full and equal dignity, they are entitled to be treated with equal respect. Much like those, including myself, who have a commitment to the human rights paradigm, Nussbaum believes that human dignity entitles each citizen to make the same set of claims against the government. In her case, she frames these entitlements as the development of a set of central human capabilities rather than rights. She argues that the capabilities in question should be pursued for each and every person, treating each as an end and none as a mere instrument of the ends of others. Although human beings have a worth that is inalienable because of their capacities for various forms of activity and striving, these capabilities must be nurtured for their full development and their conversion into actual functioning. She distinguishes between three types of capabilities: basic or untrained capacities, internal capabilities or trained capacities, and combined capabilities that combine trained capacities with suitable circumstances for their exercise (Nussbaum 2008: 357). In addition, her capability approach uses the idea of a threshold level of each capability, beneath which truly human functioning is not available to the citizen. Therefore, according to Nussbaum, the social goal should be understood in terms of the government getting citizens above this capability threshold (Nussbaum 2006: 67–71).

Nussbaum identifies a list of 10 central human capabilities, of which 5 seem particularly relevant for assessing the implications of reproductive technologies for human dignity, particularly to evaluate the dignity of the child who is born through the application of one of the technologies. They are: (1) life: being able to live to the end of a human life of normal length— that is, not dying prematurely or having a life so reduced as to be not worth living; (2) bodily health: being able to have good health, including reproductive health; (3) emotions: being able to have attachments to things and people outside ourselves; to love those who love and care for us; in general, to love, grieve, and experience longing, gratitude, and justifiable anger; (4) practical reason: being able to form a conception of the good and to engage in critical reflection about the planning of one's life; and (5) affiliations: being able to live with and toward others, to recognize and show concern for other human beings, to engage in a variety of forms of social interaction; to have the social bases of self-respect and be able to be treated as a dignified being whose worth is equal to that of others (Nussbaum 2008: 377–78).

Applying these capabilities/criteria to PGD, for example, suggests that the implications would be different for the child who is born, the wider society, and the parents of the child. The analysis also shows there may be a trade-off between capabilities, something that analysis of the implications for human dignity rarely acknowledges. It seems likely that a child who has been selected from embryos screened to eliminate or reduce the possibility of being born with severe congenital handicaps would have the possibility of a life of greater dignity and human flourishing in some ways than another child born with these physical or mental limitations. The child would also have a greater possibility of being able to live to the end of a normal human life span, of having better bodily health, and, in the case of screening for mental disabilities, to have a greater capacity for practical reason. On the other hand, widespread use of PGD could have detrimental societal implications for human dignity. Efforts to prevent handicapped persons from being born could convey a diminished sense of social dignity to those who have physical or mental handicaps. It might also make the society deficient in the capacity to recognize and show concern for disadvantaged persons, which is a significant dimension of the capacity for affiliation. There appear to be fewer dignitarian consequences for the parents of the child.

As noted above, it is important to distinguish between a particular reproductive technology and the manner in which it is applied. PGD, for example, can potentially be used in four different ways: (1) to attempt to identify embryos with severe chromosomal and genetic disorders, many of which would not be likely to survive to birth if implanted; (2) to screen for embryos with less severe or late-onset genetic conditions; (3) to engage in sex selection either for medical or social reasons; and, at some point in the future, when the genetic basis of traits are understood, (4) to attempt to enhance the characteristics of the child. The use of PGD to screen for milder or late-onset disorders or to determine the sex of the embryo that is implanted is more likely to have detrimental societal implications for human dignity than the elimination of severely abnormal embryos. It implies that significant numbers of persons in the population have a life and dignity considered to be of less value than so-called normal members. In societies that have a cultural preference for male children, the application of PGD to eliminate female embryos could further diminish women's status and dignity.

Human cloning would appear to have greater implications than PGD for the fourth dimension of human dignity identified above: subjective dignity and objective dignity. Many commentators worry that a child born through the use of cloning technology would have a reduced sense of autonomy and personal choice, but other analysts disagree. There is also apprehension that society would stigmatize a child that is a known clone. Neither of these concerns seem applicable to PGD unless and until it becomes possible to use PGD for purposes of enhancement.

The potential or actual ethical and social implications of a specific reproductive technology for human dignity raise important issues. Hopefully future discussions will be more rigorous and thereby illumine these issues better. Alternatively, given the difficulties of conceptualizing human dignity and identifying relevant evaluative criteria, it may be that human dignity is not the most appropriate lens through which to conduct an analysis of the ethical implications of a particular reproductive technology. Instead other ethical concepts and frameworks may offer more meaningful and nuanced insights.

NOTE

1. Parts of this section are based on Chapman (1999: 77–108).

11 The Language of Human Dignity in the Abortion Debate

Scott Rae

The notion that human beings possess intrinsic dignity on which fundamental rights are based is widely accepted in most versions of bioethics, and in most countries the law also reflects this critical idea. As the first section of this volume illustrates, there are a variety of ways in which the concept of human dignity is grounded, ranging from explicitly religious grounds to overtly nonreligious ones. Even those who hold that dignity is simply a human construction, or hold that it is a somewhat useless concept, nevertheless endow human beings with rights based on their moral status, autonomy, or other concept that does much of the same work in grounding rights as the concept of dignity does for its proponents (Macklin 2003, Pinker 2008).

The concept of human dignity is a critical one in the long-standing debate over the morality and legality of abortion. For some time, those who oppose abortion appealed to the idea of human dignity in support of restrictive abortion laws. The use of dignity among these groups almost always assumes or attempts to establish that embryos and fetuses are persons from conception forward. Often the use of dignity is religiously grounded, though the degree to which that is explicitly stated varies widely depending on the audience. That is, the explicit dependence on the religious grounding for the position taken can vary significantly, since it can be seen as beneficial to a public policy position to avoid such an explicit dependence on religious views.

Chief among the groups who place great weight on the concept of dignity has been the Catholic Church in its various papal encyclicals, and its appeal to dignity is both consistent and foundational to its position on abortion. It is used just as widely when extended to its views of birth control and technologically assisted procreation. Numerous Protestant commentators, namely evangelicals, affirm this view and have added their voice to the opposition to abortion. These voices have been heard in the public square most recently in the work of Leon Kass (who himself is Jewish but has much in common with many Protestants and evangelicals), former President Bush's Bioethics Commission, and the work of the evangelical Charles Colson. Though their emphasis is on the application of the idea of dignity to biotechnology, particularly enhancement technologies, the idea is repeatedly applied, or assumed to apply, to abortion.

The proponents of abortion choice laws also make good use of the concept of dignity. They are concerned about the dignity of the pregnant woman being upheld in two primary ways. First, the woman's dignity is upheld by not coercing her to carry an unwanted pregnancy. It is considered an assault on her dignity to force her to maintain an unwanted pregnancy. Second, her dignity is upheld by abortion laws ensuring that the abortion procedure is safe, thereby not forcing the desperate pregnant woman into risky procedures at the hands of less than fully qualified practitioners. Some abortion proponents acknowledge competing claims based on dignity, arguing that a woman's intuitions about abortion that give her pause are based on the dignity of the fetus but are outweighed by the dignity interests of the pregnant woman.

The courts have also made substantial use of the concept of human dignity in their decisions both to allow and to restrict abortion. The 1992 *Casey* decision attempted to balance competing versions of human dignity, though there is substantial debate on both sides of the abortion argument as to how well the Court succeeded in that task. In addition, the most recent decision of the Supreme Court in *Gonzales v. Carhart* used dignity in a way that the courts had not used it before—appealing to the dignity of the pregnant woman *to restrict abortion*, particularly in reference to what has come to be known as the partial-birth abortion procedure. The *Carhart* decision gives support for decisions that are both woman-protecting and abortion-limiting at the same time, based on the application of dignity. The use of the concept of dignity differs from the traditional ways that dignity has been used to undergird a woman's autonomy to make the abortion decision without state interference.

ABORTION OPPONENTS' USE OF THE CONCEPT OF DIGNITY

Opponents of abortion make abundant use of the rhetoric of human dignity, and the use of the concept is consistent throughout the various groups and worldviews that stand together in opposition to abortion. Both religious and nonreligious opposition to abortion appeal to human dignity as a foundational component of their ideology. I will highlight three representative uses of the concept of dignity: the Roman Catholic Church in its official encyclicals, a Protestant perspective in the writings of Charles Colson and a variety of evangelical opponents of abortion, and a Jewish perspective from the work of Leon Kass (Kass 2002). The nonreligious source I will examine comes from the volume published by the Bush administration's Council on Bioethics, *Human Dignity and Bioethics* (President's Council on Bioethics 2008).

Official Catholic Perspectives

The three Catholic encyclicals that address human dignity and bioethics are structured similarly. The first part of each addresses the foundational theological and philosophical concepts, then applies them to the bioethical issues under review. *Humana Vitae* (1968) primarily addresses birth control, *Donum Vitae* (1987) speaks principally to assisted reproduction, and *Dignitas Personae* (2008) discusses a wide variety of technologies related to procreation and the beginning of life. The language of dignity is paramount in each of the three encyclicals and forms a critical part of the foundational views of a human person. Although the application of dignity is much broader than simply to abortion, it is widely assumed that the teaching applies to abortion as well.

Perhaps the clearest exposition of human dignity comes from the 2008 encyclical *Dignitas Personae*. Here the encyclical addresses the fundamental ethical criterion necessary to evaluate any and all issues involving human embryos. Reflecting the earlier teaching in *Donum Vitae*, *Dignitas Personae* insists that from conception forward, the embryo "develops progressively according to a well-defined program with its proper finality, as is apparent in the birth of every baby" (*Dignitas Personae*: 3). The encyclical rejects the notion that embryos are simply clumps of cells and argues that embryos are human beings (and thus human persons) to be treated with unconditional respect, based on the embryo's continuity of development and identity. This gives every human being the inviolable right to life from the moment of conception—the fundamental ethical principle governing human life and procreation. What gives this anthropological notion its moral weight is the connection with human dignity. The encyclical states that, "The human embryo has, therefore, from the very beginning, the dignity proper to a person." As a result, "respect for that dignity is owed to every human being because each one carries in an indelible way his own dignity and value" (*Dignitas Personae*: 4). This is the foundational notion that governs the later application in the encyclical concerning anything that has to do with embryos and fetuses. It is assumed that destruction of embryonic and fetal life, either in the womb or in the lab, is a violation of the intrinsic dignity that is possessed by virtue of being a human being. The encyclical later grounds this dignity theologically, in the notion of human beings being made in God's image (Genesis 1:26) and human beings being the objects of God's loving care (*Dignitas Personae*: 5).

The earlier encyclical *Donum Vitae* makes repeated appeal to human dignity as part of its foundational argument for its teaching on the beginning of life. The moral criteria for evaluating a variety of medical technologies include, "the respect, defense and promotion of man, is primary and fundamental right to life, his dignity as a person who is endowed with a spiritual soul and moral responsibility" (*Donum Vitae*: 3). Further, the encyclical maintains that the moral evaluation of medicine must be done "in reference to the dignity of the human person" (*Donum Vitae*: 5). As a result, the use of human fetuses or embryos "as the object of instrument of experimentation constitutes a crime against their dignity as human beings having a right to the same respect that is due to the child already born and to every human person" (*Donum Vitae*: 11). It is assumed in these encyclicals what has been made clear in other parts of Catholic teaching—that abortion is a moral evil because it assaults the dignity of the human person. Although the direct application is to experimentation on fetuses and embryos (and in *Donum Vitae*, the way in which human persons are conceived), it is clear that abortion is problematic for the same reasons. In fact, opposition to abortion is the clear and obvious conclusion to be drawn from these statements that a human person from conception forward has fundamental dignity. The encyclicals intend that whatever applies to human embryos also applies to fetuses, that "keeping embryos alive for experimental or commercial purposes is totally opposed to human dignity" (*Donum Vitae*: 11). Further, with respect to human persons, "their dignity and right to life must be respected from the first moment of their existence" (*Donum Vitae*: 12). Equating abortion with embryo destruction, it goes on to state, "Just as the Church condemns induced abortion, so she also forbids acts against the life of these human beings (embryos)" (*Donum Vitae*: 13). The reason for this goes back to the notion of dignity—"these procedures (which destroy fetuses or embryos) are contrary to the human dignity proper to the embryo. . . . Every person must be respected for himself; in this consists the dignity and right of every human being from his or her beginning" (*Donum Vitae*: 13).

Evangelical Protestant Perspectives

Coming from outside Catholic tradition but arguing in ways consistent with it, other opponents of abortion employ the language of human dignity in defense of fetal and embryonic right to life. As is the case with the Catholic use of the concept, human dignity is a central concept that speaks to a wide variety of bioethical issues. Protestants such as Charles Colson and others highlight the centrality of human dignity, most recently applying the notion to various advances in biotechnology, but it is clear that the assumed application of the concept to abortion is never far out of their minds. Colson, for example, makes an explicit parallel between abortion and human cloning, asserting that both are to be seen as an attack on human dignity. He puts it this way, "The pro-life movement soon recognized that cloning was as much an assault on human dignity as was abortion itself" (Colson 2004: 7). Evangelical Nigel Cameron echoes this point and makes the connection between dignity and abortion explicit when he explains that,

> The point of departure for Christian reflection on medicine and biotechnology is the fact that God made us in his image, imparting to humankind a unique dignity that radically distinguishes him from other creatures and also from the inanimate creation, and declaring the sanctity of life—a safeguarding of human life by a holy God. . . . This forms the context for the challenge posed by abortion, with its notion that unborn human life is at the disposal of maternal choice. (Cameron 2004: 21)

Many of these writers see continuity in the development of biotechnology with the issue of abortion—both are different but equally concerning assaults on human dignity. Paige Cunningham, director of the Center for Bioethics and Human Dignity, and long-time attorney for Americans United for Life, calls abortion and biotechnology the "fraternal twins" of devaluation of human dignity. She suggests that,

> The pro-life movement must see the connection between the devaluing of life in the womb and its devaluation by the new biotechnologies. The issue is the worth (or dignity) of any human life, whether conceived in an experimental laboratory or discarded because of genetic profile. (Cunningham 2004: 158)

Some of the invocations of the notion of dignity are drawn from a very broad theological foundation, but not tied to any specific religious tradition. Perhaps the clearest use of the rhetoric of dignity came from former President Ronald Reagan, in his 1983 article, "Abortion and the Conscience of the Nation," written on the 10th anniversary of *Roe v. Wade*, in which he criticized the decision as well as the general phenomena of abortion in the country (Reagan 1983).[1] Reagan appealed to the tradition of civil rights in this article to undergird his criticism of abortion, drawing a parallel to the *Dred Scott* decision being overturned and the desire for *Roe v. Wade* to be similarly overturned. He put it this way in his often-repeated statement on abortion:

> The *Dred Scott* decision of 1857 was not overturned in a day, or a year, or even a decade. At first, only a minority of Americans recognized and deplored the moral crisis brought about by denying the full humanity of our black brothers and sisters; but that minority persisted in their vision and finally prevailed. They did it by appealing to the hearts and minds of their countrymen, to the truth of human dignity under God. From their example, we know that respect for the sacred value of human life is too deeply engrained in the hearts of our people to remain forever suppressed. (Reagan 1983: 2).[2]

Reagan also uses the language of the sanctity of life, life's intrinsic worth, the right to life, and the inviolability of innocent human life as parallels to the notion of dignity, similar to the way many pro-life advocates use these terms interchangeably with dignity.

A Jewish Perspective

Consistent with these is the connection made from a Jewish perspective, from Leon Kass, for whom human dignity is a central concept for his bioethic. Kass argues for a richer, more content-full bioethics that goes beyond the traditional principle-based bioethics of Beauchamp and Childress (autonomy, beneficence, nonmaleficence, and justice) that can give attention to, in his view, the more subtle ways in which human beings are dehumanized by advances in medical technology. Kass appeals to the concept of human dignity to provide a more comprehensive ethic that can be applied not only to abortion but also to other biotechnological threats to human beings. Kass puts it this way when he says,

> we are quick to notice dangers to life, threats to freedom, risks of discrimination or exploitation of the poor, and interference with anyone's pleasure. But we are slow to recognize threats to human dignity, to ways of doing and feeling and being in the world that make human life rich, deep and fulfilling. (Kass 2002: 12)[3]

Although Kass does not primarily address abortion in his key work on human dignity, he does assume that the concept applies analogous to other areas of bioethics. For example, in his section in which he addresses what he calls "eugenic abortion"—that is, ending pregnancies as a result of information obtained from genetic testing—he points out that human dignity is at stake even though the use of these testing technologies may seem benign. He puts it this way:

> For threats to human dignity can—and probably will—arise even with the free, humane and 'enlightened' use of these (testing) technologies. . . . It is, I submit, these challenges to our dignity and humanity that are at the bottom of our anxiety over genetic science and technology. (Kass 2002: 128–29)

Nonreligious Uses of the Concept of Dignity by Opponents of Abortion

The centrality of the notion of dignity for opponents of abortion also comes from those who do not argue from any specific religious tradition and do not even invoke any religious principles to ground the concept of dignity. Catholic philosopher Patrick Lee, while taking a position fully consistent with Catholic teaching, does not explicitly appeal to that tradition to support his view. His argument is that the fetus is a full member of the human community and thus entitled to all rights to life to which mature adults are entitled. He spends the majority of his work defending that idea. But in his summary statement he invokes the concept of dignity as a central underpinning for his opposition to abortion. He states plainly, "The choice to kill a child (in the womb) is a denial of the intrinsic dignity of human persons" (Lee 1997: 148).[4]

Among the opponents of abortion, even if the concept of dignity is not explicitly made, it is generally not far underneath the surface. It is frequently assumed as virtually self-evident, that all human persons, however defined, possess intrinsic dignity, such that taking his or her life would constitute the ultimate assault on such dignity. This kind of appeal to dignity is assumed to be a relatively simple step once the moral status of the fetus or embryo is established. Usually the appeal to dignity is preceded by some philosophical justification regarding fetuses and embryos as full persons, with rights to life (Lee and George 2008). Generally this involves an argument that human persons are not collections of parts and properties (a property-thing view of a human being) but, rather, what constitutes a person is something intrinsic, or essential. In other words, opponents of abortion who make this argument often distinguish between *functioning* as a person and *being* a person.

In this way, law professor Richard Stith appeals to dignity in his opposition to abortion. He is representative of many abortion opponents whose appeal to dignity begs the question of who should be included in the human community, on whom it would be appropriate to recognize intrinsic dignity. This was recognized implicitly by former President Reagan in his work mentioned above. Though not specifically invoking the language of dignity, it is clear from earlier in the article that appreciation of the value of life and recognition of the intrinsic dignity of the human person are clearly linked. He makes the claim that,

> The real question today is not when human life begins, but, "What is the value of human life?" The abortionist who reassembles the arms and legs of a tiny baby to make sure all its parts have been torn from its mother's body can hardly doubt whether it is a human being. The real question for him and for all of us is whether that tiny human life has a God-given right to be protected by the law—the same right we have. (Reagan 1983)

What Reagan refers to earlier as "the sacred value of human life" comes from "the truth of human dignity," which arises from "being made by their creator God." Although not invoking the language of dignity per se, but using the language of rights and rights-bearing status (one of which would clearly be to have one's right to have his or her essential dignity respected), philosopher Francis Beckwith argues that "the disagreement over abortion is not a dispute over moral principles per se (such as a respect for human dignity), but a clash over the question of who counts as a member of the moral community (i.e., whether it includes the unborn)" (Beckwith 2007: xiii).

Stith argues that unless one views fetuses and embryos as "developing" as opposed to "being under construction," abortion opponents' appeal to dignity seems absurd.[5] That is, unless one makes an argument for fetuses and embryos to be included in the human community, to which the ascription of rights and dignity are appropriate, their opposition to abortion is difficult to comprehend. Stith puts it this way:

> Suppose we're back in the pre-digital days and you've just taken a fabulous photo, one you know you will prize, with your Polaroid camera.

(Say it's a picture of a jaguar that has now darted back into the jungle, so that the photo is unrepeatable.) You are just starting to let the photo hang out to develop when I grab it and rip its cover off, thus destroying it. What would you think if I responded to your dismay with the assertion "Hey man, it was still in the brown-smudge stage. Why should you care about brown smudges?" I submit that you would find my defense utterly absurd. Just so for pro-lifers, *who find dignity in every human individual:* To say that killing such a prized being doesn't count if he or she is still developing in the womb strikes them as outrageously absurd. (Stith 2007: 24)

Stith correctly points out that both the pro-life invocation of dignity and the pro-choice denial of it only make sense if the prior issue of who belongs in the human community has been settled. This is why the rhetoric of dignity can often be unpersuasive to advocates of abortion choice, who view fetuses and embryos, at best, as potential persons who are currently under construction. He asserts that the concept of human dignity is something of great value and is dependent on prior agreement on who is deserving of such an ascription of dignity. He concludes that,

> pro-choice folks think pro-life claims regarding embryos to be not only wrong but absurd whenever (even unconsciously, in the back of their minds) they think that embryos are under construction in the womb. And pro-life folks find pro-choice denials of *prized human dignity* in embryos to be equally absurd whenever they think that the unborn child develops (indeed, develops itself, unlike the Polaroid photo) from the moment of fertilization. (Stith 2007: 25)

Stith links what he calls the facts of human development—that fetuses and embryos are persons who are maturing rather than beings who are under construction—with the principle of human dignity and concludes that once fetuses and embryos are included in the human community, they possess intrinsic dignity that must be respected by giving them rights to life. In addition, the invocation of dignity without establishing membership in the human community will sound illogical. He puts it this way when he writes,

> those who hold both to the truth of human development and to the truth of universal human dignity will seek to respect life from conception. But those who fall into ignorance or denial of one or the other of these truths will find our arguments against abortion to be absurd. (Stith 2007: 26)[6]

What they find absurd is not that the fetus has any dignity worthy of respect but that the dignity of the fetus nearly always is weighted more heavily than the rights and dignity of the pregnant woman.

ABORTION CHOICE PROPONENTS' USE
OF THE CONCEPT OF DIGNITY

Abortion opponents are not the only party in the abortion discussion to make appeal to the concept of dignity. Proponents of abortion choice make use of the rhetoric of dignity and tend to broaden the use of the concept. They acknowledge competing conceptions of dignity and argue that a pregnant woman's dignity must be upheld. They insist that, should the law change and abortion rights be compromised, her dignity would be undermined by being coerced into carrying an unwanted pregnancy, or having the pregnant woman being forced into unsafe procedures for ending her pregnancy. Even among those who admit that the fetus has dignity that must be protected, abortion rights advocates generally argue that the dignity of the pregnant woman should be weighted more heavily than that of the fetus or embryo. In part, these competing views of dignity reflect the debate over membership in the human community. The view that the dignity of the pregnant woman should be weighted more heavily than that of the fetus is usually, but not always, based on a view of fetuses that grants them something less than full moral status and rights to life. Often the term used is one of *respect* and the notion of dignity is often implied as that which completes the idea of respect—respect for the dignity of the fetus, but with quite a different view of what such respect involves.

Dignity and the Pregnant Woman

In attempting to balance what he considers a one-sided view of dignity by Catholic opponents of abortion, Israeli bioethicist Y. M. Barilan acknowledges a competing notion of dignity and urges the application of dignity to the pregnant woman as well. He says,

> Often life imposes harsh conflicts between fundamental interests of mothers and their unborn children; or between the value of life and the wish to avoid extreme misery. There is no reason to promote only one kind of approach to such dilemmas as respectful for the dignity of persons. (Barilan 2009: 35)

Barilan suggests that the notion of dignity goes both ways and, while not making a definitive statement on the moral status of fetuses, acknowledges that the pregnant women, especially in developing countries, can also be counted as some of the most vulnerable. He does not spell out in detail what respect for dignity of the pregnant woman involves, though his criticism of Catholic opposition to abortion in the developing world makes it clear that the pregnant woman's dignity is undermined when she is deprived of the reproductive choice. He insists that it is important to recognize that there is a need for "mutual respect for diverse traditions and ethics on the beginning of life and human dignity," which would balance the need to respect the dignity of the fetus with the dignity of the pregnant woman (Barilan 2009: 35). He argues that there are times in which the dignity of the pregnant woman would outweigh the interests of the fetus, though he does not spell out what those would be. That is, in his view, there are competing dignity interests in the choice for abortion.

Increasingly, the abortion choice discussion acknowledges this conflict and insists that even though the notion of dignity applies to some degree to the fetus, there are occasions in which the dignity of the pregnant woman can and should take precedence over the interests of the fetus. This is a more subtly nuanced view than the opponents of abortion tend to take. Abortion opponents insist that, except on rare occasions in which the mother's life is genuinely in jeopardy, the assault on the dignity of the fetus that results in its destruction is far weightier than the loss of dignity imposed on the pregnant woman in carrying an unwanted pregnancy. The advocates of abortion choice see this decision as much more complicated, due to the circumstances faced by the pregnant woman, magnified by the lack of women's rights in general in the developing world, which leaves women sometimes pregnant as a result of nonconsensual sex. Some of this complexity may also be accounted for on the basis of a more nuanced view of the moral status of the fetus. It is more rare today for abortion choice advocates to deny any moral status to fetuses, and the older justification for abortion rights that did not grant any moral status to fetuses by claiming that they are simply "products of conception" or merely "a part of the woman's body" is not nearly as common as it once was. Thus, it is possible for abortion choice advocates to invoke the concept of dignity for the fetus, and still support a woman's right to end her pregnancy.

Dignity and Moral Luck

One example of this more nuanced view involves the connection between a woman's intuitions about her pregnancy and her "bad moral luck." Hilde Lindemann observes that fetuses and embryos are increasingly being described in the "language of respect" for their dignity and moral standing. Though the specific rhetoric of dignity is not explicitly invoked, it seems clear that what is being respected is some notion of value that is either recognized or bestowed upon the developing fetus. Some would hold that dignity is to be respected when some capacity in the fetus is actualized, such as sentience, brain activity, or viability outside the womb (Warren 1997; Steinbock 1992). Others hold that dignity is bestowed, either by society or by the mother, the latter of which is the view of Lindemann, who argues that mothers "call a fetus into personhood," which she describes as

> a social practice—as, indeed, the most fundamental social practice, the one on which all other practices rest. It consists in the physical expression of human beings' intentions, emotions, beliefs, attitudes, and other manifes-

tations of personality, as recognized by other persons, who then respond by taking up an attitude toward them of the kind that's reserved for persons. (Lindemann 2009: 45)[7]

The language of respect (for dignity or some other notion denoting value) is similarly invoked with regard to embryos; at the same time, it is used by advocates of abortion rights and embryo research. For example, the NIH Human Embryo Research Panel maintained that embryos were entitled to "profound respect" but at the same time could be used in nontherapeutic research that resulted in their destruction (National Institutes of Health 1994: 50–51).[8] The same was held of fetuses, which were entitled to the same "profound respect" as nascent human life, yet the tissue of aborted fetuses could be used for a variety of research purposes (Department of Health, Education, and Welfare 1979: 35033–58).

Lindemann argues that a woman can regard the fetus she is carrying as worthy of dignity and respect and at the same time, as a result of "bad moral luck," face circumstances in which ending her pregnancy is justifiable. That is, the dignity and respect that are due to the fetus, though powerful in informing a woman's intuitions about abortion, do not necessarily mean that she is obligated to carry the pregnancy regardless of her socioeconomic and family circumstances. Lindemann cites the case of a woman who has had bad moral luck—her husband is abusive or left her when she was pregnant. Other examples of bad moral luck could include other health-related complications, lack of medical insurance, or lack of resources to assist her in caring for multiple children. Lindemann concludes that, "If her moral luck had been bad enough, the best way to discharge some or all of these responsibilities might well have been to have an abortion" (Lindemann 2009: 51). What this suggests is that the ascription of dignity and respect functions somewhat differently than it does for opponents of abortion. For opponents, it functions as something akin to a trump card, and the bad moral luck that would allow such respect to be overridden would only be medical circumstances that involve a threat to the mother's life. For the advocate of abortion choice, respect and dignity are important concepts that are the basis of a woman's sense of unease and discomfort at the prospect of abortion, but they can be trumped by a wider set of circumstances that constitute something akin to Lindemann's bad moral luck.[9]

A similar argument is made by Naomi Wolf in her widely read article "Our Bodies, Our Souls," though she does not invoke dignity specifically, nor does she use the term moral luck. For her, the notion that the fetus has dignity is just beneath the surface of her discussion. She laments the traditional pro-choice view that denied any dignity or humanity to the fetus and insists that abortion is a procedure that has a moral gravity that must be faced by pregnant women and those who advocate for abortion choice. She affirms that the fetus has value and ends up actually affirming what she calls the humanity of the fetus. This designation of humanity gives the fetus sufficient moral standing to make the abortion decision a serious one and moves Wolf to invoke the religious notions of sin and redemption to justify abortion under some circumstances. While acknowledging that the right of abortion choice should always be available, she affirms the intrinsic moral nature of the decision and criticizes her colleagues for not taking that aspect of abortion choice seriously enough. She puts it this way:

> But how, one might ask, can I square a recognition of the humanity of the fetus, and the moral gravity of destroying it, with a pro-choice position? The answer can only be found in the context of a paradigm abandoned by the left and misused by the right: the paradigm of sin and redemption. (Wolf 1995: 33–34)

One can argue that she has misunderstood the religious notion of sin and redemption—that those cannot be used as a justification for abortion since they assume that something morally wrong is being done. But what seems clear is that the fetus is seen as possessing sufficient value to make the decision a heavily considered moral decision—a value commonly captured by the term *dignity*, but, unlike the opponents of abortion, the woman's choice can still override this quality that the fetus possesses.

Although without the religious imagery, this seems to be the approach of Judith Jarvis Thomson as well in her widely read defense of abortion. She conceded that the fetus is a person from conception forward and that it possesses all that accompanies personhood, such as dignity, value, and the right to life (Thomson 1971: 48). Yet this can be overridden, in more cases than Wolf would allow, because Thomson argues that the fetus's dignity and right to life does not give it any claim on the mother for anything needed to sustain its life. Of course, this classic article has been the subject of numerous critiques, and whether these critiques hold is beyond the scope of this chapter (Beckwith 1998). Thomson does acknowledge that the fetus becomes a person well before birth, but not at conception, where it resembles a clump of cells, contrary to the later ascriptions of the embryo as deserving profound respect. Thomson acknowledges that fetuses have dignity and value by virtue of their moral status as persons, but that does not extend to a right to aid. Thus, as in the case with Wolf, the attribution of value to the fetus can be outweighed by other considerations that support a woman's abortion choice, though in Thomson's case, it would seem that the abortion choice could be justified in far more cases than in the framework set up by Wolf.

USE OF THE CONCEPT OF DIGNITY BY THE COURTS

The language of dignity is most clearly invoked in support of abortion choice in the court decisions that form the legal basis for abortion choice— namely, the 1992 *Casey* decision. Interestingly, it is here that the use of the term *dignity* by the U.S. Supreme Court has taken an unexpected turn in the recent decision concerning so-called partial-birth abortions, *Gonzales v. Carhart*. In *Casey*, the Court attempted to balance respect for life with the dignity of the pregnant woman. Both the *Casey* and *Carhart* decisions use the language of dignity in a variety of ways to refer to the dignity of life (respect for the intrinsic value of human life, often used in conjunction with the a view of the potentiality of the fetus), dignity of liberty (referring to decisional autonomy), and dignity of equality (referring to a person's right not to be subordinated or excluded as an equal member of the community) (Siegel 2008: 1737).

Dignity in the *Casey* Decision

The *Casey* decision affirmed that the state has legitimate interests in protecting both the health of the pregnant woman and the life of the fetus "that may become a child" (*Planned Parenthood v. Casey* 1992: 846). *Casey* further affirms that the pregnant woman has autonomy to make the decision about her pregnancy and appeals to the tradition of procreative liberty and its connection with dignity. *Casey*'s emphasis is on the decisional autonomy component of dignity. The Court put it this way in the majority opinion, calling the freedom to make these choices "central to personal dignity and autonomy": "Our law affords constitutional protection to personal decisions relating to marriage, procreation, contraception, family relationships, child rearing, and education" (*Carey v. Population Services International*, 431 U.S. at 685). Our cases, the Court wrote, recognize "the right of the *individual*, married or single, to be free from unwarranted governmental intrusion into matters so fundamentally affecting a person as the decision whether to bear or beget a child" (*Eisenstadt v. Baird, supra*, at 453, emphasis in original). The Court also stated that our precedents "have respected the private realm of family life which the state cannot enter" (*Prince v. Massachusetts*, 321 U.S. 158, 166, 88 L. Ed. 645, 64 S. Ct. 438 (1944)). "These matters, involving the most intimate and personal choices a person may make in a lifetime, *choices central to personal dignity and autonomy*, are central to the liberty protected by the Fourteenth Amendment" (*Casey* 1992: 851).[10] The *Casey* decision further affirms that, although the state has a significant interest in protecting potential human life, a woman's right of bodily integrity is bound up with her liberty to make the deeply personal and private abortion decision. *Casey* affirmed that decisional autonomy in such private matters is a critical part of protecting the woman's dignity. "The woman's constitutional liberty interest also involves her freedom to decide matters of the highest privacy and the most personal nature. . . . The authority to make such traumatic and yet empowering decisions is an element of basic human dignity" (*Casey* 1992: 915–16).

Further, *Casey* uses the notion of dignity to support the notion that the law cannot put undue burdens on women seeking to procure an abortion (the undue burdens requirement). In striking down the mandatory waiting period, the Court appealed to the dignity of equality in structuring the undue burdens test for the abortion decision (Siegel 2008: 1702). *Casey* put it this way:

> Part of the constitutional liberty to choose is the equal dignity to which each of us is entitled. A woman who decides to terminate her pregnancy is entitled to the same respect as a woman who decides to carry the fetus to term. The mandatory waiting period denies women that equal respect. (*Casey* 1992: 920)

Dignity is used to protect women from an undue burden in seeking abortion and from being treated unequally. *Casey* thus limits the kinds of regulations placed on abortion to those that are consistent with a woman's decisional dignity. Law professor Reva Siegel puts it this way, "Under the undue burden framework, dignity-respecting regulation of women's decisions can neither manipulate nor coerce women: the intervention must leave women in substantial control of their decision, and free to act on it" (Siegel 2008: 1753). In addition, the dignity basis for the undue burden standard protects women from understandings of women's roles that would undermine their being treated as equals under the law. Siegel insists that

> There is a further implication of the dignity-based understanding of the undue burden framework that emerges with special clarity as the Court analyzes the spousal notice provision in *Casey*. *Casey* bases the abortion right, and its application of the undue burden test, on the understanding that government cannot enforce customary or common-law understandings of women's roles. In striking down the spousal notification requirement, the Court vindicates both dignity-as-liberty and dignity-as-equality. (Siegel 2008: 1753–54)

Casey does acknowledge that these conceptions of dignity are competing, but the weighting seems clear that, until viability, the decisional autonomy and equality applications of dignity outweigh the dignity of life, since the Court has referred to the fetus not as an actual person but as "potential life" (*Casey* 1992: 871),[11] the "fetus that may become a child" (*Casey* 1992: 874). Siegel reflects this weighting when insisting that, "*Casey*'s undue burden framework insists that the state can express respect for the dignity of life only if it does so in ways that respect the dignity of women" (Siegel 2008: 1753).

Dignity in the *Carhart* Decision

The *Carhart* decision takes up the partial-birth abortion procedure and uses the language of dignity somewhat differently than in *Casey*. This decision addresses a very specific technique for late-term abortions, the intact dilation and extraction, in which the fetus is removed largely intact from the uterus, thereby decreasing the chances of dismembering the fetus and minimizing the number of times the physician must gain entry into the uterus. The fetus is removed intact with the head remaining in the cervix, at which point the physician makes an opening at the base of the skull and evacuates the skull contents (*Gonzales v. Carhart* 2007: 1621–22).

Carhart introduces an important shift in the Court's reasoning on the dignity of the woman, suggesting an openness to restrictions on abortion that are protective of women and dignity-based at the same time. *Carhart* affirms the dignity-based autonomy of the woman from *Casey* and the earlier decision concerning procreative liberty and contraception in *Eisenstadt v. Baird* ("the decision whether to bear a child is central to a woman's dignity and autonomy, to her personhood, destiny and place in society") (*Carhart* 2007: 1640). The decision also acknowledges that the law under the Court's consideration that prohibits the intact dilation and extraction procedure "expresses respect for the dignity of human life" (*Carhart* 2007: 1632). Calling the procedure "brutal and inhumane," the Court claimed that a refusal to prohibit it will further coarsen society to the humanity of not only newborns but all vulnerable and innocent human life, making it increasingly difficult to protect such life. The Court claimed that such a respect for the dignity of human life is consistent with *Casey*, though by prohibiting this particular procedure, one could argue that the weighting in this case seems reversed from the way it was done in *Casey*. *Carhart* actually presents its decision as affecting the physician's decision as to procedures, not the woman's decision concerning her pregnancy. Siegel observes that

> *Carhart* does not offer itself as limiting a woman's decision whether or when to end a pregnancy. To the contrary, the Court decides the case as if the only question in issue was the question of the medical *method* by which doctors would effectuate a woman's abortion decision; the Court authorizes regulation of the abortion procedure to the extent it does not pose an undue burden on women's decision making. (Siegel 2008: 1770–71)

It does appear on the surface that the decision is more about protecting the fetus than the woman, as the Court suggested some parallels they considered significant between the abortion procedure under consideration and infanticide. Suggesting the procedure, "had a disturbing similarity to the killing of a newborn infant," the Court affirmed that its prohibition was consistent with the dignity of the fetus (*Carhart* 2007: 1633).

However, *Carhart* also opens the door to justifications of limits on abortion that are designed to be protective of the dignity of women. Whereas in both *Casey* and *Roe v. Wade*, appeals to the woman's dignity involves protecting her decisional autonomy and fundamental equality as a member of the community, *Carhart* introduces the prospect that limits on some abortion procedures may serve to protect women. They suggest that the abortion decision has such gravity, particularly the late-term abortion decision, that it is critical that women be properly informed about its risks. This functions as an extension of the Court's linking dignity to a woman's full decisional autonomy in the abortion decision. *Carhart* introduces the prospect of the woman's regret of the decision she has made, magnified by the late-term nature of the abortion in question. The *Carhart* decision put it this way:

> While we find no reliable data to measure the phenomenon, it seems unexceptionable to conclude some women come to regret their choice to abort the infant life they once created and sustained. (See Brief for Sandra Cano et al. as *Amid Curiae* in No. 05–380, pp. 22–24.) Severe depression and loss of esteem can follow. (*Carhart* 2007: 1634)

This seems to indicate that the Court was amenable to the possibility that such restrictions on the abortion procedure could be consistent with the mandate of dignity to protect a woman's health, though the Court suggested that full informed consent was what immediately followed from the notion of maternal regret for the abortion choice. At the same time, the Court maintained that such a restriction did not "place a substantial obstacle in the path of a woman seeking an abortion, thereby not violating the undue burdens test" (*Carhart* 2007: 1635).

The criticism of the Court's reasoning here is that it indeed does undermine a woman's decisional autonomy that is dignity-based and that restrictions on the procedure do not follow from the concern about a woman's having regrets with the late-term abortion decision. Calling this a form of gender-paternalist reasoning, Siegel expresses the critique this way: "As this line of inquiry makes clear, the gender-paternalist justification for restricting abortion is in deep tension with the forms of decisional autonomy *Casey* protects" (Siegel 2008: 1701). Such decisional autonomy, as we have seen, is clearly dignity-grounded, and the gender-paternalism that Siegel suggests also serves to undermine a woman's dignity by the subtle assumptions that women are not capable of making fully informed decisions, or by the fact that they will likely face regrets in the face of decisions previously made (Siegel 2008: 1698).

CONCLUSIONS ON THE USE OF DIGNITY IN THE ABORTION DISCUSSION

The concept of human dignity is a meaningful one in the abortion debate, appealed to by both primary positions and the courts. Other terms are used somewhat interchangeably for human dignity, such as sanctity of life, intrinsic worth of life, and inviolability of life to express a similar sentiment about the moral status of the fetus. But the specific use of the notion of human dignity makes a substantial contribution to the discussion of abortion, invoked by both opponents of abortion and proponents of abortion choice.

For Roman Catholic opponents of abortion, the appeal to dignity is both explicit and clear. The encyclicals that outline the case for Catholic bioethics regularly and explicitly utilize the concept of human dignity and ultimately ground dignity in human beings' being made in the image of God, though at times, the encyclicals also appeal to something akin to a Kantian argument that human beings are to be respected as ends in themselves. By contrast, for other opponents of abortion, the appeal to dignity is not as explicit but often assumed as self-evident, given their view that fetuses are persons with full rights to life. For these groups, the central aspect of the abortion discussion revolves around the moral status of the fetus, and, for the opponent of abortion, once that is established, then invoking dignity and all of its interchangeable terms is seen to follow so naturally that it need not be explicitly invoked. It is self-evident to most abortion opponents that the dignity of a full person is violated when abortion occurs, but many of these groups find the language of rights—that the right to life is violated—to be a more persuasive and intuitively power-ful appeal. There is little doubt that abortion opponents consider the dignity of the fetus to be assaulted when abortion occurs, but the notion of dignity is most explicitly and clearly invoked in the distinctively Catholic discussion.

Other areas of the Catholic discussion that appeal to dignity are, in my judgment, not as clear. For example, the violation of dignity that Catholic teaching suggests is violated when a person is conceived via in vitro fertiliza-tion and other assisted reproductive technologies is less obvious than when abortion occurs. Bringing a new person into the world by assisted reproduc-tive technologies is considerably different than taking a fetus's life through abortion when it comes to respecting their dignity. I would suggest that similar ambiguity in terms of the violation of human dignity also applies to human cloning and enhancement biotechnology. Though those who have reservations about some of these practices often conclude that dignity is violated when these practices occur, the argument that this is so is far more nuanced and less intuitively obvious than in the case of abortion.

The use of the concept of dignity on both sides of the abortion discus-sion reflects the broader debate on the issue. Abortion opponents concede that a pregnant woman's dignity must be respected but often argue that it is outweighed by respect for the dignity of the fetus. Similarly, proponents of abortion choice argue that even though the fetus has value, often expressed in terms of "potential life," a woman's dignity should be weighted more heavily. Opponents of abortion maintain that the notion of potential life is incoherent and argue that fetuses and embryos are "persons with poten-tial" who mature according to what they already are through gestation, as opposed to becoming something of a different kind in the process. This disagreement over the relative weighting of dignity points toward the deep division about the moral status of the fetus. For if the fetus is indeed a full person with attendant rights to life, then its dignity is clearly violated when abortion ends its life. Weighting the dignity of the fetus more heavily than the woman's right to her own body and her decisional autonomy seems intuitively obvious to most opponents of abortion. By contrast, if the fetus is not a person, but something less than, in terms of moral status, then weighting the woman's dignity more heavily seems comparably clear. The specter of a woman being forced to carry a pregnancy that she does not desire strikes many people as a deep violation of her dignity.

The exception to this weighting would be those supporters of abortion choice, such as Thomson and Wolf, who concede that the fetus is a person and still weight the woman's dignity more heavily than that of the fetus. They argue that the dignity of the fetus does not give it the right to the aid of another (Thomson) or that it can be overridden in some circumstances by the woman's compelling interests (Wolf). Abortion opponents counterargue that once intrinsic dignity of the fetus as a person is conceded, then the weighting of dignity is clearly in favor of the fetus. Some further press that concession by insisting that if the fetus is a person with intrinsic dignity, then the fetus also has a claim on the mother for the resources it needs to flourish, analogous to a newborn (Beckwith 2007: 172–201). Others further press the concession to draw conclusions about paternal responsibilities. That is, if the woman, through her decisional autonomy, can justifiably refuse to make substantial sacrifices and assume responsibility for an unwanted child, why would that not also apply to the father when it comes to child support (Pavlischek 1998)? In general, however, the ascription of dignity in this discussion reflects the broader conversation about the moral status of fetuses and how to weight the interests of fetuses against the interests of the pregnant woman.

The Courts claim to balance these competing conceptions of dignity. The *Casey* Court, reflecting the *Roe v. Wade* decision, maintained that the dignity of the fetus was important, a part of the state's compelling interest in upholding the value of life and protecting life in the final stages of pregnancy. That is balanced by the woman's decisional dignity to make life's most value-laden choices, particularly those about procreation and child rearing, apart from state interference. Most abortion opponents argue, correctly in my view, that this not a balance at all, but that in the Court's decision, the dignity and rights of the woman clearly outweigh that of the fetus. It may be that respect for the dignity of the fetus shown by the Court is primarily *symbolic* and thus allows for the eventual weighting of the woman's interests ahead of the fetus's. In my judgment, appeals to this kind of symbolic dignity are not helpful and are often justifiably seen by opponents of abortion as a meaningless gesture at best, and insulting at worst.

Although the use of the rhetoric of human dignity is generally used well in the abortion debate, it is not always clear how the notion of dignity is grounded philosophically and how competing claims based on dignity are resolved. As a result, it can appear that the application of dignity is either somewhat ambiguous (because it can be appealed to for either side of the abortion debate) or that dignity is being used as a rhetorical device with the expectation that invoking basically ends the debate—that it has something akin to trump-card status in the discussion. It functions to cut off debate rather than facilitate it. That is, the rhetoric of dignity can be used as something resembling a philosophical club with which one can hammer one's opponent. It can actually be used as the virtual equivalent of an ad hominem argument, in which the person is attacked by virtue of being accused of undermining someone's dignity. I would suggest that these are abuses of the concept of dignity and are not positive contributions to the overall discussion on abortion.

To more effectively make use of the concept of dignity would seem to involve being clear about both the grounding for dignity and the process by which competing claims to dignity are resolved. It does seem that the respect for dignity enjoys a broad consensus as a general moral principle, but for it to move beyond a general moral norm in the abortion discussion would involve being clear about one's metaphysical and anthropological commitments that form the basis for human dignity. Our use of dignity reflects our broader philosophical commitments, which are sometimes smuggled in to the discussion without being explicitly identified. Those who argue from a distinctly religious framework tend to be clearer about these underlying commitments but would need to articulate them with more publicly accessible reasons if they expect them to be taken seriously as a basis for public policy.

In addition, for dignity to be a more helpful concept in this debate, both sides need to recognize that there are legitimate competing claims to dignity. Further, they need to be clear about the process they use to resolve these competing claims. Failure to recognize that there are indeed competing claims here sets the discussion back. At times, it appears that neither side even acknowledges the dignity claims of the other. Although the respective weighting of these competing claims may seem self-evident to both sides, a lack of clarity on either side often makes the use of the term *dignity* confusing. Again, one runs the risk of smuggling in a highly debatable way of addressing these contrasting claims.

In my assessment, appeals to dignity are helpful in this discussion but not without the assumptions underlying the use of the concept being clear. What it means for a fetus to have dignity depends on whether the claim for its moral status can be sustained. If so, then weighting the life of the fetus more heavily than the decisional autonomy of the woman seems clear. If those claims cannot be sustained, then the dignity of the fetus involves some sort of symbolic value or its dignity as a potential, but not actual, person. By contrast, the weighting of the woman's decisional dignity vis-à-vis the fetus depends largely on what is meant by the dignity of the fetus.

NOTES

1. Reagan also published a book by the same title (Nashville: Thomas Nelson, 1984).
2. For some of the repetitions of this statement, see Brownback 2004 and Clark 2004. Clark put it like this, "Ronald Reagan's record reveals that no issue was

of more importance to him than the dignity and sanctity of human life. . . . He emphasized the truth of human dignity under God" (Clark 2004: 69).

3. For additional reading that illustrates the centrality of the concept of human dignity to Kass's thinking, see President's Council on Bioethics 2008, which he chaired, *Human Dignity and Bioethics*, the application of which was primarily to matters of biotechnology, not abortion.

4. Lee also uses concepts such as intrinsic value, sanctity of life, and sacredness of life interchangeably with intrinsic dignity.

5. When describing fetal growth in the womb, I would use the term *maturing* instead of *developing*. That is, fetuses should be described as maturing from the completion of conception onward, based on the biological fact that embryos have all the genetic information and capacities they need from that point. They simply mature or, as Stith puts it, develop. Stith's distinction between "developing" and "being under construction" would be clearer if he substituted the word *maturing* for *developing*.

6. For further discussion of arguments that fetuses and embryos should be included in the human community, see Lee 1997, Lee 2004, Beckwith 2007, and Moreland and Rae 2000.

7. See also Green 1974 and Green 2001.

8. See also the statement of the National Bioethics Advisory Commission 1999 for a similar view.

9. For more on the concept of moral luck, see Nagel 1993.

10. The italicized portion in the text of the citation reflects the author's emphasis and is not in the original citation.

11. The Court put it this way, "On the other side of the equation is the interest of the State in the protection of potential life. . . . Yet it must be remembered that Roe v. Wade speaks with clarity in establishing not only the woman's liberty but also the State's 'important and legitimate interest in potential life.'" Numerous other instances in the opinion echo this notion.

12 Human Dignity and the Debate over Early Human Embryos

Nathan J. Palpant and Suzanne Holland

Perhaps in no other aspect of science is the meaning of human dignity as contested and as little agreed upon as it is in the science of early human embryos, especially in human embryonic stem cell research and its uses. Although human dignity is invoked with regularity and richness, in our view it has added little to constructive rhetorical engagement on the ethical limits and uses of early embryos. In this chapter we examine the literature of human dignity as it pertains to the embryonic stem cell debate. We conclude that, although the appeals to human dignity are numerous and multivalent, the rhetoric itself hinders rather than helps the crafting of workable public policy on human embryonic stem cell research.

PRIMER ON HUMAN EMBRYONIC STEM CELLS

The historical precedence for research on early embryos began in the development of techniques used currently in the context of in vitro fertilization (IVF). In their seminal 1970 publication in the journal *Nature* (Edwards et al. 1970), Edwards, Steptoe, and Purdy described the first successful IVF procedure nearly 8 years before Louise Brown was born. However, the research by Robert Edwards on IVF techniques began 20 years prior, providing a history of research on early human embryos dating back to the 1950s. In their *Nature* report, Edwards and colleagues state, "We fertilized many more eggs and were able to make detailed examinations of the successive stages of fertilization. We also took care to photograph everything because we would have to persuade colleagues of the truth of our discoveries" (Edwards et al. 1970: 1308). These experiments at the inception of IVF technology in the 1950s provided a long precedence for generating spare embryos for research purposes without evidence of the kind of controversy currently seen with human embryonic stem (hES) cell research. Paul Lauritzen and Carol Tauer are the only scholars to our knowledge who have really accentuated this observation (Lauritzen 2004; Tauer 2001). Although there is some opposition to IVF, particularly by the Roman Catholic Church, it has garnered significant support due to the opportunity it provides for couples with various reproductive deficiencies to have children of their own lineage. Interestingly, the Catholic Church does not cite opposition to these early seminal findings using embryos solely for research purposes as reason for opposing IVF technology.

In addition to IVF, the work on culturing human embryos in vitro also provided the basis for current practices involving preimplantation genetic diagnosis (PGD), which became a clinical practice in the 1990s (Handyside et al. 1990). Although PGD is controversial, the prominent social issue concerns the genetic information derived from the test and not the work required to establish the technical practicality of single cell biopsy of early human embryos. We note that the problem of moral complicity is equally problematic in the context of IVF and PGD as it is with the derivation of hES cells from early human embryos. The desire to (1) provide children to infertile couples through IVF, (2) provide genetic information to inform reproductive decisions regarding the genetic status of a preimplantation embryo, and (3) generate cell therapies for heart attack patients are all equally complicit in that they all depend upon research on early human embryos (Holland 2007). Social opinions and ethical dialogue reveal a significant inconsistency in attitudes and ideas about these three practices. Paul Lauritzen has made this point, saying,

> there is something of an irony in the fact that so much attention has been devoted to developing and defending the distinctions between embryos created solely for research and embryos left over from IVF procedures, because there would be no embryos left over from IVF procedures had there not been embryos created solely for research purposes to develop IVF in the first place. (Lauritzen 2004: 18)

Similarly, Bartha Alvarez Manninen has noted that

> it is not at all clear . . . that embryos may be instrumentalized in order to treat the disease of infertility, but that they cannot be instrumentalized in order to treat Alzheimer's disease, Parkinson's disease, spinal cord injuries, retinal deterioration, or diabetes. (Manninen 2008: 13)

In parallel with the work of Edwards and others, the principles of embryonic stem cell research were developed with the establishment of mouse embryonic stem cells in 1981 (Evans and Kaufman 1981; Martin 1981). Nearly two full decades later in the late 1990s, James Thomson at the Wisconsin Regional Primate Research Center successfully cultured the first human embryonic stem cell line (Thomson et al. 1998). These cells were derived from blastocyst-stage preimplantation embryos produced by in vitro fertilization. With the work of mouse embryonic stem cells as the background for this breakthrough, Thomson's work became a precedent for establishing cultures of undifferentiated hES cell lines, which are able to continuously self-renew under appropriate in vitro culturing conditions.

Science of the Human Embryo

Human embryonic stem cells are derived during the early stages of embryo-genesis. Approximately five days after fertilization, the developing embryo differentiates into two cell types: the trophectoderm, which gives rise to the supportive structures of the pregnancy including the placenta, and the polarized inner cell mass (ICM) cells, which give rise to the body proper. Appropriately understood, these ICM cells are not stem cells but rather pro-genitor cells. Progenitor cells represent cells in a precursor state but inher-ently destined toward a particular lineage, which, in the case of the early embryo, includes all mature cell types of the body. As development pro-gresses, these lineage decisions of ICM cells give rise to more mature cell types with increasingly restricted fates. If allowed to develop normally, these ICM cells have the intrinsic potential to become a mature human being. When taken out of the context of the developing embryo and cultured in vitro, these ICM cells are what we define as embryonic *stem cells* and, as noted, can be manipulated to give rise to all types of cells that comprise the three germ layers of the adult human body. In vitro human embryonic stem cells are not embryos and cannot naturally develop into mature organisms *in themselves.* That is, if hES cells were placed back into the uterus *alone* they could not survive. Their capacity to contribute to organismal development requires the protective milieu of the blastocyst (cumulatively including the intact trophectoderm and associated ICM cells). The placement of embry-onic stem cells into a blastomere is common in the development of murine models wherein genetically (or otherwise) modified mouse embryonic stem cells (mES) cells are injected into a blastocyst. These chimeric blastocysts containing an ICM with both normal and modified ES cells results in devel-opment of chimeric progeny.

The cellular flexibility of embryonic stem cells is known as pluripotency and is the reason why embryonic stem cells are so attractive for research and therapeutic purposes. This is in contrast to many types of adult tissue–derived stem cells that are considered multipotent, able to give rise to sev-eral, but not all, cell types. The blastomere cells of the early zygote (prior to compaction and development of the trophectoderm and ICM) are set apart from all other cell types as totipotent, because they are genuinely undiffer-entiated and thus can give rise to both the embryo and the extraembryonic tissues. Although separable based on their potentiality, stem cells (but not progenitor cells as with ICM cells in vivo) have one singular characteristic in common: they undergo asynchronous cell division. Specifically, progeny from a stem cell can include (1) additional undifferentiated stem cells to renew the stem cell population and/or (2) differentiated organ-specific cell types that develop in a directed fashion in vitro or in vivo toward mature cellular phenotypes. In the intact human embryo in utero, the lineage decisions of ICM cells as progenitor cells ultimately give rise to the full mature human being without repopulating any kind of embryonic stem cell (thus, they do not fall under the canonical definition of a stem cell). In contrast, in vitro ICM-derived embryonic *stem cells* can be maintained in an undifferentiated fashion with a normal karyotype for extended times (even years), well beyond the cell culture capacity of any other kind of primary somatic cell type. To date, no hES cell line has been reported to stop replicating. These cells can also be stimulated to differentiate into any cell type of the body (albeit with varying success inherent to current cell culture techniques). In vivo transplantation of undifferentiated hES cells results in teratoma formation, defined as the differentiation into mature cells from all three germ layers. The observation that hES cells become teratomas supports their pluripotent phenotype. In the adult body, tissue-specific stem cells are necessary to maintain normal physiological function and retain cellular homeostasis in the context of cell loss due to injury or normal cellular turnover. For example, stem cells are necessary in the maintenance of skin, the hematopoietic system, intestinal epithelium, and elsewhere to maintain both the stem cell niche as well as to repopulate mature cell types required for normal physiological functions.

Developments in manipulating the cellular genome have enabled scientists to generate cells with an embryonic phenotype from differentiated somatic cells. Specifically, fibroblasts have been reengineered to an embryonic state by overexpression of key transcription factors. These cells are called induced pluripotent stem (iPS) cells. Although alternative sources of pluripotent stem cells are attractive scientifically and ethically, their derivation is based almost entirely on the extensive work performed on human embryo research, and it is open to debate about the extent to which iPS cells remove ethical issues.

Human Embryonic Stem Cells in Regenerative Medicine

The historical paradigm for therapeutic management of acute and chronic diseases has been based on both preventative measures (stop it before it shows up) or postonset disease management (treat when able or don't let it get any worse). Until recently, the possibility of reversing disease progression to reestablish original physiological function has remained elusive, merely a remote hope. Although the manifestation of actual cell-based therapeutics has not come to full fruition in everyday medicine, significant advances in animal models and some clinical trials have established the basis for an ever-growing field known as regenerative medicine (Anversa et al. 2002; Fukuda and Yuasa 2006; Laflamme and Murry 2005). The fundamental goal of regenerative medicine is to restore the original function of a damaged organ by introducing cells (or genes that influence cellular phenotype) that correct the inciting pathology. Many chronic diseases that affect millions of individuals worldwide, including Alzheimer's disease, diabetes, Parkinson's disease, spinal cord injuries, myocardial infarction, and many others, are being studied to determine the potential for cell- and gene-based therapies.

Human embryonic stem cells are considered an attractive approach for tissue engineering and regeneration because of their intrinsic nature as pluripotent cells. In other words, embryonic stem cells have the potential to be differentiated in a directed way to generate cells that comprise any cell derived from any organ of any germ layer. This happens flawlessly during healthy embryogenesis in vivo, and scientists have made significant progress toward generating embryonic stem cell–derived organ-specific cell types in vitro (Laflamme et al. 2007; Laflamme et al. 2005). Although scientists are beginning to define the conditions for directed differentiation of various cell types, the ability to reach *maturity* of adult cell types has been difficult. This is likely due to the requirement for complicated signaling patterns and heterogeneous cellular interactions needed to reach these late stages of cellular maturity. Even as work is being done to address these matters, it is widely held that the extensive replicative capacity of hES cells in vitro makes it feasible to generate massive numbers of cells needed for tissue patches or artificial tissue development. These cells can then be optimized for delivery to injured organs in vivo. Extensive research in tissue engineering looking at different types of scaffolds and cellular manipulations seeks to determine the optimal methods for delivery of *in vitro*–derived cells to regenerate or repair organs.

BIOETHICS AND THE HUMAN EMBRYO CONTROVERSY

Although the scientific context for the therapeutic utility of hES cells is emerging, social, political, and academic discourse continues to wrangle over more fundamental questions of whether this research should be done in the first place. Since Thomson's first report on the topic, the rhetorical strategies arguing for or against this type of research have been largely influenced by the abortion controversy. Numerous statements put forth by prominent private, public, and religious bodies in the last 15 years have suggested that defining the moral status of the embryo is *the* fundamental ethical variable in determining the legitimacy of research using human ES cells, although others have disagreed (Holland 2001). These discussions are framed in such a way as to highlight the properties of embryological developmental instances or stages to inform moral decision making. For example, arguments opposed to research on human embryos focus on the event of conception as the initiating event of human life (George and Tollefsen 2008). Those in favor of research using embryos suggest, for example, that individuation does not occur until after 14 days have passed and the embryo is no longer capable of dividing into twins (Shell 2008). So far, the literature on the matter is heavily weighted on the side of determining embryo status as the ground for moral decision making. This essentially makes dialogue on the issue a binary matter. With an either/or outlook on the topic, as Lauritzen explains, "either the embryo is a person or it is essentially a kind of property" (Lauritzen 2004: 8). The result is a rhetorical and ethical stalemate over the moral status of the early embryo.

This incipient question of the embryo's moral status has hinged, at least in part, on the issue of human dignity. Indeed, the term *human dignity* has become fashionable in writings about the moral status of the embryo. Although the concept has a long history dating back centuries, it emerged in bioethical discourse largely after the atrocities of the Second World War. Since then, its utility in bioethics has been invoked in numerous issues, including cloning, stem cell research, abortion, genetic engineering, jurisprudence and patenting of human biological material, genetic testing, and issues of death and dying. In an effort to stimulate dialogue on what human dignity actually means and whether it has any utility at all in bioethics, Ruth Macklin wrote that

> a close inspection of leading examples shows that appeals to dignity are either vague restatements of other more precise notions or mere slogans that add nothing to an understanding of the topic. . . . Dignity seems to have no meaning beyond what is implied by the principle of medical ethics, respect for persons: the need to obtain voluntary, informed consent; the requirement to protect confidentiality; and the need to avoid discrimination and abusive practices. (Macklin 2003: 1419)

Indeed, it seems that many arguments invoking human dignity rhetoric use it as an all-purpose justification for social unease, a term of general condemnation about technological advances that must therefore require regulatory restraint. Caulfield and Brownsword have followed Macklin at the policy level, writing,

> it is a pre-condition of effective regulation that the rationale and purpose of the regulation are clearly stated. However, so long as human dignity is a contested concept open to different interpretations, regulatory references to "respect for human dignity" cannot possibly give a clear steer to regulators and regulatees. (Caulfield and Brownsword 2006: 75)

The language of human dignity seems ubiquitous in bioethics literature on issues related to the impact of science and technology on human beings, but it is too often rhetoric without clarity of content. What, actually, does human dignity mean, and for whom, especially when applied to embryos?

In the next section, we seek to determine whether the current issues in bioethics about hES cell research can be excavated without invoking the term *human dignity*. Is language of autonomy, respect for persons, and human rights sufficient in the absence of human dignity to dialogue about the issues with more clarity, and/or are some kinds of human dignity rhetoric necessary and informative? We combine the illustrations that follow with suggestions for new avenues of deliberation that might be more fruitful than the rhetoric of human dignity, which we believe has contributed to the constrained dialogue about embryos and hES cells.

AN EVALUATION OF HUMAN DIGNITY
AND ITS VARIOUS MEANINGS

Given our assessment of this literature, we will outline our discussion of the relevance of the term *human dignity* in the human embryo controversy in light of Lennart Nordenfelt's assessment of three varieties of dignity (Nordenfelt 2004). Although Nordenfelt outlines four concepts of human dignity, only three of the four are especially pertinent to the topic at hand:

1. The *dignity of moral stature* is the result of the moral deeds of the subject.
2. The *dignity of identity* is tied to the integrity of the subject's body and mind.
3. The *dignity of Menschenwürde* (intrinsic dignity) pertains to all human beings to the same extent and cannot be lost as long as the subject exists—this human dignity is inviolable.

To some extent, all of these various forms of human dignity rhetoric have been invoked in the human embryo controversy. Their association with these categories has been at times explicitly discussed, but often reference to human dignity is not entirely clarified and the language framing the concept is only mildly informative as to the authors' purpose. We will take each of these as single understandings of dignity pertaining to the lived human experience and address how each of these is used in the literature and therefore reveals the relevance (or lack thereof) of human dignity rhetoric in the context of the hES cell controversy.

The Dignity of Moral Stature

According to Nordenfelt, the dignity of moral stature is understood as the character or conduct of an individual with particular emphasis on the moral value of an individual's *actions*. The dignity of moral stature is dimensional in that it can be attributed to an individual in varying degrees—high or low, depending on the moral value of one's actions. Nordenfelt points out that this dignity is unique in that it does not confer any rights on the subject. In fact, rights given in response to moral actions diminish the value of the action and the associated attributed dignity.

The dignity of moral stature, certainly underrepresented in the literature, is an important component of discussions about research on human embryos for two reasons. In the first case, it is recognized that we as moral agents have an obligation to respect the inherent dignity of our neighbor (*Menschenwürde*). In other words, those who attribute rights (even limited rights such as the right to be respected even in the absence of the right to life) would support the view that we gain dignity when we respect the *Menschenwürde* of the early human embryo. In the second case, as rational moral agents, we are obligated to act with dignity (make the moral choice) such that the implications of our decision support our sense of self-respect. This is significant because most of the literature on human dignity revolves around our conceptions of the human embryo as the subject. Notwithstanding the plethora of discussion on the topic, the embryo, lacking any sense of self-interest, cannot be said to care what we decide about its rights or moral status. That is, the dignity (of whatever sort) of the embryo is only important as it pertains to and informs this critical point of an acting agent's dignity of moral stature.

One manifestation of the dignity of moral stature is personal sacrifice. Leon Kass uses the image of the crucified Christ as exemplifying the dignity of moral stature. He writes,

> To be sure, for Christians, Christ on the Cross may be regarded as the supreme exemplar of human dignity, notwithstanding the fact that the image of the crucified man-God is, deliberately, a complete inversion of what would ordinarily and everywhere be regarded as "dignified" or "elevated." (Kass 2008: 318)

Nordenfelt expands on the dignity of moral stature by drawing attention to its various relations with the concept of respect. First, moral agents gain dignity by paying respect to other people based on attributed (*dignitas*) or inherent (*Menschenwürde*) rights. In this case, the subject of the dignity is the one *giving* recognition to another's attributed or inherent rights. Second, intrinsic to the dignity of moral stature, there is reciprocity of respect given among moral agents in response to dignifying moral actions *irrespective of rights*. Kass's example of the sacrificial Christ fits this paradigm of dignity. Similarly, this is the kind of respect we give, as witnessing moral agents, to Janusz Korczak for his sacrificial devotion as he accompanied children into Nazi gas chambers. Nordenfelt describes the third kind of respect attributed to the dignity of moral stature as self-respect. This kind of dignity comes about when the moral agent *chooses* the moral action. Korczak, who ran an orphanage and spent his life writing children's books, could not have retained self-respect (the dignity of moral stature) had he merely led the children to the chambers but, dismissing his compassion, not actively and sacrificially comforted the singing children in the gas-filled chambers. These types of respect are clearly independent from the inherent respect due persons as beings, discussed in the dignity of *Menschenwürde*. Overall, Nordenfelt makes clear that the dignity of moral stature associated with moral decision making is closely related with our understanding of personal and *relational* varieties of respect.

What are the implications on the dignity of moral stature if we determine that the embryo is (or is not) worthy of rights? Although conservative ethicists may balk at this statement, we maintain that the subtle differences in the derivation of alternative sources of pluripotent stem cells such as embryos generated only for research purposes, supernumerary embryos, organismically dead embryos (Landry and Zucker 2004), induced pluripotent stem cells (Takahashi and Yamanaka 2006), or embryos derived by altered nuclear transfer (Condic 2008) each require a unique ethical analysis. Use of embryos created solely for research purposes might have, for example, very different implications on the dignity of moral stature than iPS cells. These distinctions are critical in determining an informed ethic about pluripotent stem cell technologies.

The corollary perspective is equally important. That is, what are the implications to our self-respect if we deprive sentient rational beings of treatments generated by human ES cells? Our historical response to research on IVF and PGD techniques may inform this issue to some extent. If one values consistency in ethical decision making, then those maintaining opposition to early human embryo research need to reevaluate their social and political views on IVF and preimplantation genetic diagnosis technology. Unfortunately, this point as a whole has only gained passing attention in the dialogue to date. It would seem that moral consistency in particular matters such as human embryos (pertaining to hES cell research *and* IVF *and* PGD etc.) is a necessary, if not sufficient, prerequisite to determining the impact of these choices on the dignity of moral stature.

The dignity of moral stature has been threaded into various discussions on human embryo research, albeit not directly the focus of any given article that we have examined. As a matter of general applicability, there is discussion of the obligations of virtuous agents (those capable of moral decision making) in the treatment of marginal or nonagents, those who have limited or no autonomy or the capacity for autonomy (Ashcroft 2005; Beyleveld and Brownsward 2001). These works may provide starting points for discussion of the dignity of moral stature pertaining to human embryo research.

Fuat Oduncu argues that "only the rejection of research which destroys embryos and the promotion and support of [adult stem] cell research will help society keep its humanity, which again is deeply rooted in the respect of human dignity" (Oduncu 2003: 14). Oduncu's reference to humanitarianism, like human dignity is perhaps a good starting point, yet remains an ambiguous term for which he provides no further clarification. If we consider that humanitarianism is an ideology that influences moral decision making, what are the more fundamental components that influence our dignity of moral stature or self respect? Some scholars have provided ideas to this end such as the concept of vulnerability. That is, our ability to make the moral choice is inextricably associated with our capacity to respect the vulnerable nature of our humanness or the humanness of others. Quoting Gabriel Marcel, Gilbert Meilaender notes, "we will gain insight into this mystery [of human dignity] chiefly . . . when we are moved by a spirit of compassion that recognizes our shared vulnerability; hence . . . 'dignity must be sought at the antipodes of pretension and . . . on the side of weakness'" (Meilaender 2008: 264). Presumably, attributing weakness and vulnerability to a subject assumes the subject is relationally integrated into humanity. This type of rhetoric may be clearer in talking about the poor or disabled but becomes markedly more complex in discussions about human embryos, where terms such as *personhood* and *potentiality* and putatively morally significant embryological stages are not agreed upon and thus add dimensions of discord among commentators. To what extent does compassion toward the vulnerable and weak play a part in how we understand the dignity of moral stature in this controversy? In addition to vulnerability, what else contributes to our understanding of humanitarianism?

Another perhaps more common reference on the part of those who oppose human embryo research is the argument of utilitarianism, generally an appeal to bring about the greatest good for the greatest number of persons in any given situation. In this sense there is the concern that an attitude of utilitarianism negatively impacts our self-respect as it pertains to the dignity of moral stature. Robert George and Christopher Tollefsen make this point in their book *Embryo*:

> And this brings us to what is perhaps the most crucial point concerning all forms of utilitarianism and consequentialism: within any such ethic, there will always be human beings who are dispensable, who must be sacrificed for the greater good. Utilitarianism fails to respect, in a radical way, the dignity and rights of individual human beings. For it treats the greater good, a mere aggregate of all the interests or pleasures or preferences of individuals, as the good of supreme worth and value, and demands that nothing stand in the way of its pursuit. (George and Tollefsen 2008: 92)

George and Tollefsen broadly consider utilitarianism as leading to the most significant abasement to humanity with historical examples being Nazi extermination of Jews and former eugenics programs in the United States. Kant's categorical imperative from the *Groundwork of the Metaphysics of Morals* often accompanies this type of argument. Although discussed in greater detail later, Kant's imperative states, "So act that you use humanity, whether in your own person, or in the person of any other, always at the same time as an end, never merely as a means" (Kant 1997). Invoking Kant in a strict sense to support an antiutilitarian position is not entirely legitimate if one considers the broader context and meaning of Kant's use of the term *humanity*, which is based in rationality. Furthermore, arguments of utilitarianism in this debate are only mildly useful because the position assumes that the subject of the utilitarian agenda is a rights bearer (e.g., the human embryo). For commentators who support research on early human embryos, the embryo is not a rights bearer, not even of the most basic right to life. A lack of consensus on this fundamental premise of human rights precludes any substantive dialogue on the matter. Caulfield and Brownsword have addressed this point in saying,

> As a matter of principle, those regulators who take a permissive approach to . . . stem cell research see no negative impingement on human dignity. From a human-rights perspective this is an understandable position. There is no direct compromising of human dignity because the human embryo is not yet a bearer of human rights; eggs are collected and embryos are derived from rights-holders on the basis of their free and informed consent. Neither is there evidence of indirect threats to rights-holders or of researchers becoming more casual about respect for the rights of fellow rights-holders. (Caulfield and Brownsword 2006: 75)

These juxtaposed viewpoints illustrate that the issues that influence our dignity of moral stature (self-respect) are integrally tied to highly contested matters in this debate: rights, autonomy, personhood, humanity, rationality, sentience, dignity. The pillars of pro- and antihuman embryo research arguments seem frail in light of significant rhetorical ambiguity. For some, the moral choice recognizes a right to life of the embryo and the desire to protect its *dignity of Menschenwürde* matched with Kantian opposition to so-called utilitarian attitudes. Antiutilitarianism is pitted against beneficence. For others, the moral choice is beneficence and the healing imperative seeking to protect the *dignity of identity* of our rational, sentient brethren, attributing limited rights (or none) to the early human embryo. Thus, we see that the dignity of moral stature is inextricably tied to our interpretations of the dignity of *Menschenwürde* and the dignity of identity.

The Dignity of Identity

The second kind of dignity described by Nordenfelt is that of the dignity of identity, which is "attached to ourselves as integrated and autonomous persons, persons with a history and persons with a future with all our relationships to other human beings" (Nordenfelt 2004). This kind of dignity is closely associated with the integrity of an individual in a very general sense, including biological, psychological, and relational elements. This type of dignity is perhaps best illustrated by situations that infringe on such things as personal autonomy (e.g., imprisonment) or physical and mental attributes or capacities (e.g., mental or physical deficiencies associated with disease or aging) that necessarily impose restrictions on personal freedom and social relations. The dignity of identity is best understood biographically. That is, this dignity is only understood in the context of changing experiences of a single individual across his or her life. For example, beauty or intelligence are attributes of personal dignity that are vulnerable to the influence of external factors such as aging, injury, or other kinds of violation that may impinge on these dignifying characteristics. Juxtaposing the status of personal attributes or relationships in a biographical sense gives context for variations that occur in an individual's dignity of identity. As such, the dignity of identity can come and go in conjunction with changes in the integrity of an individual's physical or mental capacities or changes in one's communal relationships and is relevant only in the context of comparison. The dignity of identity directly correlates with developmental stages (e.g., biological development, cognitive awareness, levels of rationality and autonomy), the human embryo having the least (if any) compared to rational, autonomous, moral agents.

The connection between the dignity of identity and the human embryo controversy is aligned primarily on the side of those in favor of human embryo research. The main argument of proponents is that ES cell research serves highly ethical goals that preserve the dignity of identity, for example, by providing adequate cell and tissue replacement therapies for the treatment of incurable diseases such as Alzheimer's disease, dementia, multiple sclerosis, and so on. Hence, those who are opposed to this kind of research are guilty of impeding promising chances for medical progress and defying the "therapeutic imperative" and the "ethics of healing." From this perspective, the dignity of identity of the suffering patient is held paramount to any putative dignity of the embryo.

Theologian Ted Peters, in his essay, "Embryonic Stem Cells and the Theology of Dignity," claimed this vantage point when he wrote,

> we might side with proponents to say, because embryonic stem cells derive from excess fertilized ova at in vitro fertilization clinics, and would never under any circumstances reach implantation or have the natural environment necessary to become a human being, and because hES cells show such enormous potential for developing new therapies that could dramatically enhance human health and well being, these cells may be used to serve the dignity of future persons who will benefit. Because pluripotent stem cells do not have the actual potential for becoming a human being, they do not have dignity to be compromised. In this case, the dignity honored is that of future beneficiaries of this medical research. (Peters 2001: 136)

Generally, society's endeavors, including biomedical research, medicine, education, social work, and finance directly attend to an individual's dignity of identity. Human embryos clearly have no participation in this given their lack of self-interest or involvement in the shared vulnerability associated with biographical changes in this kind of dignity. In contrast, disease, aging, weakness, limitations, suffering, and mortality put a person's existence at risk, and rational beings with self-interest are acutely aware of this, especially in the face of disease. In the given context, the interests of those who could benefit from hES cells (for therapeutic applications) or human embryo research (e.g., for advancements in IVF or PGD technology) are set in contrast to those arguing in favor of the dignity (*Menschenwürde*) of the embryo. As alluded to previously, the dignity of identity is one instance where we are faced with evaluating values on the basis of some kind of hierarchy. In other words, moral agents have an obligation to make a moral decision about human embryo research. Is the higher obligation toward the dignity of identity, which impacts the dignity of friends and neighbors in our community, or an obligation toward the dignity of *Menschenwürde*, where protection of the embryo is tantamount? And is it even a question of either/or where discrete lines of right and wrong demarcate the ethical way?

The Dignity of *Menschenwürde*

The dignity of moral stature and identity are types of dignity understood in terms of degrees. For the most part, these forms of dignity are independent such that an individual can simultaneously experience different scales of each variety of dignity. Although we certainly recognize the validity of aristocratic notions of dignity (those that can vary in degree), there remains a final characteristic of dignity which is generally held as common among all human beings regardless of merit, moral actions, or identity.

The dignity of *Menschenwürde* is the kind of dignity often understood as inextricably tied to being human. It is a shared feature of humanity. In other words, by virtue of being human, you have *Menschenwürde*. In Dieter Birnbacher's "Ambiguities in the Concept of *Menschenwürde*" (Birnbacher 1996), this dignity is defined as the "sphere of unforfeitable individual minimal rights to be respected irrespective of the considerations of merit and quality." Interestingly, as with Macklin's "dignity is a useless concept" (Macklin 2003: 1420) and Horton's statement that dignity is "a linguistic currency that will buy a basketful of extraordinary meanings" (Horton 2004: 1081), Birnbacher expresses similar sentiments in his introduction about the German term *Menschenwürde*:

> The use made of the concept of *Menschenwürde* . . . is an irritating one . . . due to the unclarities and ambiguities of the concept itself . . . with no fixed content of its own, lending itself to merely rhetorical and opportunistic application. . . . It typically functions as a conversation stopper. . . . There are several reasons to reject an extensive recourse to the concept of *Menschenwürde* in bioethics. One reason is that one gets the impression that the inherent emphasis and the inherent pathos of the concept is exploited simply in order to eschew the difficulties of giving rational arguments for moral and legal injunctions against unwelcome practices. . . . [The] extensive use of the argument of *Menschenwürde* necessarily weakens the authority and moral emphasis of the concept.

> This is deplorable because the concept has still an important role to play. It would be a pity if by importing subjective and fashionable contents into its meaning the concept loses its normative force and ends up as a mere expressive gesture, a piece of empty rhetoric full of connotation but devoid of denotation. (Birnbacher 1996: 112)

The seeming nonchalant references to human dignity may be, at least in part, the reason for such negative sentiments about the term and its use in bioethical and political dialogue. For example, in the first article of their 1948 Universal Declaration of Human Rights, the General Assembly of the United Nations stated: "All human beings are born free, *equal* in dignity and human rights. They are endowed with reason and conscience and should act towards one another in a spirit of brotherhood." More recently, the executive order given by President George W. Bush to expand approved stem cell lines in 2007 stated in article 2.b., "it is critical to establish moral and ethical boundaries to allow the Nation to move forward vigorously with medical research, while also maintaining the highest ethical standards and respecting human life and human dignity." There are significant assumptions about what dignity means in these contexts.

Regardless of the responses such as those issued by Macklin and others, there is no doubt that phrases such as "respecting human dignity" have a powerful appeal. George and Tollefsen illustrate this point using a very general example. "Our dignity," they write, "is violated when the basic goods are deliberately damaged or destroyed in our person, as when someone intentionally takes another human being's life. That action, as an assault on human life, is an assault on human dignity" (George and Tollefsen 2008: 96). Yet the actual basis for the assumptions behind the moral compass guiding this type of ethic is challenging to pinpoint. In David Badcott's evaluation of the matter,

> with *Menschenwürde* we encounter a serious problem that might be termed the *problem of essentialism*. Just what is the essential property or element possessed by all human beings beyond a certain embryonic state through infancy and childhood, from adulthood to old age, but is nevertheless retained by the dead body? (Badcott 2003: 123)

What are basic human rights? What rights do we have a duty to respect and protect, and who holds these rights?

Given the current topic, we must determine whether the normative influence of the term *dignity* has any substantive power in the rhetorical strategies over the moral worth and rights of the early human embryo. This question is important given the effort scholars have made toward determining whether the dignity of *Menschenwürde* is attributable to early human embryos. The significance of the terms or phrases such as "personhood," "human being," "membership in the species *Homo sapiens*," and putatively morally relevant points of embryological development are all points of departure for determining the validity of attributing dignity to the early human embryo. In the following discussion, we examine a number of scholars' efforts to excavate the validity of dignity, or lack thereof, in evaluating the moral status of the early human embryo.

Two Ontological Evaluations of Dignity and Personhood

In response to David Badcott's problem of essentialism, there are two major viewpoints that influence people's *perceptions* about what gives a human being dignity. The first, a theistic view, holds that, in contrast to all other animals and beings on the earth, humans are created in the image of the Almighty (*imago Dei*). Among others (Jones and Whitaker 2009), Richard Neuhaus has written about this:

> A human being is a person possessed of a dignity we are obliged to respect at every point of development, debilitation, or decline by virtue of being created in the image and likeness of God. Endowed with the spiritual principle of the soul, with reason, and with free will, the destiny of the person who acts in accord with moral conscience in obedience to the truth is nothing less than eternal union with God. This is the dignity of the human person that is to be respected, defended, and indeed revered. (Neuhaus 2008: 227)

Indeed, for some, the divine blessing is one that provides reason to respect human life regardless of its stage of development, capacity for rational thought, or autonomy.

The secular humanist correlate to Neuhaus is that human beings are worthy of the dignity of *Menschenwürde* because we belong to humanity, or, as others see it, we belong to the species *Homo sapiens*. Two scholars presented here show marked differences in the opinion of using species evaluations as a framework for determining the validity of attributing dignity to early human embryos. In the first case, Fuat Oduncu holds the view that our common genetic constitution sets us apart as unique:

> The mere membership of humanity creates and preserves the fundamental value of human dignity until death. So, even if the embryo has never developed reason or mental powers because of severe damage or a lack of the material neural conditions, the embryo, nevertheless, will unconditionally be accorded the value of human dignity. . . . The widely accepted value of human dignity is intrinsically connected with the substantial existence of a human being whose starting point is the genetical uniqueness resulting from the process of conception. (Oduncu, 2003: 12)

In contrast to this, Lisa Bortolotti and John Harris have written that species boundaries or demarcating dignity-worthy beings based on genetic constitution is an arbitrary demarcation point that cannot act as a legitimate ontology for attributing dignity to persons. They write:

> Notice that . . . the objection relies on dignity being an attribute of *human* life as such. But . . . there is nothing intrinsically valuable about belonging to the species *Homo sapiens*. Granting rights and interests on the basis of species membership alone seems to be totally arbitrary and it is comparable, as a practice, to granting rights and interests on the basis of race or sex. (Bortolotti and Harris 2005: 74)

It is clear that arbitrating between these viewpoints is a difficult endeavor where fundamental philosophical differences about the ontology of dignity prevent these camps from achieving an epistemological praxis.

The Kantian Person and Arguments on the Dignity of *Menschenwürde*

The invocation of Kant's second principle formulation of the categorical imperative is perhaps the most often cited reference designed to support the concept of personhood in the early human embryo. The basic framework for this philosophy is that a person ought not to be used merely as a means but always also as an end. George W. Bush's 2001 speech regarding federal funding of hES cells illustrates this point. "Even the most noble ends," he says, "do not justify any means . . . the fact that a living being is going to die does not justify experimenting on it or exploiting it as a natural resource" (Bush 2001). Viewpoints such as this suggest an inexorable dignity attributed to persons. In the *Groundwork of the Metaphysics of Morals*, Immanuel Kant states this imperative as follows:

> Now I say that the human being and in general every rational being exists as an end in itself, *not merely* as a means to be used by this or that will at its discretion; instead he must in all his actions, whether directed to himself or also to other rational beings, always be regarded *at the same time as an end*. The practical imperative will therefore be the following: *So act that you use humanity, whether in your own person, or in the person of any other, always at the same time as an end, never merely as a means.* (Kant 1997)

At first glance this statement supports the notion that all persons are worthy of dignity and its associated respected rights. Many scholars have used Kant's imperative as the groundwork for arguments against exploiting human embryos. Scholars of this persuasion, such as Fuat Oduncu, have interpreted the core principles of Kant's categorical imperative as follows:

> The human embryo is looked upon as a human being from the moment of its conception and thus attributed the fundamental principle of

human dignity that guarantees the right to life of the embryo. Accord-ing to Kant, human dignity forbids and even condemns instrumental-ization and reduction of a human being to a mere means and object. Human beings are persons and as such they are ends in themselves. . . . *Vulnerability* is to be considered as a condition for the respect of the specific value of *human dignity*. Existing in this way includes specific anthropological features—vulnerability, human dignity, integrity, cor-poreality—aspects which forbid any actions to instrumentalize, exploit and reduce humans to a mere means in order to reach some ends. Con-sequently, slavery, exploitation of the vulnerable nature of man, or the use of embryos for research are to be prohibited, as this kind of behav-ior rigorously offends against the fundamental principle of human dig-nity. Here again, the Kantian categorical imperative forces us to look upon humans not only as mere means, but always as ends in themselves. Embryos are vulnerable living human corporealities, and their integrity ought to be held in respect. (Oduncu 2003: 14)

However, taking a Kantian interpretation of the categorical imperative, as Fuat Oduncu and Michael Novak have done, has come under significant scru-tiny in recent years given that Kant's evaluation of personhood or humanity does not, strictly speaking, apply to characteristics of the early human embryo.

The most direct opposition to using a Kantian argument in favor of pro-tecting the putative dignity of the early human embryo has been presented by Bartha Alvarez Manninen (Manninen 2008). Making issue of the use of a strict interpretation of Kant's categorical imperative, Manninen deconstructs the assumption that the early human embryo is an instantiation of a new Kantian person. Kant, she points out, describes his definition of a person in his *Metaphysics of Morals*, where he states that a person is "a subject whose actions can be *imputed* to him. *Moral* personality is therefore nothing other than the freedom of a rational being under moral law." Based on this, it seems strict Kantian personhood pertains to beings with certain capacities—specifically reason and autonomy. Given that many individuals have varying degrees of reason and autonomy, this interpretation casts Kantian determina-tions of the dignity of *Menschenwürde* into categories of degree and varia-tion attributed to the dignities of moral stature and identity described above. Thus, Manninen argues that a strict evaluation of Kant's philosophy does not support a position wherein this inexorable dignity of *Menschenwürde* can be attributed to the early human embryo. Reflecting on this, she writes,

if Kant's theory really endorses the claim that only actually rational or self-conscious beings have moral status, we would have no more duties toward infants and the severely mentally disabled than we have toward non-human animals. This would not license us to treat them cruelly, as it does not license us to treat animals cruelly, but certainly the formula of humanity would not apply to [human embryos, fetuses, infants, the senile, or the severely mentally disabled], as it does not apply to non-human animals. Moreover, they would not be regarded as persons with intrinsic dignity, but rather they would be regarded as mere things. . . . [As such], Kant himself misinterpreted the very spirit of his own moral theory; that, if interpreted correctly, some nonrational human beings certainly should be treated as ends in themselves even if they technically lack Kantian personhood. (Manninen 2008: 10)

We agree with Manninen and others that Kantian philosophy, in the strict sense, cannot be applied logically to the determination of the moral worth or any attribute of dignity of the human embryo. Some scholars who support embryo research agree with this. Bortolotti and Harris, for example, state that

> Personhood begins in humans when human beings start having a sense of self and responding to standards of rationality. . . . Although there is no ultimate test for the capacities involved in self-consciousness, these empirical studies rule out that human embryos and fetuses are persons. As they do not satisfy the requirements for personhood (rationality and self-consciousness), they do not qualify as persons and, in our view, do not have an interest in their own continued existence. (Bortolotti and Harris 2005: 73)

Bortolotti and Harris, it seems, understand dignity as a characteristic that is transient and vulnerable, like rationality or autonomy, and thus can be attributed in units of degree. Perhaps not wanting to go the distance like Bortolotti and Harris, Manninen builds on scholars such as Allen Wood who provide an "extended" interpretation of Kant where the relation of the being in question is set in reference to its *capacity* for rational nature. That is, having the potential, or having previously had the potential, or having some of the essential parts for rationality may be starting points for evaluating a being's worthiness of inherent dignity. This brings us to another contentious point, the issue of potentiality.

Potentiality of Personhood in Determining Human Dignity

As with many issues in this debate, the matter of potentiality is as divisive as any other given that scholars use different starting points to evaluate the capacity for potential personhood. In this case, dignity of *Menschenwürde* is dependent on one's interpretation of potentiality and so is not seen as a matter of gradation but as an either/or variable. Manninen and Woods provide one persuasion of this issue, pointing out that, although the human embryo does have the intrinsic capacity for development, one key feature—the womb—is not in place and this removes its "potential" development into a human life (Manninen 2008). For Manninen and others

(Peters 2001), in utero embryonic and extraembryonic components must be in place for the argument of potentiality to hold up. Thus, an embryo in a Petri dish has no more capacity for natural development than a human ES cell even though, in both cases, the integral capacity (genetic predisposition) for human life exists. Given this philosophical basis, Manninen writes, "If we must decide between discarding surplus embryos in fertility clinics . . . or using them for very promising research, which of these two options most expresses due respect for rational nature" (Manninen 2008: 13)? It is not clear whether Manninen holds that the early embryo does not have dignity as such, but some scholars of this persuasion place human embryos in what Birnbacher calls the extended meaning of *Menschenwürde*. Birnbacher fills out this vantage point in more detail as follows:

> While *Menschenwürde* in its core meaning needs an individual subject as bearer, this is not necessary with the extended concepts [pertaining to early and the residual stages of human life]. With them, there need not be a real subject to correspond to the grammatical subject. While *Menschenwürde* in its core meaning postulates minimal rights, *Menschenwürde* in its extended meaning postulates obligations without corresponding rights since there may be no bearer of rights. With *Menschenwürde* in its core meaning the object of respect and protection is the concrete human being. With *Menschenwürde* in its extended meaning it is something more abstract: humanity, human life, or the identity and dignity of the human species defined by its specific potentialities. My thesis is that the distinction between core concepts and extended concepts is of the utmost importance both in ethics and in law. . . . *Menschenwürde* in the abstract . . . is generally thought of as a much weaker principle that does not rule out, as the strong principle does, being given up in favor of other values such as individual autonomy or scientific and medical progress. The respect due to a human embryo or a human corpse is a *weak* form of respect, much weaker in any case than that due to a human person with its capabilities of consciousness and self-consciousness. (Birnbacher 1996: 113)

We note that Manninen, like Birnbacher, holds that human embryos are interpreted as having an extended version of the dignity of *Menschenwürde* so that hierarchical arrangements of the varieties of dignity are manifest where the dignity of identity of "concrete human beings" suffering from debilitating diseases is of more import than nascent embryonic life. This perspective suggests that human embryos in vitro are not Kantian persons, nor do they have the "potential" for natural development into human life. In vitro embryos, by this interpretation, have dignity in the extended manner, which carries less moral weight than dignity in its core meaning. From this view, the greatest respect we can provide to the human quality of rationality is to use spare IVF embryos for research and therapeutic purposes as opposed to disposing of them. Opponents to this interpretation of dignity and respect of the embryo, such as Richard Callahan, have countered by asking, "What in the world can that kind of respect mean? It is an odd form of esteem at once high-minded and altogether lethal" (Callahan 1995).

In contrast to the vantage point held by Manninen, there are other more conservative interpretations of potentiality held by scholars such as George and Tollefsen. In their chapter, "Human Dignity," from their 2008 book, *Embryo*, the protection for human life from the moment of fertilization is outlined as follows:

> It seems to us that the natural human capacities for reason and freedom are fundamental to the dignity of human beings—the dignity that is protected by human rights. The basic goods of human nature are the goods of a rational creature, a creature who, unless impaired or prevented from doing so, naturally develops capacities for deliberation, judgment, and choice. . . . We become subjects of human dignity, in other words, from the point at which we begin to exist as human beings, and we are, for the same reasons, the subjects of absolute human rights from precisely that point as well. . . . [In] the vast majority of cases, excluding only cases of monozygotic twinning, human beings come to be by the completion of fertilization, when there is a single-celled zygotic member of the human species *able to direct its own integral organic functioning and development toward maturity*. It follows that at that point there exists a subject of human dignity and human rights, and that any choice to deliberately damage or destroy such a subject is a violation of an inviolable right, the human right to life. (George and Tollefsen 2008: 106–9)

Thus, George and Tollefsen hold that the ability to orchestrate one's own "integral organic functioning and development to maturity," which is initiated at the moment of fertilization, is sufficient to merit the early human embryo dignity of *Menschenwürde* and worthy of protection based on its equal right to life. From this viewpoint, the human ES cell is unique from somatic or germ cells in that the ES cell has the intrinsic potential for development into a complete human person. Neither Manninen, nor anyone who knows the biology of the human embryo, would argue with this last point. However, as opposed to focusing on potentiality as existing at the level of the intact embryo in utero, George and Tollefsen hold that the *singular entity capable of full development into a human individual*—that is, unequivocally the hES cell—is the final arbiter of determining potentiality.

CAN WE MOVE BEYOND HUMAN DIGNITY RHETORIC?

Where does all of this analysis on the rhetoric of human dignity in the stem cell debate leave us regarding the possibilities for genuine dialogue among stakeholders and forging constructive public policy on embryonic stem cell research?

We are not particularly sanguine about those prospects for it seems that the issue of embryonic stem cell research has become as toxic to constructive dialogue and public policy as has the abortion issue. In fact, our research has persuaded us that stem cell research has become a toxic topic to the precise extent that the rhetoric of human dignity is invoked in the debate. In other words, when human dignity is used as a qualifier or a descriptor of the early embryo from which embryonic stem cells are derived, we observe that the public temperature rises to a near-fever pitch among the general public, and sometimes among academics as well. Unwittingly perhaps, this shrouds the prospects for clarity in a cloud of confusion: on the one hand, the term *human dignity* suggests what is owed to human beings—dignity—and on the other hand, *human dignity* itself is a nebulous term—what kind of dignity is owed to what kind of being, and at what stage, and on the basis of which commitments?

It is especially ironic that what is arguably the simplest biological state of human life (the early blastocyst) has become mired in convoluted language and tortuous debates about its moral status. Embryonic stem cell research in this country needs coherent public policy, and because public policy exists in the realm of the practical, it is perhaps unwise to attempt to construct it on the basis of polarizing rhetoric. Concepts such as dignity, humanity, and personhood undergird opposition to embryo research, as we have seen. The same concepts can also be used to muster support for embryo research when the embryonic stem cells are regarded to be in service of enhanced dignity, humanity, and personhood for suffering and ill corporeal human beings. So where and how does this cash out? That the same rhetoric can be used on both sides in this debate (with a polarizing outcome), and that we are no further toward a resolution, leads us to urge that such language be jettisoned in favor of appeals to more fruitful concepts. This is not to say the term is useless *altogether*, but rather, narrowing discussions of human dignity to appropriate contexts will help preserve its rhetorical power. However, we maintain that in situations such as the human embryonic stem cell controversy, the term *human dignity* is insufficient (if not counterproductive) to function as an argumentative premise.

What makes a concept or rhetorical appeal more fruitful than another? Usually it is sound argument, but on this issue we have reached an argumentative stalemate. The yawning chasm that seems to exist between two camps (pro- and antiembryo research) needs a rhetorical bridge, and the usual ethical appeals—appeals to utility, or to duty, or to autonomy—seem only to increase the divide. Just as in the colloquialism that "one person's trash is another one's treasure," an ethical appeal to human dignity or autonomy, for instance, that might seem useless to one person might be of highest relevance and clarity to another. Whether it is or is not futile to attempt to ground the moral case for or against embryonic stem cells in appeals to human dignity, this much has become clear to us: the time has come for a new kind of rhetoric about early embryonic research. In that spirit, we offer two modest suggestions.

Healing

Perhaps it is not too bold a statement to claim that the imperative to heal is something of a common human denominator. It crosses over and transcends virtually all cultures and religions and is something to which governments and even corporations have dedicated themselves, to say nothing of the obvious orientation of health care researchers, scientists, and health care practitioners across the globe. Indeed, the imperative to heal is a concept that is as firmly rooted in secular approaches to science and medicine as it is in religious and theological approaches; it is the premise of both medical-craft and soul-craft. The duty to heal is both the grounding for the Hippocratic Oath and a foundation of the Abrahamic religions. In rabbinic Judaism, for example, the imperative to heal is so strong that it is one of the only legitimate reasons for breaking the Sabbath. So, too, in both Christianity and Islam, there are numerous examples of the divine mandate to heal the sick and broken among us.

Thus, it is tempting to consider whether the concept of the imperative to heal might provide a rhetorical and ethical opening—an approach to the ethics of the early embryo that could be more unitive and cohesive than it is divisive. What materials may we use in fulfilling the obligation we have as humans, whether religious or secular, to heal one another? May we use pieces of ourselves? Clearly, health care ethics permits rational, autonomous persons, free from coercion, to consent to donate blood, tissues, kidneys, eggs, and even embryos for the healing of other such persons in need. Although we might not realize it, the impulse to heal is the basis of both fertility medicine (also known as assisted reproductive technologies or ARTs) and the new field of regenerative medicine. In assisted reproduction, couples who suffer from infertility issues are able to make use of technology (developed as a result of research on human embryos) that assists them in producing a child, thus "healing" the condition of infertility and the suffering that attends it. In regenerative medicine, it is hoped that patients will one day be able to regenerate diseased or failing organs, blood, tissues, and so on using human ES cells so that a chronically ill or dying person may experience healing from the disease that ravages her body. It is ironic that in both these arenas the animator of healing—a good we affirm cross-culturally—is the embryo. Because the rhetoric of "duty to heal" appears across theological and secular ethical traditions, we offer it as food for thought that could have the potential to move the debate away from the stalemate produced by appeals to human dignity. Another framework that merits consideration is the concept of moral complicity.

Complicity and Agency

The question of the complicity of the moral agent in evil deeds has long occupied the Roman Catholic moral tradition and, in fact, is the central issue that animates the Catholic tradition of just war theory. There is a long history of scholarly debate on issues of moral complicity in the Catholic tradition, to be sure, but one also finds a great deal of discussion concerning complicity in secular as well as theological bioethics. However, complicity has not been the subject of much scholarly focus in the embryonic stem cell debate; though it has been addressed, it has not had a central place in the literature.[1] We suggest that complicity may be a fruitful avenue for exploration in the stem cell debate precisely because it forces us to take seriously objections to early embryo destruction, and perhaps opens a new pathway for traversing this seemingly intractable issue. The question is, when, if ever, is a moral agent complicit in the destruction of embryos that are used for stem cell research and therapeutics, and what does this do to an agent's self-respect?

Ron Green, a Kantian bioethicist, in a 2002 stem cell article, explored the pertinent question of whether an agent may ever benefit from evil deeds (Green 2002). Green considers concerns about both "moral contagion" and the "moral encouragement" of an agent, dismissing the former as "primarily an emotional concern." The latter issue of moral encouragement, however, Green finds relevant insofar as at the level of public scrutiny, we can agree that we generally do not want to support "forms of benefiting from evil that encourage the further commission of evil deeds." Thus, Green scrutinizes three ways in which we might encourage cooperation or might benefit from evil: (1) "direct encouragement through agency" (I benefit from another's having done the wrongful act that I encouraged but did not perform); (2) "direct encouragement through the acceptance of benefit" (a second person, B, benefits from the evil deed person A committed, though B has no connection to A, and B does not forego the benefit); (3) "indirect encouragement through the legitimization of a practice, or practice encouragement" (one accepts benefit from an evil deed having been done, though no doer of the deed is known, such that acceptance of wrongdoing opens the door to future wrongdoing) (Green 2002: 550).

Those opposed to embryonic stem cell research might raise concerns similar to Green's "practice encouragement" concerns about a moral coarsening effect of embryo destruction over time. Green concludes that even for those who worry about being complicit in embryo destruction by making use of embryonic stem cell lines that they did not create (or destroy), it may be possible to find a moral warrant allowing such research. He points out that since there is no clear causal connection between embryo destruction in fertility clinics, for instance, which makes stem cell research possible, then "none of our three types of encouragement need occur when someone uses an hESC line" (Green 2002: 554). And if none of the three kinds of encouragement with evil deeds needs to occur, then "although one may abhor the deeds that led to this biological material becoming available, one can use it without incurring any encouragement responsibility" (Green 2002: 554–55). Crucially, he claims, embryonic stem cell research does not *cause* the destruction of embryos; the destruction of embryos is caused by the decision to discard them (which means destroy) once stored in fertility clinics. "Research," he claims, "causes only the *manner* of their destruction" [and this is] "morally unimportant" (2002: 555). Need embryo destruction (believed by many to be an evil act) entail benefiting from the act such that it encourages evil? Green's argument shows that it does not.

There is still the fact that embryo destruction is an integral part of the way in which assisted reproduction is practiced and of the procedures that fertility clinics routinely employ. Unless those procedures change (and there is no reason to think that they will), those who hold that there is no moral warrant for being involved in embryo destruction (complicit in an immoral act) presumably will conclude that neither ought they to participate in the assisted reproduction industry nor advocate for any enterprise involving use of the early human embryo, whether directly or indirectly.

We noted that the embryo is the root of both fertility and regenerative medicines. There are many who oppose the destruction of the embryo that takes place in both these arenas, and yet it is almost a truism that objections to embryo destruction in stem cell research are both more vociferous and more numerous than are objections to (1) the practice of discarding embryos by the assisted reproduction industry or (2) the fact that all assisted reproductive technology is built on the basis of human embryos derived and manipulated solely for research purposes. If one strives for ethical consistency and objects to the killing of embryos, consistency would seem to demand that one object at every juncture and in every arena. Further, if one has concerns about being morally complicit with embryo destruction by benefiting from an evil act (destruction of human life), then complicity concerns in one arena cannot be ethically cordoned off from another. More to the point, it seems worth pointing out that if one objects to benefiting from evil, then one has to look closely at other forms of stem cell derivation as well.

Induced pluripotent stem cell (iPSC) research, which anti–embryonic stem cell forces have sanctioned because iPSCs do not originate from embryonic stem cells, is entirely dependent on knowledge and practices derived from embryo destruction in embryonic stem cell research. To be clear, we do not mean to suggest, nor do we think that iPSC research is "evil"; we merely point out that (1) all of the research is dependent upon the embryo one way or another, and (2) ethical consistency seems to demand acknowledgement that if complicity is a moral concern, it is a thread that runs through the whole garment. As such, neither blanket opposition nor categorical support of embryonic stem cell research is a useful approach. This brings us again to the point that nuanced ethical arguments—regarding, for example, notions of healing and moral complicity—on specific kinds of pluripotent stem cells and their related technologies require further ethical scrutiny.

A FINAL NOTE

In conclusion, we urge readers to the awareness that what is sincerely needed is something that has eluded scientists and scholars thus far—if not a common language with which to speak, then a common respect for the differences in ethical persuasion among us that have led good people to passionate and sometimes unalterably opposing conclusions. The poet Adrienne Rich once dared us to "dream of a common language." If we cannot find such a language on this issue, let us at the least foster a climate of mutual respect for views that are as incisive as they are divergent. And perhaps, not unlike the national debate that still rages over what constitutes brain death, we will at least respectfully begin to agree to disagree and finally to set a line of demarcation for the legal, if not always ethically comfortable, uses for embryonic stem cell research.

NOTE

1. For exceptions to this, see Brown 2009, Green 2002, and Guenin 2004.

13 Human Dignity in End-of-Life Issues
From Palliative Care to Euthanasia

Thomas R. McCormick

In this chapter I will show that claims about normative ethical duties in the provision of medical treatment to human individuals in the terminal phase of life are often anchored in the concept of human dignity. Interestingly, opposing arguments about society's duty to patients nearing the end of life may also appeal to the concept of dignity while recommending medical choices that are quite disparate. For example, both advocates and opponents for withdrawing life support (or hastening death) ground their arguments in the notion of human dignity. Similarly, patients may speak of a desire to "die with dignity," yet the meanings they attach to this concept are very different, as seen in the differing clinical decisions they choose. For patients such as Helga Wanglie in Minnesota, dying with dignity meant using all relevant medical technology that could reasonably be expected to prolong her life, even in an unconscious state (Capron 1991). Others, suffering a terminal illness where curative measures are no longer effective, make a transition to palliative care believing that the management of symptoms and the provision of excellent pain control will help provide a dignified death. Some patients receiving care in the intensive care unit may recognize that curative efforts are no longer possible and may elect to stop the technologies that are keeping them alive, allowing death with dignity. Another cohort of persons who are nearing the end of life from a terminal condition believe strongly that their dignity will best be preserved by taking matte rs into their own hands and planning the date and hour of death by taking a medication that will end their life at a time and place of their own choosing. While one can perhaps justify these differing approaches from the principle of respect for persons, I argue that such diverse actions cannot be justified solely by an appeal to dignity. It appears that the meaning of the term in such instances is too imprecise, standing in need of reinterpretation in this generation. Although all might agree with the general concept that we would not want to be treated in ways that demean our human dignity, it is clear that the meaning of dignity in common clinical endeavors needs specification.

HUMAN DIGNITY IN EVOLVING PRACTICES
OF CARING FOR DYING PATIENTS

Popular usage of the term dignity in recent years is often associated with notions about dying with dignity. This will be pointed out in the three cases that follow in the next section. Our current preoccupation with death with dignity is ironic in that, for the greater part of human history, most people have struggled to live with dignity, with adequate food, water, clothing, and shelter. For much of the world's population, this is still the case.

In 1900, the average life expectancy in the United States was only 48 years. One-third of all babies died before the age of 18. Epidemics such as the flu and infectious diseases caused the premature death of many adults. Most adults died quickly from accidents or infections. In the first half of the 20th century, life expectancy was enhanced largely by public health measures that provided clean drinking water and improved sanitation and by improvements in the supply and distribution of food.

In the second half of the 20th century, startling advances in medicine and technology allowed physicians to prevent many premature deaths. Scientific breakthroughs abounded, such as Fleming's discovery of penicillin (1943), isoniazide treatment for tuberculosis (1951), dialysis for chronic kidney failure (1960), and vaccinations preventing the spread of infectious diseases. These innovations, along with many new treatments for heart disease and cancer led to the prolongation of lives. With the development of interventions such as ventilators and the subsequent rise of intensive care, more lives were prolonged and an increasing number of seriously ill patients spent their last days in hospital or in an intensive care unit. Early in the 21st century, the average life expectancy in the United States had grown to 77.9 years (U.S. National Center for Health Statistics 2011). Such advances in prolonging lives gave rise to a new kind of problem when some began to claim they feared being robbed of their dignity by a prolonged period of debilitation in the dying process.

With more people living longer in developed countries, an array of complex ethical issues arose surrounding the nature, extent, and duration of care near the end of life. Today's patients want more control over treatment choices and demand respect for their wishes to die with dignity. This demand for death with dignity occurs in a time when the process of dying is lengthened for many. Over the past three decades, both public and professional attention has shifted to questions about the proper management of the dying process. Currently, dying with dignity involves concern for quality of life, the appropriate use of aggressive life-prolonging treatment, whether or when to forego life sustaining treatment, and how to manage pain and symptoms related to the disease and its treatment when death appears to be inevitable. (Some call this living with dignity until death occurs.) In addition to these medical-ethical issues, other psychological, social, and economic issues arise. Dying patients face further threats to dignity from protracted suffering, becoming a burden upon their families, and the impact of the cost of care in the final period of life. Such questions are not going to diminish but are likely to grow in relevance as the cohort of baby boomers now reaching retirement age begins to engage with these important considerations and demand dignity in dying.

Only a few years ago, dying with dignity for many meant that all medical resources appropriate to prolonging survival were employed in patients' care. Most people wish to live as long as possible, and some are willing to purchase very expensive medications that may only prolong their lives by a few weeks or a few months at best. Yet many others, having observed the tortuous dying of friends or family members, consider prolonged suffering a "fate worse than death" and wish for themselves a more peaceable ending. Increasing rates of Alzheimer's disease and other forms of dementia in the elderly lead some to fear that their bodies might outlive their minds. The loss of cognitive function is considered a severe threat to the dignity of personhood by many. Therefore, the timing of death and the use of life-prolonging technologies are important considerations.

The rising popularity of the home hospice movement marks a significant shift in patients wanting to avoid hospitalization near the end of life. Hospice patients believe that dying with dignity means dying at home, with good pain control and symptom management, psychological and spiritual support, accompanied by family and friends when death is inevitable and imminent. Such factors are intended to enhance the quality of life in the end days, allow a natural death, and preserve the patient's dignity.

Many very ill patients are hospitalized near the end of life and receive critical care in the intensive care unit, where death is common even though life can be prolonged despite severe organ system dysfunction through the most technologically advanced treatment methods. In most cases, patients, or family, make decisions to halt treatments that are sustaining life. Thus, the timing of death is affected by decisions to withhold or withdraw key medications or technologies. In such cases, families find ethical grounding for these actions in the notion of allowing a dignified death for their loved one.

In still other cases, terminally ill patients appeal to their physicians to provide a lethal medication that will help them end their lives quickly in the face of suffering. In prior years, health care providers in the United States could decline on the basis that such actions were forbidden by law or by the ethics of medicine. Currently, however, three states—Oregon, Washington, and Montana—have legalized physician-assisted death under certain circumstances that will be examined later in this chapter. Such patients often claim that such a step is necessary under the circumstances to allow them to die with dignity. Thus, we have seen a sort of evolution in which some patients choose a natural death with palliative care support; others choose to hasten death by removing life-prolonging technologies; still others request physician aid in dying; and all have appealed to the notion of dignity in support of their differing choices. It is clear that there has been an evolution in health care practices related to end-of-life care. Exerting control in the final stage of life reflects patients' notions about what it means to die with dignity.

While many patients and families use the language of dignity, it is not the language that is commonly used when controversial cases are reviewed by the hospital ethics committee. In the clinical setting, ethics committees (or consultants) are more likely to couch their analysis of the issues in light of principles such as autonomy, nonmaleficence, beneficence, and justice. All things considered, a competent patient who is informed of treatment alternatives and consequences has the right to decide. Disparate choices of patients are considered morally acceptable under the principle of respect for persons. The power of patients and their stories to shape policy and practice may be seen in the case studies that follow.

THREE LANDMARK CASES IN THE DEATH WITH DIGNITY DISCUSSION

The Case of Karen Ann Quinlan

Questions related to dying with dignity and appropriate end-of-life care burst dramatically into national awareness in recent years surrounding a few tragic cases that gained notoriety in the public press and caused many to question how they would want to be treated in similar circumstances. In 1975, 21-year-old Karen Ann Quinlan survived an anoxic event that left her alive but in a persistent vegetative state. She lived in this unconscious state, not knowing herself or others, for a decade until her death in 1985. Lacking the ability to breathe on her own, her death was at first prevented through the use of a ventilator. When her parents wished to discontinue the ventilator due to her profound neurologic impairment, their decision was challenged and the case went eventually to the New Jersey Supreme Court, where a ruling was made recognizing the parents of Karen as appropriate surrogate decision makers as well as their right to discontinue the use of a ventilator. Later, although legally authorized to turn off the ventilator, Karen's parents were persuaded to try weaning her from the ventilator. She was successfully weaned, and her life was prolonged by the employment of tube feedings and skilled nursing care. Although unconscious, in a persistent vegetative state, her biological life continued for 10 years, and she died in 1985. Her nationally prominent case raised the awareness of millions that one's physical body could be sustained even in the face of irreversible loss of consciousness. Many individuals claimed to family and friends they would not want to live in such a state, believing that biologic life without conscious awareness was life without human dignity. The right to discontinue the use of a ventilator was established in the court; the mandatory employment of a feeding tube for an irreversibly unconscious patient was the next issue to be tested.

The Case of Nancy Cruzan

In 1983, 25-year-old Nancy Cruzan suffered a closed head injury in a single-car accident and survived, after stabilization, in a persistent vegetative state. Although she was not ventilator dependent, her body was maintained through the use of tube feedings, antibiotic therapy, and skilled nursing care. She died of dehydration in 1990 following the removal of a feeding tube after her parents provided testimony in Missouri circuit court that there was "clear and convincing evidence" that Nancy would not have wanted her life prolonged in this way. Missouri state law required the employment of life-prolonging therapies in cases where patients were unable to speak for themselves, unless there was clear and convincing evidence that they would not want to live in such a condition. This case, following on the heels of the Quinlan case, led to an even greater awareness in the public that biologic life could be sustained by machines long after what many considered meaningful life had ended. Discussions of both the Quinlan case and the Cruzan case were often headlined as "death with dignity" cases. Increasingly, persons began making their wishes known about treatment preferences to family and friends in fear of having their bodies sustained with no hope of returning to a conscious and sapient state. In sharp contrast to earlier years, removing life support to allow death with dignity was increasingly considered by many to be a right, and the court decisions in Quinlan and Cruzan were widely cited in support.

The Case of Harold Glucksberg

In 1997, the case of Harold Glucksberg, who claimed he wanted "death with dignity," was heard before the U.S. Supreme Court. Glucksberg, along with three terminally ill patients (who have since died), four physicians, and the Compassion in Dying organization, challenged the law in Washington State forbidding physician-assisted suicide as an infringement on due process as protected under the Fourteenth amendment. Glucksberg claimed that some dying patients could choose to end their lives by ordering that a ventilator or feeding tube be withdrawn, while others were slowly dying without such technologic support. Since there was no technology to withdraw to speed the onset of death, Glucksberg argued that patients like him needed the assistance of their physician to help them die in a timely manner, in order to have equal opportunity. The U.S. Supreme Court found in favor of Washington State, that the state's law forbidding physician-assisted suicide was not unconstitutional. The court went on to claim that these matters were better left in the hands of the states and determined by state statute. Interestingly, about a decade later, voters in Washington State on November 4, 2008, passed a referendum, I-1000, entitled the Death with Dignity Act, codified as RCW 70.245 and becoming effective March 5, 2009. This act allows terminally ill adults the legal right to request lethal doses of medication from medical and osteopathic physicians that they may self-administer to hasten their demise (Death with Dignity Act of Washington 2009).

These three cases are noteworthy. The Quinlan case established the legal right of a patient, or the patient's surrogate, to discontinue life support in the form of a ventilator. The Cruzan case established the legal right of a patient to discontinue the use of artificially administered nutrition and hydration. The Glucksberg case, although dismissed by the Supreme Court, eventually led to the passage of a law that provides legal permission for physicians to assist terminally ill patients by providing a lethal medication that causes the death of the patient shortly after being ingested. In all three cases, arguments were built upon the notion of respect for the dignity of the patient and his or her right to a dignified death. Interestingly, the factors that contribute to or are deemed necessary for a dignified death are not defined. It might be assumed by inference that hastening the moment of an imminent death was a desirable contributing factor.

In all three cases, there were others who voiced opposition to withdrawing a ventilator, withdrawing tube feedings, or providing a lethal prescription. These opponents also based their arguments upon an appeal to respect for human dignity, arguing that ventilator support or tube feedings are appropriate means to support human life and ought to be employed out of respect for the dignity of the unconscious patient, who remains a human being even though permanently unconscious. Opponents of physician-assisted death also argued against this policy change on the basis that it mitigates against human dignity in general and poses a threat to disadvantaged individuals, the elderly, and others whom society already tends to devalue. Opponents saw a threat to the dignity of the individual who ended life by taking a lethal medication and a potential threat to the dignity of classes of disadvantaged persons without clearly establishing any basis in reality for such perceived threats. This could be interpreted as an emotive claim signifying strong feelings of disapproval rather than a substantive argument based upon empirical data.

SOCIAL POLICY CHANGES RELATED TO HUMAN DIGNITY IN END-OF-LIFE CONSIDERATIONS

In retrospect, it is clear that many Americans were deeply affected by these three cases. Social policy changes by way of court decisions and by legislation have changed the climate of support for patients, allowing patients to have a greater voice in their own care and treatment as the end of life approaches. It may be argued that such advocacy for patient autonomy supports the dignity of patients. With a population of approximately 284 million and nearly 2.5 million deaths per year, many in the United States will be affected by social policy changes. Such legislation warrants a closer look as to how it does or does not support human dignity.

Washington's Death with Dignity Act of 2009

Regarding Washington's Death with Dignity Act, only incurable patients who are expected to die within six months are eligible to make such a request. Supporters of this legislation argue that allowing dying patients to end their lives up to six months before they might otherwise die of their underlying illness prevents needless suffering and preserves the dignity of such patients. Dignity is supported by allowing the patient to choose the moment of death, shortening the dying trajectory, ending suffering or the fear of future suffering, and reducing the burden of care from the family and the cost of treatment in the final months of life. In the nine months after the law went into effect, the Washington State Department of Health statistics showed that 63 qualified patients received a prescription, and by the end of 2009, 47 had died (7 died without having ingested the medication); 82 percent of the patients cited a concern for the loss of dignity as a factor in choosing to die in this way (Washington State Department of Health Death with Dignity Report 2009). In the second year, 2010, 87 requested and received the lethal prescription from 68 physicians and 40 pharmacists; 72 patients died, 51 after taking the medication; 15 died without taking the medication, and the status of 6 was not known at the time of the report as to whether or not they had ingested the medication (Washington State Department of Health Death with Dignity Report 2010).

The widow of a Washington State patient who chose to take the lethal prescription to end his life in the first year the law went into effect described his gradual decline from metastatic cancer. Her husband had been an active, take-charge person. He loved being the skipper of his boat, choosing the course, setting sail, and directing the journey. She saw this as a metaphor for how he lived his life, relishing control. Near the end, however, he suffered chronic pain with acute painful episodes, shortness of breath, loss of appetite, weight loss, energy loss, and fatigue. It was important for him to end his life before experiencing further deterioration and the risk that he might not be capable of self-administering the medication when the time came. He put his affairs in order, said good-bye to family members, and, a few days later, took the medication in the company of his wife. He said his final good-bye and expressed his love to her and died within minutes. She reflected that he died as he had lived, wanting to be in full control until the end. For this man, dignity was found in living and dying as an autonomous being.[1]

Over a decade earlier, the state of Oregon passed a similar referendum. On October 27, 1997, Oregon enacted its Death with Dignity Act (Death with Dignity Act of Oregon 2010). Over the first 13 years since its passage, there have been 525 deaths in Oregon attributed to persons using the Death with Dignity Act (DWDA). To put this into perspective, it is estimated there have been 20.9 DWDA deaths per 10,000 total deaths during that period. Oregon reports that, "As in previous years, the most frequently mentioned end-of-life concerns were: loss of autonomy (93.8%), decreasing ability to participate in activities that made life enjoyable (93.8%), and loss of dignity (78.5%)" (Oregon Health Authority Death with Dignity Act 2010). Although the notion of dignity played a prominent role, in 2010 it was overshadowed by concern for loss of autonomy and the loss of enjoyable activities.

An Oregon woman who requested the prescription from her physician and took it to end her life was described by a family member as formerly an active and involved contributor to her family and community. However, during the previous 24 months as chemotherapy lost its effectiveness, the cancer continued to spread while she grew weaker and more dependent upon her family for care. Near the end, her summative comment was, "I'm sick and tired of being sick and tired, and I just can't take this any longer."[2] Her family believed she had taken the medication to end her physical suffering as well as the suffering from losing her role as a productive and giving person. "The existential suffering was as powerful as the physical suffering."[3]

It seems noteworthy that only slightly more than 20 persons out of 10,000 who died in Oregon chose to utilize the DWDA. Why such a small number? Many of these apparently were motivated by a concern about the potential for their loss of dignity in the dying process. What did loss of dignity mean to these individuals? One Oregon physician echoes the view of many in claiming that the patients who had asked him for a prescription under the law were usually people who liked being in charge and were in positions of control in their work or professional lives and in their family systems. He surmised that rather than simply having death overtake them, this particular cohort found meaning and dignity in taking control of this final situation and ending life at a time of their own choosing.

This gives rise to the questions: what about the other 9,980 of the 10,000 who died, what were they thinking, were they not concerned about the loss of dignity in the dying process? Perhaps others felt that their dignity would best be preserved in other ways, like some terminally ill patients I have talked with who chose hospice and palliative care as a way of preserving their dignity in the dying process. One hospice patient said, "my condition is incurable and I know I will die fairly soon, but I will die with dignity if I can live as fully as possible until that day comes."[4] This patient felt that her relationships and visits from her beloved family members made her remaining days valuable to her as well as to her loved ones up to the very end. Some dying patients I have counseled felt strongly that attending to spiritual issues was an important component in maintaining their dignity at the end of life as the focus moved from curative measures to quality-of-life concerns. Important spiritual tasks before dying include asking for or granting forgiveness, reaffirming love for family and friends, and saying good-bye. Depending upon the care team to help alleviate pain and other symptoms frees patients to carry out spiritual practices that bring comfort and provide dignity as death draws near (Ai and McCormick 2010: 24–41).

The Rise of Advance Directives in 1976

The Quinlan and Cruzan cases, because of their widespread reporting, succeeded in raising public awareness that decisions must be made concerning the nature and extent of medical care that is desired near the end of life.

Individuals began to document their concept of a dignified death. It is now commonly accepted today that these choices must be memorialized through a living will, also called an advance directive, or by designating a trusted person with a durable power of attorney for health care decisions.

California was the first state to pass Natural Death Act legislation, taking effect in 1976, with the stated intent of protecting the dignity of individuals in the final stage of life by securing their right to make choices about limiting treatments that might prolong the dying process.

Surveys have found that some physicians are still uncertain about the legality of stopping life-prolonging procedures, even though this is clearly permitted by law. All 50 states and the District of Columbia now have legislation that allows people to create a document intended to instruct care providers about their wishes in the event they become cognitively incapacitated. In Washington State, for example, the rights of patients are explicitly expressed in the Natural Death Act (1979). The language of the legislators noted their intent to protect and preserve the dignity of patients nearing death in the light of two well-known cases, Quinlan and Cruzan. It is instructive to examine the conceptual language of one such legislature. Washington State legislators wrote: "The legislature finds that adult persons have the fundamental right to control the decisions relating to the rendering of their own health care, including the decision to have life-sustaining treatment withheld or withdrawn in instances of a terminal condition or permanent unconscious condition" (Washington Natural Death Act 1992). Clearly, under this legislation, patients who fear ending up with conditions such as Quinlan or Cruzan could sign, while cognitively capable, an advance directive instructing their families and care providers about how they would want to be cared for in the event of permanent loss of consciousness in order that their dignity might be preserved.

The legislature further found that, "in the interest of protecting individual autonomy, such prolongation of the process of dying for persons with a terminal condition or permanent unconscious condition may cause loss of patient *dignity*, and unnecessary pain and suffering, while providing nothing medically necessary or beneficial to the patient" (RCW 70.122 1992). A key concept here is respect for the dignity of the patient, a notion that is closely associated with the autonomy of the patient as discussed above. Further, the concept that medical care can prolong life at the biological level, but without benefit to the dying or unconscious person, is recognized in the law. The legislature noted that "there exists considerable uncertainty in the medical and legal professions as to the legality of terminating the use or application of life-sustaining treatment" even when the patient has provided evidence of such a desire (RCW 70.122 1992). In the light of these findings, the legislature went on to lift up the right to dignity and privacy of patients and enacted the law giving adult patients the right to an advance directive and the right to designate another to serve as the patient's durable power of attorney for health care (RCW 70.122 1992). The original 1979 legislation was revised in 1992, first, to include patients who were permanently unconscious as well as patients who were terminally ill and, second, to specifically allow the withholding or withdrawal of artificially provided nutrition or hydration in addition to mechanical ventilation (Washington Natural Death Act 1992]). Clearly, in this revised legislation, we see the effects of the Quinlan and Cruzan decisions shaping the language of statutory laws to preserve patient autonomy and dignity at the end of life.

Washington State was the third state, following California and Idaho, to pass a law making advance directives a legal right for its citizens. Currently, all 50 states and the District of Columbia have some form of law providing for advance directives. President Obama was the first president to publicly declare that he and his wife have advance directives. There seems to be growing acceptance of the rationale behind these living wills, and more citizens are expressing their preferences by writing advance directives while they are cognitively able. The concept of preserving dignity in dying by enabling free and autonomous treatment choices for patients gained a major foothold in U.S. society.

Patient Self-Determination Act of 1990

Over a decade after the first Natural Death Act legislation went into effect, senators Danforth and Moynihan became concerned that too few were aware of their rights under the law to stop or refuse unwanted medical treatment. They believed that institutions and physicians needed to do more to educate patients about their autonomy in treatment decisions that were most likely to preserve their dignity at the end of life.

In 1990, the Patient Self-Determination Act, a federal law, was passed and went into effect in December 1991. This law requires any health facility receiving federal funds to inform patients about advance directives and their right to make decisions about the nature of their health care, including the right to obtain and file an advance directive with their care givers.

Shortly after the Patient Self-Determination Act went into effect, only about 21 percent of seriously ill elderly patients needing hospitalization had advance directives (Teno et al. 1994: 23–30). More recently, a study by Silveira and colleagues reported that among subjects who needed surrogates to make decisions on their behalf, 67.6 percent had an advance directive (Silveira et al. 2010: 1211–18). This is a considerable increase in those choosing to have an advance directive and may indicate that the concept is finding more widespread acceptance, especially in the population over 60 years of age. The study by Silveira and colleagues suggests that about one-quarter of all seriously ill older patients will lack the capacity to make decisions concerning their care at the end of life. The best-case scenario in preserving patients' concepts of dignity in dying would equip the patient with both an advance directive and a person holding durable power of attorney for health care in order to respond appropriately to unique situations that couldn't be anticipated in writing in the living will, thus allowing surrogates to act in the best interest of the patient whom they represent. Silveira and colleagues recommend that our health care system provide the necessary support to allow these important conversations between physicians and patients to take place in a timely manner. A number of recent studies have shown the need for improved communication between physicians and patients regarding end-of-life care. As a result of this communication deficit, the following problems have been identified by Larson and Tobin (2000: 1573):

1. Patients are dying after prolonged hospitalizations or intensive care, often in unrelieved pain.
2. Preferences concerning life-sustaining treatments are not adequately discussed, documented, or adhered to.
3. Referrals to hospice and home care, which could address these shortcomings, occur late or not at all.

THE CONCEPT OF HUMAN DIGNITY IN END-OF-LIFE DECISION MAKING

Here, I turn directly to an examination of the concept of dignity and a critique of its usefulness in justifying ethical decisions about the nature, extent, and duration of treatment at the end of life.

Opponents of the DWDA have argued that hastening death by taking a lethal medication is a threat to intrinsic human dignity and have asserted the claim that human dignity is best preserved within a society by valuing every human life and cherishing it until the breath of life is overwhelmed by the force of disease or accident. They contend that it is consistent with our human nature to seek survival. In both Judaism and Christianity, life is considered a good, a gift from God. The commandment, "Thou shalt not kill" (Exodus 20:13) has had a centuries-long influence on the thought of Jews and Christians and analogues exist in other religions. This threat to dignity seems to stem from a belief that death is a natural process and should not be taken into human hands.

Yet, while human life is very valuable, it is not an absolute good, and death is recognized as inevitable for all. What are appropriate limits upon the obligation to preserve life? Distinctions between allowing death and causing death are important, as are the motives and intentions behind our actions. How may we best take the unique circumstances and values of the patient into consideration?

As we have seen, both supporters of the DWDA and those who oppose the involvement of physicians hastening death through providing a prescription for a lethal dose of medication appeal to the concept of dignity as one justification for their positions. How do we account for this apparent discontinuity in clinical decision making in which various parties base their actions on an appeal to dignity? The answer seems to reside in the fact that the word *dignity* may be interpreted differently or may be nuanced differently in its various usages. Webster's dictionary defines dignity as "the quality or state of being worthy, honored, or esteemed" (*Webster's New Collegiate Dictionary* 1973: 319). In general, it might be argued as intuitive that persons universally hold a desire to be esteemed and treated as worthy, hence the broad appeal to the general notion of dignity.

However, to be used in a meaningful way, the various meanings and uses of the term *dignity* must be identified. Once specified, it may be more clearly evident when differing meanings of the term come into play. For example, Lennart Nordenfelt offers his view that there are four kinds of dignity and spells out their differences, from the commonly held notion of universal dignity to more particular distinctions associated with individuals. Nordenfelt's concise definition of the four types of dignity is as follows:

> *Menschenwürde* pertains to all human beings to the same extent and cannot be lost as long as the persons exist. The dignity of merit depends on social rank and position. There are many species of this kind of dignity and it is very unevenly distributed among human beings. The dignity of merit exists in degrees and it can come and go. The dignity of moral stature is the result of the moral deeds of the subject; likewise it can be reduced or lost through his or her immoral deeds. This kind of dignity is tied to the idea of a dignified character and of dignity as a virtue. The dignity of moral stature is a dignity of degree and it is also unevenly distributed. The dignity of identity is tied to the integrity of the subject's body and mind, and in many instances, although not always, also dependent on the subject's self-image. This dignity can come and go as a result of the deeds of fellow human beings and also as a result of changes in the subject's body and mind. (Nordenfelt 2004: 69)

Perhaps Nordenfelt's distinctions can illuminate the various appeals to dignity that seem confusing in application. What does it mean for the patient nearing death to possess dignity, to be "honored or esteemed?" For the sake of clarity, I make a distinction between the terms *dignity* and *undignified*. I note that some, in making a claim for dignity in dying seem to use the term as meaning the opposite of undignified and speak about wanting to avoid an embarrassing, undignified death. In Nordenfelt's terminology, this is closely tied to the dignity of identity. Emil J. Freireich, a noted oncologist at the Texas Medical Center's MD Anderson Cancer Research Center was a guest speaker for this author's medical ethics class at Baylor Medical School in 1973 on the topic of death with dignity. His opening statement was an emphatic exclamation, "There is nothing dignified about dying!"[5] He claimed to have been a witness to hundreds of deaths during his career and bluntly made the point that some patients, suffering a sudden heart attack or stroke, lose consciousness and fall flat on their faces; some dying patients lose control of bladder and bowels; some vomit; some lapse into a coma in the hours or days before death occurs. He insisted that we make a clear distinction between the commonly used term *undignified* and the more serious notion of dignity.[6]

Freireich's concept of intrinsic human dignity was based on the claim that there is worth and value in every human life even when the ravages of disease may cause one's appearance to become undignified. This notion seems to equate with Nordenfelt's concept of "universal" human dignity, which persists even in the face of an altered "dignity of identity." Freireich went on to explain that his high regard for the worth and dignity of every human being was the cornerstone of his worldview and the chief motivator that led him to devote his life to medicine and the care of the sick. For Freireich, the value of human life was not diminished by illness or the threat of death. He recognized that many adult patients often feel undignified or embarrassed when they can no longer care for themselves independently, when they need assistance with the tasks of everyday living such as eating, bathing, or toileting. He emphasized the fact that feelings of embarrassment can be overcome as patients adjust to the realities of physical impairments that make their former independence impossible. He echoes what many nurses have observed: that it is quite culturally acceptable for infants to be fed and assisted with bowel and bladder hygiene in the first few years of life. Likewise, it is culturally acceptable in assisted living facilities to assist residents with all aspects of daily living in the final stages of life. Formerly independent adults must go through a psychological transition when circumstances force them into a more dependent state that allows them to more gracefully accept the kinds of care that must be provided in their current condition. Thus, it seems imperative that we make a distinction in our public discourse between situations that are at first blush embarrassing or seem undignified for persons who were formerly independent (the dignity of identity) and the universal dignity that we ascribe to all humans and that is not lost due to illness or disfigurement.

Some critics of practices such as physician-assisted death or voluntary positive euthanasia frame their opposition in theological terms. Many whose worldview is shaped by a monotheistic religion believe that human life is a gift from God. As a gift, it is therefore not up to the individual to choose the time and manner of death. These critics claim that we are not the creators of our own lives but rather that the very processes that allow human life to spring forth are attributable to the ongoing creative work of God, in whose image we are created, the *imago Dei*. Thus, for such believers, the source of intrinsic human dignity stems from the fact that human life is a gift from God—and thus of inestimable value. The corresponding human duty, or moral obligation, is to treat the self and all other humans as equals and to show them the respect due to a fellow creature of God. As creatures of a creator God, a part of our stewardship is to protect and preserve not only our own dignity but the dignity of our fellow creatures as well. This is a radical claim in that it implies a duty to provide those essentials of life that are the foundation of human dignity, such as food, shelter, and health care. It goes much further than the claim of Mill, who argues for the primacy of individual liberty and the like liberty of others. Many religions claim that our duties to patients nearing the end of life are to provide comfort and care in the dying process but not to practice physician-assisted death nor euthanasia. Theological concepts may be cherished and practiced among the like-minded within a community of faith but may not be persuasive to nonbelievers. We need a language suitable to the public square.

In secular society, dignity is often associated with the exercise of *autonomy*. This word is derived from the Greek language, *auto*, meaning "self," and *nomos*, meaning "law"; thus, self-governing. Although dignity cannot be reduced simply to respect for autonomy, it seems to require such respect as an essential dimension of the concept. We have seen considerable development in the concept of the autonomous patient in the field of clinical medicine. Medical historians such as James Whorton at the University of Washington, claim this development has grown in significance since World War II. Prior to the early 1940s, medical historians claim that one of the hallmarks of health care was medical paternalism, the notion that the doctor knows best about what is good for the patient. It was generally assumed in that period that promoting the usual goals of medicine was equivalent to promoting the highest good for patients (Jonsen et al. 2006: 16). In the current climate of health care, it is generally held that the dignity of dying patients is best preserved by protecting their individual values and helping them to die in ways that are consistent with those values. Earlier in this chapter, I traced the development of patient autonomy in end-of-life considerations through examining the cases of Quinlan, Cruzan, and Glucksberg and the subsequent policy changes that have evolved in the United States.

Beyond the borders of this country, an even stronger step in underscoring patient autonomy is the practice of voluntary euthanasia as seen in the Netherlands and in Switzerland, where patients have the right to request the physician to take an action that will end the patient's life. In the Netherlands, involuntary euthanasia is also practiced in some cases where the family requests euthanasia for an incompetent elderly family member or in the case of a child born with multiple handicaps where it is thought a swift and merciful death through an act of the physician would be better than dying of natural causes while withholding nutrition and hydration (Gomez 1991: 137). The practice of involuntary euthanasia raises concerns among many that such a practice might lead to abuses such as targeting the elderly, the mentally ill, or other vulnerable groups within the society.

In Switzerland, an organization founded in 1998 known as Dignitas promotes voluntary euthanasia for the terminally ill and for incurable mentally ill persons who have decisional capacity, following the guidelines of the Court of Switzerland. The name of the organization seems to imply that dignity may be promoted by the right of individuals to active assistance in ending their lives at a time of their choosing. The causes of suffering from which they seek an escape may be physical, mental, or emotional. From this perspective, it appears that the notion of dignity has been conflated with that of autonomy—for example, if one has autonomy, one also has dignity; if one loses autonomy, one experiences a corresponding loss of dignity (Select Committee on Assisted Dying for the Terminally Ill 2005). Those who advocate for death with dignity and groups such as Dignitas seem to primarily emphasize the notion of dignity most closely associated with Nordenfelt's dignity of identity and to promote policies that protect the rights of individuals to end their lives so as to avoid continuing threats to their sense of themselves and their identities.

Many religious groups articulate a strong belief in the worth and dignity of human persons in an appeal that seems consistent with Nordenfelt's *Menschenwürde*, or universal dignity. Since health care decisions involve human values and human needs, such decisions are not merely technical, but are ethical in nature. Choices about the nature, extent, and duration of medical treatment make life better or worse for human individuals, so the practice of medicine always has ethical implications. Humans can use their intelligence and freedom to act and live creatively in fulfilling needs and pursuing values. Humans live in community, so it is not only important to secure the rights of individuals but to secure equal rights for disadvantaged individuals within that community. Social policies often create tension between the desires of individuals and the value of maintaining a safe and secure environment for society as a whole. Such reasoning provides a strong argument against policy shifts that would allow practices such as involuntary active euthanasia.

CONCLUSION

Dignity is a concept widely used in current bioethics discussions in the United States and frequently referenced in international bioethics discussions in agencies such as UNESCO, the European Commission, and the Council of Europe. Bioethics scholars differ widely in their positions about the usefulness of this concept. Ashcroft (2005) points to divisions of opinion that sort out to four groups among scholars:

> One group, along with Macklin, regards all "dignity-talk" as incoherent and at best unhelpful, at worst misleading. Another group finds dignity talk illuminating in some respects, but strictly reducible to autonomy as extended to cover some marginal cases (Deryck Beyleveld and Roger Brownsword). The third group considers dignity to be a concept in a family of concepts about capabilities, functionings and social interactions. This group is exemplified by the authors in a recent suite of articles in *The Lancet*, inspired by the writings of Amartya Sen and Martha Nussbaum on development and freedom. The final group considers dig-

nity as a metaphysical property possessed by all and only human beings, and which serves as a foundation for moral philosophy and human rights (Leon Kass and much of European bioethics and theological writings). (Ashcroft 2005: 679)

In addressing the ethical issues that arise in end-of-life care, I am particularly interested in whether dignity can provide practical guidance in sorting out the alternative courses of action that should be available to dying patients. Can dignity alone provide guidance, not only for individual choices, but for shaping social policies as well? I have suggested that *respect for persons* has played a powerful role in the development of current public policies surrounding the care of patients at the end of life, offering patients a great deal more freedom of choice than existed four decades earlier. Thus, an implication of the principle of respect for persons is that of preserving a frame in which a variety of choices are available to persons nearing the end of life. Such an approach shows respect for those valuing individual autonomy and honors as well the choices of those with a communitarian view of dignity that is centered on membership in the extended family and allows diversity in opinions about human flourishing. Dignity must be enriched by additional concepts if it is to shape policy. Particular religious communities may have a high level of agreement about a particular vision of human flourishing and the implications of dignity in supporting such; however, outside of this context, conflict is inevitable in a pluralistic culture. As shown in this chapter, the concept of dignity is used both in support of and in opposition to such policies as the Death with Dignity Act.

Nordenfelt makes a very useful contribution by showing four distinctive types of dignity, and it appears that two types in particular relate to the dignity of dying persons: the notion of *Menschenwürde*, or universal dignity, and the dignity of a unique identity. Concepts of dignity have originated from both secular and religious sources. Those from secular sources often associate the notion of dignity more closely with respect for liberty and individual autonomy. Those from Judaism, Islam, and Christianity seem to embrace the broader term of respect for persons as having intrinsic value and also associate dignity with the concept of the stewardship of life as authored by a Creator-God.

I suggest that in place of arguments from dignity alone, a common ground can be found by fostering the notion of *respect for persons* as a normative obligation in the moral community as a way of protecting the dignity (*Menschenwürde*) of all human beings. Gustafson (1974) reminds us of the moral ambiguity of human experience and the fact that moral considerations begin with the unique human person. I hold that there are a variety of ethical options in providing care for patients nearing the end of life. Such options should allow for individuality while protecting the society as a whole. Good medical care in the dying process treats suffering as defined by the individual patient and seeks to bring comfort in ways that are commensurate with the patient's values. For example, even in those extreme situations beyond usual palliation, where patients suffer from severe escalating pain, medicine should provide comfort. Respect for the patient's dignity might involve sedation at such levels that the patient may not awaken during the last hours or days of life but will die peacefully while sedated. This form of pain management, when chosen by the patient, has been carefully considered by ethicists and is justified under the principle of double-effect, whereby one aims at the mitigation of pain while acknowledging that life may be foreshortened. The alternative of allowing grievous suffering to continue is not tolerable. This approach has been affirmed not only by ethicists but indirectly by the U.S. Supreme Court. In the case of *Washington v. Glucksberg* (1997), the court instructed states to examine state laws so that they do not impede palliative care and do not obstruct the alleviation of pain. Justice Sandra O'Connor claimed in this case, " a patient who is suffering from a terminal illness and who is experiencing great pain has no legal barriers to obtaining medication from qualified physicians to alleviate that suffering, even to the point of causing unconsciousness and hastening death" (*Washington v. Glucksberg* 1997). Clearly, the highest court upholds the concept of respect for persons as a high value by supporting the appropriate treatment of pain, even if it foreshortens the life of the patient.

Great strides have been made in providing a more comfortable and humane death for many. I argue that this has evolved from the principle of respect for persons. Some dying patients wish to avail themselves of nearly every technology that will prolong their life in the face of death. The natural drive for survival is strong. Some prefer palliative care near the end of life when no cure is possible. Hospice has increasingly been utilized to assist dying patients by promoting the best quality of life possible under the circumstances of their illness. Recognizing that too many patients were dying painful deaths, Oregon now has the highest utilization rate per capita of morphine. Thus far, in both Oregon and Washington, following the passage of Death with Dignity Acts, physicians have paid greater attention to the needs of dying patients. There has been a relatively small proportion of terminally ill patients seeking a lethal prescription. Rather, follow-up studies seem to indicate that a minority of patients with an exceptionally high desire for control and personal autonomy are more likely to choose to act within the framework of the DWDA. Following the notion of theologian James Gustafson, who commented on the "moral ambiguity of the human experience," I believe that, as a society, we should allow various notions of dying with dignity to coexist (Gustafson 1974). By doing so, we are on the way toward creating an environment that shows respect for persons, recognizes the uniqueness of individuals who are facing death, and allows for a variety of choices among these moral alternatives in support of human dignity at the end of life. Human dignity is best supported by an overarching commitment to revere the life of human individuals and by providing medical care in support of a high quality of life until the end. Human dignity and respect for persons is also supported by social policies that recognize the pluralism of our society and allow for autonomy in patient choices at the end of life, following the caveat of John Stuart Mill, that our liberty is restricted where its expression threatens to abridge the liberty of others (Mill 1869). Respect for persons is at work in the framing of the DWDA in that patients have a legal right to request the lethal medication, yet physicians who disagree may opt out as prescribers of this medication on the basis of conscience. The rights of both parties are respected. Further, hospitals and other health care institutions may opt in, or opt out, thus allowing the rights of all to be protected.

In the Netherlands, the practice of voluntary euthanasia appears to have led to instances of involuntary euthanasia (Kass 2002: 208). I argue that any practice of nonvoluntary active killing is contrary to the goals of medicine and opens the door to the use of medical technology to end the life of nonconsenting individuals. Such a practice, or a social policy allowing such a practice, is contrary to our understandings of respect for persons. On the whole, while our society has developed and put into place a variety of effective and ethical ways to support the dignity and individuality of persons facing the end of life, controversy remains regarding the limits that might best protect those most vulnerable within our society as we face the future together. I recognize that people will continue to use the term *dignity* in support of their values about care at the end of life yet argue that the principle of respect for persons is a more useful concept. Agich (2007) points out the lack of specific content in many conceptions of dignity but sees a broad utility in dignity as a "concept expressive of shared social understanding of the status of . . . people." I agree, and suggest that even though a theoretical appeal to dignity alone is inadequate to resolve differing opinions about clinical choices at the bedside or policy disputes about the range of options that should be allowed, we may nonetheless intuitively agree and tacitly accept that dying patients possess (universal) dignity and deserve respectful care at the end of life. One aspect of respectful care is an openness to support patient choices about the nature of the care they desire as life comes to an end.

NOTES

1. Personal conversation with a family member of a patient who utilized Washington's Death with Dignity Act, 2009.
2. Personal conversation with a family member of a patient who utilized Oregon's Death with Dignity Act, 2008.
3. Ibid.
4. Personal conversation with hospice patient.
5. E. J. Freireich, MD, Personal communication, MD Anderson Cancer Research Center, Houston, TX, 1973.
6. Ibid.

14 The Evolving Bioethical Landscape of Human–Animal Chimeras

John Loike

The history of chimeras goes back to ancient Greek mythology. Originally, a chimera referred to a mythical animal with the head of a lion, the body of a goat, and the tail of a serpent. In the current world of scientific research, chimeras are organisms composed of cells or genes obtained from two or more different species. Human–animal chimeras can be generated by either transplanting human stem cells into animal fetuses or human genes into the genome of animal fetuses. The first modern interspecies chimeras were engineered in the 1980s with little ethical debate. However, in the last decade, new and challenging bioethical dilemmas regarding human–animal chimeras have emerged, creating considerable debate on the appropriateness of such research.

In this chapter, I first characterize human–animal chimeras and outline their use in scientific and medical research (Huther 2009). I then focus on two bioethical challenges emerging from human–animal chimeric research that attempt to enhance animal cognitive and reproductive capacities with humanlike properties. Ethical concerns, such as respecting species identity and respect for human dignity, are presented from both a secular and religious vantage point. Finally, I propose that the principles of respect for species identity and human dignity raise a broader bioethical challenge—whether there are limits to the type of research that scientists should pursue in generating human–animal chimeras.

HUMAN–ANIMAL CHIMERAS

To begin, it is important to distinguish between a chimera and a hybrid. Hybrid animals are usually created when an ovum from one species is fertilized by the sperm of another species. Thus, every cell in a hybrid organism contains the same genetic information that was derived from two different parental species. Examples of hybrid organisms include the mule (a genetic cross between a male donkey, *Equus asinus*, and a female horse, *Equus caballus*) and the more exotic liger (a cross between a male lion, *Panthera leo*, and a female tiger, *Panthera tigris*). Most hybrids are sterile and usually exhibit characteristics not seen in either parent species.

Chimeras, in contrast, are generated by introducing stem cells from one species into the fetus of another species. Thus, chimeras contain two distinct populations of cells. One cell type contains a complete genome of one species, and the second cell type contains a complete genome of the other species. Although there are a wide variety of chimeras,[1] this chapter focuses on chimeras that are generated by transferring cellular or genetic material obtained from a human being into an animal fetus.[2]

The scientific utility of human–animal chimeras research is well documented (Huther 2009). Human–animal chimeras have been used in medical centers around the world for several decades, leading to significant medical applications (Nagy and Rossant 2001). In the 1980s, for example, Irving Weissman and his research team created one of the first human-mouse chimeras (McCune et al. 1991). They transplanted human bone marrow, which contains hematopoietic stem cells, into a strain of mice that lacked its own immune system and triggered the formation of a nearly complete human immune system in these mice. These mice have served as valuable animal models for studying a variety of human diseases, such as AIDS. Because the HIV virus that causes AIDS does not normally infect mouse cells, these human-mouse chimeras enable scientists to learn how the virus infects and replicates in human immune cells within a live animal. These chimeras also provide an animal model of AIDS to test potential antiviral drugs. Further, human-mouse chimeras generated to contain human liver cells serve as another valuable animal model to study the biology of the virus that causes hepatitis (Turrini et al. 2006; Chayama et al. 2011), liver development, hematological cancers (Shizuru et al. 2005), and drug metabolism (Meuleman et al. 2005).

Chimeras may eventually be used to reproduce organs of the human body. Researchers at the University of Nevada are attempting to reconstitute human organs in sheep by transplanting human stem cells into specific organ areas in sheep embryos (Almeida-Porada et al. 2007; Chamberlain et al. 2007). Their research is based on observations that foreign stem cells are not immunologically rejected by the sheep fetus during embryological development and that transplanted human stem cells can emerge as the dominant cell type of the sheep's organ (Almeida-Porada et al. 2010; Ersek et al. 2010). As of 2010, they have engineered human-sheep chimeras with livers composed primarily of human cells that are genetically compatible with the human stem cell donor. A human liver developed in such a human-sheep chimera can then be transplanted into an individual needing a liver transplant. Potentially, this transplant technology could increase the number of available transplantable organs and reduce the need for antirejection drugs normally administered to most organ transplant recipients. Human-sheep chimeras are also being examined as potential sources for other cells, such as the human hematopoietic cells (Chamberlain et al. 2007) and human beta cells of the pancreas (Balaban 1997; Almeida-Porada et al. 2010; Ersek et al. 2010).

The potential medical risks associated with chimeras must be addressed before this research can enter clinical trials. Chimeras risk the transmission of unknown viruses, oncogenes, or diseases between species (Hug 2009). In addition, stem cell lines derived from human–animal hybrid embryos created by somatic cell nuclear transfer could transmit animal mitochondrial diseases to humans if used therapeutically in people (Hug 2009).

Until recently, transplanting human stem cells into animal fetuses to generate human–animal chimeras has rarely elicited an ethical outcry (DeGrazia 2007). Yet, now there are at least two areas of research associated with human–animal chimeras that raise bioethical concerns (Bobrow 2011). The first area is neural chimeric technology, which involves reconstituting parts of an animal's brain with human neurons that may enhance cognitive capacities (White 2007). Included in this research area is the transplantation of human genes, such as those that regulate human speech, into animal fetuses (Reiner et al. 2007). The second area is reproductive chimeric research, in which transplanted human precursor stem cells are used to generate animals that produce functional human sperm or eggs (Narayan et al. 2006; Leake and Templeton 2008). It is important to note that, as of 2011, there are no reports showing that human sperm or eggs generated by these transgenic human-mice chimeras can be used successfully in human fertilization protocols. Included in this second research area is the use of human stem cells to generate a cow or sheep with a human uterus (Lupton 1997; Rosen 2003; Alghrani 2009). Because the normal gestational period of a calf is nine months, a human–cow chimera containing a human uterus could theoretically serve as a bioincubator for human embryological development.

While the above-mentioned areas of chimeric research raise ethical concerns, there are sound scientific rationales for these research scenarios. For example, human–animal neural chimeras containing human neurons could provide novel model systems for identifying various human cell types that are necessary for human embryonic brain development. Human–animal neural chimeras could also be used to study human neurological diseases for which no appropriate animal models exist. For example, one could transplant human stem cells from an individual with a genetic predisposition to develop Alzheimer's disease or Huntington's disease into a murine fetus. These human-mice chimeras would offer the advantage of having only a two- to three-year life span, allowing scientists to follow neural degeneration, characteristic of these diseases, from birth until old age in a relatively short time frame. Similarly, human–animal reproductive chimeras that produce human gametes may be valuable animal models to understand sperm and oocyte embryological development and to enhance our understanding and treatment of infertility. Japanese scientists have used laboratory-made sperm, using embryonic stem cells, to restore fertility in sterile mice (Sato et al. 2011). This may also open up new avenues to reprogram human cells from the skin to act like sperm and to treat infertility in people. Human–animal reproductive chimeras containing human uteri may provide a useful model to develop human uterine transplant technologies or may serve as an artificial incubator for premature infants. Finally, transplanting human genes, such as those that regulate speech or other behaviors, would help scientists understand how these genes function and the effects of behavioral gene mutations in an animal model.

As stated above, transplanting human genes into animal fetuses is a viable approach to study how human behavioral genes function. There have been many recent articles, for example, on identifying and characterizing the Forkhead-box protein P2 (FOXP2) gene and its protein product in human language development (Hug 2009; Rees 2011). This gene was mapped from genetic studies of a family with language dysfunction and found to encode a transcription factor. This is the first gene, as yet, implicated in Mendelian forms of human speech and language dysfunction (Fisher and Scharff 2009). From an evolutionary perspective, the human form of this transcription factor differs in only two amino acids from FOXP2 transcription factors found in nonhuman primates (Enard et al. 2002). In addition, this amino-acid change appears to occur around the time that language emerges in human evolution. Other studies in both humans and animals reveal that this gene is present in both vocal and nonvocal animals and is important for muscle coordination in vocalization and in many other motor functions. For example, mutations in the human form of FOXP2 gene cause region-specific cellular changes in the developing human cortex and striatum resulting in functional disturbances in language comprehension, grammar, and syntax (Newbury et al. 2010; Watkins 2011). Studies using human-mouse chimeras where the human form of FOXP2 was transplanted and expressed in mice reveal changes in vocal output and cellular changes in the neural circuitry in their brains (Enard et al. 2009; Kurt et al. 2009). Thus, this gene not only affects motor functions controlling voice but neural circuitry as well.

BIOETHICAL CHALLENGES EMERGING FROM HUMAN–ANIMAL CHIMERA TECHNOLOGY

Respecting Species Identity

Several bioethicists (Loike and Tendler 2002, 2003, 2008; Baylis and Robert 2003; Elmer and Meyer 2010) argue that generating chimeras that display human characteristics or human reproductive capacities will blur the boundaries of species. Defining species boundaries is not unique to chimeras. This issue of species identity arose during our long history of using breeding or genetic technology to develop new species or hybrid species (Goble et al. 2006) of plants and animals. There is, however, no consensus on how to define species identity. In many basic biology texts, a species is defined in reproductive terms as a group of organisms capable of interbreeding and producing fertile offspring (Elmer and Meyer 2010). Aside from the reproductive definition of species, there are more than 20 other definitions or characteristics of species (Mayden 1997, 1999). For example, some biologists claim that unique genome sequences or genomic similarity represent the identity of that species or its blueprint (Robert 2006; Loike and Tendler 2008; Elmer and Meyer 2010). One problem, however, in using a genomic approach to define species is that the human male genome is genetically more similar to the male chimpanzee than to the human female because the human female lacks a Y chromosome (Marks 2002). Thus, DNA sequence similarities may not be a valid approach to defining a species. Other scientists (see Orr et al. 2004) are attempting to identify unique genes that are thought to underlie the origin of a specific species and differentiate one species from another. When it comes to humans, other scholars prefer to avoid a genomic approach and include criteria such as similarity of morphology, behavioral characteristics, self-awareness, complex language skills, or ecological niche in their definition of *Homo sapiens* (Loike and Tendler 2002, 2008; Baylis and Robert 2003; Elmer and Meyer 2010). Baylis and Robert (2003) conclude that, while species identity is difficult to ascertain biologically, notwithstanding, there may be moral or bioethical concerns attendant on efforts to address speciation.

> There is, however, no one authoritative definition of species. Biologists typically make do with a plurality of species concepts, invoking one or the other depending on the particular explanatory or investigative context. . . . This fact of biology, however, in no way undermines the reality that fixed species exist independently as moral constructs. That is, notwithstanding the claim that biologically species are fluid, people believe that species identities and boundaries are indeed fixed and in fact make everyday moral decisions on the basis of this belief. (Bayliss and Robert 2003: 6)

While establishing a biological definition of species with respect to human–animal chimeras may be difficult, the moral consequences of speciation have profound scientific, social, political, and legal ramifications. How society differentiates plant, animal, and human species influences how human beings treat other creatures, what we eat, what we patent, and what kinds of organisms are legally required to follow societies' moral and civil laws. From a scientific perspective, Coors et al. propose that the same evolutionary proximity that makes humanizing ape research strategies attractive renders it ethically unacceptable, because apes have the greatest potential to produce humanlike phenotypes and carry a unique potential for harm (Coors et al. 2010).

From a political and social vantage, enhancing human characteristics using chimeric technology can confer specific social rights. This belief is reflected in the legal systems of several European countries that ban the use of nonhuman primates for animal experimentation (Murphy 2010; Vatican 2011). These laws are based on the belief that nonhuman primates exhibit specific humanlike characteristics and should be granted special humanlike rights (DeGrazia 2007). In addition, nonhuman primates may not be critical animal models for studying diseases since a great deal of neuroscience knowledge can be learned from rodents. However, further development of chimera technologies may create the need for new laws: for example, should neural human-mouse chimeras that display humanlike characteristics be treated much as nonhuman primates and possibly be exempted from certain types of invasive research?

An anthropological ramification of maintaining species boundaries emerges from the Neanderthal Genome Project. In 2005, research was initiated to sequence the Neanderthal genome from fossilized Neanderthal bones that are tens of thousands of years old (Noonan et al. 2006; Noonan 2010). Various new technologies, such as metagenomic approaches for characterizing complex DNA mixtures, are being used to sequence Neanderthal DNA. Using this method, it will be possible to recover significant amounts of Neanderthal genomic DNA with high-throughput sequencing. This will allow scientists to characterize comprehensively all sequences present in a sample, and the origins of each sequence, by comparison to known genomes. Moreover, multiplex automated genome engineering (MAGE) is being improved to package the DNA into chromosomes and make the millions of changes necessary to transform a human genome into a Neanderthal genome. Scientists developing MAGE estimate that about 10 million changes are needed to make a modern human genome match the Neanderthal genome (Zorich 2010). Once this task is accomplished, the moral issue is whether scientists should differentiate genetically generated embryonic stem cells containing the Neanderthal genome into gametelike cells to produce a Neanderthal zygote for implantation into a human or nonhuman surrogate. Even though there may be anthropological benefits to studying Neanderthal humans, attempts to generate a Neanderthal-like organism raise bioethical concerns about whether such research activity crosses the boundaries of species integrity.

While there is no scientific consensus on how to establish species identity, there is an intrinsic belief that species demarcation is an important biological dogma. Historically, human beings always have been attuned to the prevailing philosophy that they are a unique species and that this uniqueness, in part, gives our species certain governing rights over all organisms. Creating human–animal neural chimeras with enhanced humanlike cognitive or language capabilities may blur how we identify a species and could challenge these philosophical concepts.

Respecting Human Dignity

The other primary ethical concern related to human–animal chimeras is the principle of respecting human dignity. Respecting human dignity has been invoked in contemporary bioethics regarding human genetic enhancement as well as generating human-nonhuman chimeras (de Melo-Martin 2008; Loike and Tendler 2008). Before addressing the ethical relevance of safeguarding human dignity in human–animal chimeras, two controversial parameters must be delineated. The first is how to define human dignity, a concept that has been notoriously elusive. The second is if, and when, the principle of respecting human dignity should play a role in scientific research.

Human dignity can be viewed either within a secular or religious perspective. Immanuel Kant proposed a secular definition that human dignity is associated with the capacity to think for oneself and direct one's actions. Using a Kantian moral framework of human dignity, human beings possess an unconditional and incomparable worth that is independent of metaphysical or religious precepts (Paton 1971; Macklin 2003; Karpowicz et al. 2005). According to Kant, human beings have dignity because of their reasoning faculties, which give them the freedom and ability to distinguish moral from immoral actions. Using this Kantian definition, however, some scholars have argued that not all human beings have dignity. Patients in a permanent vegetative state, for example, have irreversibly lost their autonomy and may no longer have dignity (Schroeder 2010).

In contrast to this secular definition of human dignity, a Judaic-Christian–based definition formulates human dignity as an inviolable right invested by God in all human beings, including fetuses, comatose patients, and patients in a permanent vegetative state (Kass 2004; Loike and Tendler 2011). In its simplest religious formulation, human dignity can be equated with the sanctity or infinite worth of human life and assumes that there is something uniquely valuable about human life. From a religious view, human dignity emanates from the first chapter of Genesis, which records how human beings were uniquely fashioned and divinely created (Soloveitchik 1993). Several biblical scholars comment that the Bible describes how God created human beings using two different processes (Soloveitchik 1983). The first process was biological/genetic as indicated by the fact that human beings were created on the same day as other animals. The second process was metaphysical as God infused into human beings a spiritual entity that differentiates human beings from all other creatures. This metaphysical, and almost divine, quality of human beings confers a sanctity that exists within each human being from the beginning of life as an embryo until natural death (Soloveitchik 1993).

From a Judaic perspective, all creatures deserve a certain degree of respect that prohibits the unnecessary destruction of all organisms and animal suffering (Loike and Tendler 2011). There is, however, a God-given human right to govern plants and animals, enabling human beings to use nonhuman life forms for food, clothes, and even for scientific experimentation (Bekoff 2007; Loike and Tendler 2011).

Thus, respecting human dignity, according to Judaism and Christianity, is in part based on divinely created characteristics, such as language and cognitive capacity to engage in rational thought. One outcome of these characteristics is reflected in their moral virtues. Moral virtues that people often consider as being good are courage, compassion, and altruism. Without such virtues and human cooperation of its members, a society cannot survive.

A second outcome of respecting human dignity, discussed in primarily Judaic literature and law is an intrinsic moral right to be born from a woman and, whenever possible, to be raised by two parents. The Talmud in Tractate Niddah (p. 31b) states that there should be three partners in creating and bringing up a human being: man, woman, and God. This means that all human embryos have the dignified and moral right to be gestated within a woman. If, in the future, artificial incubators were developed that could gestate a human embryo from fertilization until term, under Jewish law, they would not be permissible unless there were an immediate medical crisis.

If one accepts the principle and outcomes of human dignity, then it is appropriate to examine the role human dignity may play in bioethics. On the one hand, bioethicists such as Ruth Macklin argue against applying the principle of respecting human dignity to bioethics; they do so by using arguments that weaken the validity of this principle.

> [Human] dignity is a useless concept. . . . It means no more than respect for persons or their autonomy. . . . A close inspection of leading examples shows that appeals to dignity are either vague restatements of other, more precise, notions or mere slogans that add nothing to an understanding of the topic. (Macklin 2003: 1419)

Other scholars and bioethicists argue from a secular (Fukuyama 2003; Kass 2004) and religious perspective (Loike and Tendler 2011) against Macklin and advocate the paramount importance of applying the principle of respecting human dignity in bioethical matters. Mahmudur Rahman Bhuiyan summarizes the views of bioethicist Francis Fukuyama, who uniquely blends a secular approach of species identity with human dignity:

> Fukuyama fears that human nature is the most valuable thing that may be affected by recent advances in human biotechnology. He defines human nature as "species-typical traits" of human beings (such as language and cognition, which provide the grounds for feelings such as pride, anger, shame, and sympathy), arising from genetic factors. According to Fukuyama, these species-specific traits of humans differentiate us from all other nonhuman species, and this differentiation constitutes the basis of human dignity. The reduction of shared traits among humans will result, according to Fukuyama, in the degradation of human dignity . . . thus makes human nature uncertain. (Bhuiyan 2009: 831)

APPLYING THE PRINCIPLES OF RESPECTING SPECIATION AND HUMAN DIGNITY TO HUMAN–ANIMAL NEURAL CHIMERAS

Scientists generally are not interested in transferring behavioral characteristics from humans to animals or from animals to humans. Nonetheless, there is a fear that generating human–animal neural chimeras may inadvertently trigger the transfer of human behavioral characteristics (such as intelligence, consciousness, or speech) into an animal. This notion has some scientific foundation. Evan Balaban (Balaban et al. 1988; Balaban 1997, 2005) reported that transplanting small sections of brain from developing quail into developing brains of chickens resulted in chicken chimeras that exhibited vocal trills and head bobs unique to quail. Balaban clarified the importance of his research:

> This is the first experimental demonstration that species differences in a complex behavior are built up from separate changes to distinct cell groups in different parts of the brain and that these cell groups have independent effects on individual behavioral components. (Balaban 1997: 2001)

Importantly, these data suggest that neurological studies involving cell transplantation from one species to another can result in the transmission of behavioral phenotypes found in the donor organism and not in the host.

The fear of engineering human–animal chimeras that express human behavioral characteristics, especially with regard to creating humanlike rational cognitive thought in animals, has triggered debates in the public and academic press (DeGrazia 2007; Loike and Tendler 2008). One assumes that scientists using this technology could create a human–animal chimera with a significantly higher cognitive capacity than that of the original animal species but not possessing the same rational intellect or reasoning that is normally present in human beings. Enhanced cognitive capacity is defined here to include the ability to act intentionally, engage in complex communication and speech, and act for moral reasons. Given this assumption and definition, it is important to discuss the bioethical concerns of speciation and human dignity with respect to human–animal neural chimeras.

Speciation

If one accepts the reproductive definition of a species, then clearly a human–animal chimera with reconstituted human neural circuitry and enhanced humanlike cognitive capacities is not human, because it lacks human gametes and therefore cannot successfully fertilize with human sperm or eggs. Since a human-mouse neural chimera will only possess sperm or eggs that encode the genome of the mouse, only normal mice will be its progeny. Using a reproductive or genetic criterion for speciation, generating human–animal chimeras with human cognitive capacities would *not* violate any bioethical issues related to this definition of speciation.

If human speciation is defined using cognitive parameters, then one might propose that a human–animal neural chimera with enhanced cognitive capacity should be viewed as a humanlike species with the same rights as cognitively comparable human beings. At the very least, such neural chimeras may deserve enhanced rights, such as those given by European countries to nonhuman primates. Yet this issue is further complicated in considering how human beings view themselves as a unique and superior species (Atran 1999; de Melo-Martin 2008; Eberl and Ballard 2009). What are the minimal enhanced cognitive functions that differentiate animals from humans? Some ethicists (see Kure 2009) argue that a human–animal chimera with a partially reconstituted human brain will appear animal-like and should be no different than an animal that contains a transplanted human heart valve. But others would view such neural chimeras as humanlike (Kirksey and Helmreich 2010).

In conclusion, whatever criteria are used to define a species, scholars must provide clear criteria in deciding whether to classify neural human–animal chimeras as human, animal, or as a new species. From an ethical perspective, those scholars who believe that species demarcations are not determined by genetic or reproductive criteria would not object to generating human–animal neural chimeras. In contrast, scholars who adopt a behavioral perspective on speciation might argue against engineering any human–animal neural chimera unless, perhaps, there is strong scientific or medical justification for such research.

Human Dignity

The other bioethical concern related to human–animal neural chimeras that requires further clarification is the issue of respecting human dignity. The relevance of respecting human dignity in scientific experimentation depends on whether scholars adopt a religious or secular definition. According to a religious definition, creating a human–animal chimera with humanlike cognitive capacities may violate respecting human dignity for at least two reasons. First, transferring human stem cells or genetic material into an animal fetus to produce a human–animal chimera with human behavioral characteristics or human language skills will be viewed as an attempt to create a humanlike being in an animal shell. If this neural human–animal chimera is viewed as humanlike, then it differs from other human beings in not being gestated, born from a woman, or raised by human parents. Creating such a humanlike organism would deprive it of the right of being born from a woman or being raised by other human beings and would be viewed religiously as disrespecting human dignity. Second, a human–animal neural chimera that displays humanlike behavioral characteristics may also exhibit some enhanced autonomous capacity. Assuming we have a scientific method to assess autonomous capacity in these chimeras and that they may somehow communicate their autonomy, it would seem unethical to use such chimeras for scientific experimentation without their consent. Even if their autonomous capacity cannot be assessed scientifically, there is still a possibility that these chimeras may possess enhanced self-awareness and understanding rending these concerns as sufficient grounds to restrict their use in animal experimentation.

It is important to further characterize what the violation of human dignity really means. Leon Kass uses the term *human degradation* as a violation of human dignity as he describes his opposition to creating a humanlike organism within an animal shell (Kass 2004). Moreover, a human–animal neural chimera with enhanced cognitive capacities might recognize itself as a physical anomaly with restricted social acceptability with other human beings or with its species of origin. Will such a chimera express a kind of humanlike consciousness or soul that can be tormented? There are many fictional examples in literature of humanlike creatures, such as the Golem of Jewish legend and Frankenstein's monster, who feel tormented because of their monstrous physical nature, belonging nowhere, and being socially cut off from "normal human beings." Jewish law is quite sensitive to respecting human dignity, instituting moral guidelines that avoid actions leading to human degradation and embarrassment, and promoting activities to ensure that all individuals be socially accepted in the community (Loike and Tendler 2011). Even though there may be sound scientific reasons to create human–animal neural chimeras, the potential to violate human dignity would prohibit such research activity according to Jewish law (Loike and Tendler 2011).

Would scientists/bioethicists who accept a secular definition of human dignity sanction the creation of a human–animal chimera with a partially reconstituted human brain? The answer is probably no since the issue of respecting human dignity was also in the ethical debates regarding human cloning:

> To be a legislative member in the [human] kingdom of ends is to possess autonomy, i.e. to be able to give oneself one's own moral law that is universal and unconditional. To be treated as a means or commodity is to be denied one's own autonomy; and because autonomy gives support to human dignity, it is also to undermine human dignity. (Shuster 2003: 523)

In any scenario that might possibly emerge in the foreseeable future, human–animal neural chimeras would most likely be quarantined and treated as laboratory animals for scientific experimentation. Under these circumstances, generating such human–animal chimeras with enhanced cognitive and autonomous capacity would violate even a secular interpretation of human dignity.

DeGrazia (2007) presents a different secular reason against creating human–animal chimeras with enhanced cognitive function. Such experimentation should be prohibited but not because of the issue of human dignity. First, he states,

> (1) such experiments involving Great Apes should be prohibited out of respect for the research subjects and (2) such experiments involving

rodents may or may not be morally permissible, depending on answers to unresolved questions regarding rodents' moral status. In neither case do concerns about human dignity prove significant? (DeGrazia 2007: 311)

Then, DeGrazia combines moral and species identity issues to argue against creating human–animal chimeras.

> As borderline persons, Great Apes have full moral status. . . . Possessing full moral status, Great Apes should not be used in research unless (1) their participation is realistically expected to pose no more than minimal risk to them or (2) greater risks are justified by the prospect of direct veterinary benefit to them and the absence of alternatives offering a better benefit/risk ratio. . . . Meanwhile, the purposes of such studies would primarily be human centered: (1) to advance biology by better understanding how ESCs and neural stem cells work and (2) to advance the gradual development of treatments for such diseases as Alzheimer's and Parkinson's. In other words, the ape subjects' interests would be subordinated to social utility. (DeGrazia 2007: 324–325)

Clearly, DeGrazia believes that research to modify the neural capacity of animals does not come close to satisfying those ethical standards and should not be allowed.

Irrespective of a religious or secular view of human dignity, many other ethical and legal questions remain to be explored and addressed. For example, who will serve as the parental proxy of these human–animal chimeras? Is there an ethical conflict of interest of the research scientist who engineers these human–animal neural chimeras and decides their fate as research animals? Do these scientists have an ethical right to use human–animal reproductive chimeras as a source of human sperm or eggs?

These ethical concerns are in some ways similar to one of the most controversial uses of preimplantation genetic diagnosis: the creation of so-called savior siblings. Savior embryos are produced to serve as a source of stem cells to be used in potentially life-saving therapy for an existing child. To put this hypothetical ethical issue into more realistic perspective, society today would find it troubling when parents conceive a child primarily to be a potential future donor for an extant loved child with a serious illness. One of the primary unanswered concerns is whether a savior child will receive the same sort of love and parental protection as other children (Murphy 2010). Similarly, if one assumes that human–animal neural chimeras should be treated as humans rather than experimental subjects, will they receive the same protection or care that other human subjects receive?

From both religious and secular perspectives, there are serious ethical concerns related to human dignity in generating neural human–animal neural chimeras. Moreover, issues such as the autonomous rights of such chimeras need to be addressed and debated. Since there are no clear ethical solutions to these issues, a cautionary approach is what has been instituted by the U.S. government in discouraging or even prohibiting research involving neural human–animal chimeras (National Academy of Sciences 2005). In addition, other countries, such as Great Britain, are following the lead of the United States in establishing guidelines that discourage such research (Academy of Medical Sciences 2011).

While the issue of respecting human dignity will continue to be debated in academic circles, there are practical political and scientific ramifications. The guidelines proposed by the National Academy of Sciences, for example, presented a viable platform for research institutions and corporations to evaluate the ethics and directions that human–animal neural chimera research should take (National Academy of Sciences 2005). While the National Academy of Sciences never defined human dignity (either religiously or secularly), it used human dignity as a criterion to influence its recommendations. Specifically, the National Academy of Sciences' recommendation is that research should be limited to in vitro experiments without actually generating a human–animal chimera possessing enhanced cognitive function (National Academy of Sciences 2005). On July 22, 2011, the Academy of Medical Sciences in London published a report that recommended similar ethical guidelines on human–animal chimeric research (Academy of Medical Sciences 2011). The Academy of Medical Sciences also included as criteria for its recommendations respect for human dignity, as defined from a Kantian perspective, and the concerns of crossing the boundaries of speciation. Thus, international scientists are using the principle of respecting human dignity as an important bioethical consideration for drafting their recommendations with regard to human–animal chimera research.

APPLYING THE PRINCIPLES OF RESPECTING HUMAN DIGNITY AND SPECIATION TO HUMAN–ANIMAL REPRODUCTIVE CHIMERAS

Apart from the case discussed above in partially reconstituting a human brain in animals, several other scenarios exist in which the issue of respecting human dignity and speciation should be considered. One futuristic case involves the application of embryonic stem cell technology to genetically engineer a cow that contains a human uterus (Loike and Tendler 2008). Such a human–animal chimera may be able to serve as a bioincubator for human fetal development and may be useful for women who do not desire to carry their own child or who choose not to use a human surrogate for delivering their genetic child. Another case involves using technology to create a human–animal chimera that produces human sperm or eggs. These animals might provide a replenishable source of human gametes that, when applied to in vitro fertilization (IVF) protocols, will lead to a decreased incidence of genetic mutations and birth defects. As older men or women have children, there is an increased rate of birth defects commonly observed in their children (Gourbin 2005; Tartaglia et al. 2010). In these reproductive-medical scenarios, there will be no enhancement of cognitive capacity in the human–animal chimera created. Rather, these chimeras will only contain human organs and may be viewed, bioethically, as being no different than creating a sheep with a human liver, pancreas, or hematopoietic system.

For the secular bioethicist or scientist who supports a Kantian position of human dignity, these human–animal reproductive chimeras do not exhibit any change in cognitive status and therefore would not be in violation of human dignity. From a Jewish or Catholic perspective, however, creating such human–animal chimeras may infringe upon the principle of respecting human dignity. As discussed above, in both Judaism and Catholicism, important elements in human dignity are (1) the right for all human babies to be born from a woman rather than from an animal and (2) to engage in parental relationships.

Filial and parental relationships are guiding principles in the Catholic Church's positions regarding IVF (Weissmann 2010; Vatican 2011; Shea 2003). According to these religious scholars, IVF itself violates human dignity of children because it deprives these children of their filial relationships with their parents and hinders the maturing of their personalities (Shea 2003). This position on IVF was clearly presented when the International Federation of Catholic Medical Associations declared its disagreement with Robert Edwards being awarded the Nobel Prize for Medicine for his work in developing in vitro fertilization (Weissmann 2010; Vatican 2011). The problems of infertility, the group said in an official statement, must be solved within an ethical framework that respects the dignity of the embryo as a human being. Noting the "enormous cost"—that of undermining human dignity—with which IVF has "brought happiness" to couples who have conceived through this method, the International Federation of Catholic Physicians (FIAMC) decried the use of millions of embryos, thus human beings, created and discarded "as experimental animals destined for destruction." This use of human embryos "has led to a culture where they are regarded as commodities rather than the precious individuals which they are" (Benagiano et al. 2011:)

Similarly, in Judaism, sperm or egg donation is not merely viewed as a one-time event without ensuing moral responsibilities (Fischbach and Loike 2008). The use of gametes to generate children elicits a moral responsibility to care for them. Thus, any technology that dissociates these responsibilities from the gamete donor and the child must be viewed cautiously. Thus, the use of human–animal reproductive chimeras would have to be considered within a framework of medical intervention and question whether creating such chimeras would violate respecting human dignity.

With regard to the principle of respecting human speciation, the capacity to use human gametes of a human–animal reproductive chimera for human reproduction blurs the boundaries of speciation. As repulsive as it may sound, the theoretical possibility exists that a male human–animal chimera could mate with a human female or another human–animal female chimera to generate fertile human offspring.

LEGAL ASPECTS OF CHIMERA RESEARCH

The United States has witnessed various legislative attempts to prevent the use of chimeras in research that uses both public and private funding (National Academy of Sciences 2005; Academy of Medical Sciences 2011). For example, On April 29, 2010, the Arizona legislature approved a law banning the creation of "chimeras"—human–animal hybrids. The new law prohibits any resident of Arizona from "creating or attempting to create an in vitro human embryo by any means other than fertilization of a human egg by a human sperm." No legal prohibitions have yet been enacted on the use of private funding to sponsor the creation of human–animal chimeras (Streiffer 2005). Canada has adopted a set of guidelines similar to that of the United States (Bordet et al. 2010). For example, it is an offense knowingly to create a chimera or transplant a chimera into either a human being or a nonhuman life form (Canada Department of Justice 2004: Section 5(l) (i) of the Act, as cited in Bordet, Feldman, and Knoppers 2007)). This law defines a chimera as either (a) an embryo into which a cell of any nonhuman life form has been introduced or (b) an embryo that consists of cells of more than one embryo, fetus, or human being. Uniform regulation for chimera research has also been achieved by the United Kingdom in choosing a fairly permissive but controlled approach in which research is monitored through a licensing system. Other countries such as Germany, Denmark, the Netherlands, Spain, Switzerland, and Australia have laws that ban human cloning, experimentation on human embryos, eugenic selection, and research on human chimeras (Payne 2010).

ARE THERE LIMITS TO SCIENTIFIC EXPERIMENTATION?

Even if governmental agencies accept the concept that respecting human dignity and speciation are relevant issues to consider with regard to chimeric research, it is unclear whether laws restricting or banning human reproductive cloning or chimeric research will actually stop or restrict such scientific experimentation. This raises a broader issue of whether society should impose limits on scientific experimentation and whether such limits will be effective. From a secular perspective, the main deterrent to pursuing any technology is the potential harm it may bring upon plant life, animal life, human life, or the environment. From a religious perspective, these questions relate to a long-standing religious debate about whether human beings have the right to interfere with "God's creations." Many religious groups believe that human beings do not have an unlimited license to "play God" and modify species through hybridization or technologies that generate chimeras (Dabrock 2009). This ethical concern about humans playing God or interfering with God's creations came to the secular forefront in the 1970s during debates about the possible limits on recombinant-DNA technology (Goodfield 1977; Coady 2009). Religious scholars believe that God created all animals as individual species and that it is biblically prohibited for human beings to manipulate and cross-breed God-given plant and animal species. Therefore, creating hybrids or chimeras might violate this biblical principle (Loike and Tendler 2008; Coady 2009; Dabrock 2009).

Baylis and Robert (2003) link speciation and the concern of playing God from a slightly different perspective:

> crossing species boundaries is about human beings playing God and in so doing challenging the very existence of God as infallible, all-powerful, and all-knowing. . . . The creation of new creatures—hybrids or chimeras—would confirm that there are possible creatures that are not currently found in the world, in which case the world cannot be perfect; therefore God, who made the world, cannot be perfect; but God, by definition is perfect; therefore God could not exist. This view of the world, as perfect and complete, grounds one sort of opposition to the creation of human-to-animal chimeras. (Baylis and Robert 2003: 7)

In contrast, other religious scholars believe that human beings serve as co-creators, partnering with God in the creation process, and are directed to use any technology to improve the world (Loike and Tendler 2008; Dabrock 2009). From this perspective, there is moral justification to engineer human–animal chimeras and to create new species when the goal is to understand disease or develop new therapies.

Whatever view is adopted, history has shown how difficult it is for any society to limit scientific experimentation. This issue is clearly seen within the current legal and ethical debates concerning human embryonic stem cell research (Cornetta and Meslin 2011). The attempts by the U.S. government to restrict embryonic stem cell research have centered on banning federal funds to be used for such research. The outcome of these attempts has, at best, only resulted in a slowing of the research, because other funding agencies are providing funding. It is not difficult to predict that, in the case of human–animal neural or reproductive chimera research, funding from nongovernmental agencies will be provided if this research shows therapeutic potential.

CONCLUSIONS

The ethical concerns related to human–animal neural and reproductive chimeras will continue to be debated in the future and will depend, in part, on how neural human–animal chimeras display human cognitive functions or language. Once again, history can provide an important lesson. The reality of creating and visualizing the first mammal, Dolly, generated from adult cells, triggered debates on human cloning that resonated around the globe. Most likely, the bioethical debates about the ethics of generating human–animal neural chimeras will follow a similar course of stimulating debate, when scientists actually engineer the first neural human–animal chimera that exhibits enhanced cognitive functions. At that time, when it may be possible to evaluate the chimera's human capacities, the issues of respecting human dignity and respecting human speciation will come to the forefront.

With respect to human–animal reproductive chimeras, science has come a long way. Research is being conducted to produce chimeric mice that possess human gametes. No human has been produced using mouse-derived human gametes, because this would be viewed as a violation of human dignity. Nonetheless, it may be difficult, if not impossible, to limit or restrict this research in reproduction. Once again, the history of IVF provides an important lesson. IVF, despite its ethical conundrums, continues to be utilized across the globe. If the lesson of history in IVF repeats itself, society will recognize that research generating meaningful health care benefits will more likely be approved and actively pursued than research that, even inadvertently, causes human suffering or death.

Human beings are often mesmerized by new biotechnologies and are driven to develop these technologies in order to perfect the human experience in their limitless pursuit of new knowledge. There is great academic and intellectual concern about "forbidden knowledge" in limiting the type of research that scientists should purse. Leon Kass's words are insightful: "The forbidding aspects of science are, by and large, wholly separate from the forbidding of science. It in no way follows a priori that all—or indeed any—forbidding knowledge deserves to be forbidden—proscribed or disallowed" (Kass 2009: 274). In other words, it is difficult to restrict technological development when society believes there are potential medical benefits.

In conclusion, reconstituting a human hematopoietic system in an irradiated mouse does not conjure up the same ethical concerns as producing neural or reproductive chimeras. This chapter considers two issues—respecting human dignity and respecting human speciation—that underscore the ethical concerns of human–animal chimera research. From a secular perspective, one could view these concerns as distinct. Defining human dignity and speciation from a Judaic perspective, in contrast, provides a way to intimately interconnect and link these issues. Respecting human dignity, according to the Jewish perspective, confers both a sanctity and infinite value to human beings. It is this unique sanctity, emanating from being singularly created by God, that demarcates human beings as a unique species that deserves special moral considerations. Moreover, this sanctity, in large measure but not absolutely, defines the human species identity by highlighting humans' unique cognitive qualities, such as complex reasoning, self-awareness, and complex language. Human sanctity is further linked to the criterion of human identity of being born from a human being. In this manner, human dignity and speciation are intimately interconnected. Adopting this integrated approach suggests that chimeric research that crosses species boundaries or violates human dignity should not be pursued.

NOTES

1. This chapter will focus on prenatal generation of human-animal chimeras as opposed to postnatal generation. Postnatal generation of chimeras involves the introduction of human stem cells, tissue, or organs into postnatal animals.
2. This chapter does not discuss the use or ethics of human-embryonic hybrids, where a human nucleus is transplanted into an enucleated animal oocyte. Arizona is joining Louisiana in prohibiting scientists from creating such "human-animal" hybrids. The Arizona statute also prohibits human cloning through various means, including fertilizing human eggs with nonhuman sperm and vice versa. As of 2009, there has been no successful attempt to maintain human embryonic hybrids for more than a few days (Skene et al. 2009).

15 Psychotropic Drugs and the Brain
A Neurological Perspective on Human Dignity

William P. Cheshire, Jr.

> The desire to take medicine is perhaps the greatest feature which distinguishes man from animals.
>
> —Sir William Osler (Cushing 1925)

The phrase "human dignity" arises most often in discourse on ethical decisions at life's margins. Accordingly, previous chapters in this volume have addressed the language and meaning of human dignity in regard to the moral principles guiding decisions about the application of biotechnology at the beginning and end of life. Where technology touches on life at its boundaries, ethical dilemmas confront deeply held beliefs about what it means to be human. Philosophers continue to grapple with how or even whether humans have natures and whether the term *dignity* defines human nature accurately, sufficiently, or usefully. Despite its unresolved and often contested meaning, human dignity has not gone gently into the good night of obsolete language. Rather, it persists in the lexicon of bioethics even as new insights into the brain challenge us to rethink what dignity might mean.

If the language of dignity has significance at the edges of the human life span, then how much more meaning might it have throughout a person's life? In between life's delicate beginning and fragile end are matters of what it means not only to be a human but also to live as one. Central to this question is how biotechnologies affect the organ of thought, the brain. For how people perceive, think, act, reflect, and relate to others influences judgments about the worth (Latin *dignitas*) of human life and its purpose. The ways of living in the world involve mental activities within the brain, about which neuroscience offers increasingly detailed and informative descriptions. Neuropharmacology, meanwhile, provides an ever-enlarging array of psychotropic drugs capable of modifying brain function to treat a wide range of neurological and psychiatric disorders. If how one thinks and feels and acts is relevant to the meaning of human nature, then decisions about the appropriate use of drugs that affect thought, emotion, and behavior are germane to an exploration of human dignity.

The phrase "human dignity" has appeared infrequently in the bioethics literature concerning the use of drugs that modify brain chemistry. One reason for this may be custom, in that this language is conventional parlance in discussions of life issues. Another reason is that ideas about brain and mind are in transition. Advances in neuroscience combined with emerging pharmacologic capabilities are raising fresh questions about the wise stewardship of the neurochemistry that underlies human thought, mood, and behavior. Neuroethics is the new neighbor on the bioethics block. Now that neuroscience is making it possible to illustrate questions about human dignity by use of brain images, language must adapt if it is to keep up with increasingly detailed scientific portraits of human nature and their interpretations.

THE NEUROPHYSIOLOGIC BRAINSCAPE

Advances in neuroscience are opening new windows into the human brain, which is arguably the most highly complex structure in the known universe. The adult cerebral cortex contains approximately 100 billion neurons and a greater number of nonneuronal glial cells (Pakkenberg and Gundersen 1997; Williams and Herrup 1988; Azevedo et al. 2009). Cerebral neurons communicate with one another by way of more than 100 types of neurotransmitters and neuropeptides. The outpouring of these signaling molecules occurs at the junction between neurons known as the synapse (Cheshire 2008b). The human brain contains approximately 160 trillion synapses—a number larger than the 200 to 400 billion stars in the Milky Way galaxy (Tang et al. 2001).

Prior to the development of neurosurgery and antisepsis, study of brain function was limited to postmortem descriptions and, in rare cases such as that of Phineas Gage (Damasio et al. 1994), clinical observations on survivors of focal brain injuries. With the advent of sophisticated neuroimaging techniques, it is now possible to study the brain during life both structurally and functionally. Brain scanners routinely capture breathtaking images of the gyral folds of the cerebral cortex that underlie human thought and personal identity (Cheshire 2007a). Functional magnetic resonance imaging permits precise, noninvasive, spatial, and temporal resolution of psychological processes in the living brain. When metabolic activity increases in the brain regions involved in a specific cognitive activity, those areas will light up on functional brain scans, which apply computer algorithms to compare signals from the brain during a defined activity to the brain at baseline. These correlations tell what brain regions are involved in the exercise of such capacities as vision, language, and movement. Within the last decade, functional brain imaging has turned to investigating the neural substrates of moral reasoning, agency, decision making, altruism, emotional bias, fear, deception, belief, and spirituality—all of which have profound implications for the conceptualization of human dignity (Cheshire 2007b; Jones 2010).

PSYCHOTROPIC DRUGS AND THE BRAIN

Medical science intervenes in brain processes most commonly by use of psychotropic drugs. A psychotropic drug is an administered chemical substance that acts upon the central nervous system. Drugs may affect brain function in a number of ways. Some act on macromolecular structures or signaling molecules within the cell, while others bind to specific receptors at the synapse, where they may inhibit presynaptic release, postsynaptic binding, or presynaptic reuptake of neurotransmitters. The resulting effect at the cellular level may be altered excitability of nerve membranes or modified flow of neuroeffectors or neurotransmitters such as serotonin, dopamine, or norepinephrine. These complex chemical changes within the brain can translate to alterations in perception, consciousness, mood, memory, or other aspects of cognition or behavior. The clinical responses and side effects may be profound or subtle.

Psychotropic drugs are widely used in medicine. A study by the U.S. Centers for Disease Control and Prevention found that of 2.4 billion prescriptions in 2005, the most frequently prescribed class of drug was antidepressants, which amounted to 118 million prescriptions (Burt et al. 2007). Antidepressant prescriptions to brighten or stabilize mood have tripled over the past two decades. One reason may be that the selective serotonin reuptake inhibitors (SSRIs) have a more favorable side effect profile than earlier antidepressants, and their introduction has been followed by an increased rate of prescribing. Also, aggressive pharmaceutical advertising is not without market influence. Along with benzodiazepines, SSRIs are also used to treat chronic anxiety. SSRIs have been prescribed to an estimated 10 percent of the U.S. population (Olfson and Marcus 2009; Pirraglia et al. 2003; Fukuyama 2002).

Farther down the list were pain relievers, with 108 million prescriptions for nonnarcotic and 94 million for narcotic analgesics in 2005. A broad range of pharmacologic measures is available for the alleviation of acute and chronic pain. These include the opioids; nonsteroidal anti-inflammatory agents; antiepileptics, which also have nerve membrane stabilizing properties; antidepressants, which also enhance pain inhibitory pathways; and others that block or modify the perception of pain. Additionally, local anesthetic agents such as lidocaine are used to block pain transmission during medical procedures, and general anesthetic agents are used to induce temporary unconsciousness during invasive surgery.

In contrast to anesthetics, stimulants such as caffeine, modafinil, amphetamine, methylphenidate, and atomoxetine are used to heighten the level of alertness or mental focus. Recent interest has focused on the potential use of stimulants as cognitive performance-enhancing agents, and debates continue over how to guide their appropriate use. Licensed off-label as well as illicit use of stimulants and other drugs that sharpen mental focus, sustain wakefulness, boost alertness, and improve information retention and learning has increased among healthy students and professionals (Cheshire 2006b). Some pharmaceutical companies have targeted "lifestyle drugs" as a potential market for lucrative growth with emphasis on improving quality of life, as such drugs would not necessarily alleviate disease (Cheshire 2010a). Current and foreseeable applications of psychotropic drugs are also moving beyond the scope of medical practice into academic and military (Russo 2007) applications for purposes of cognitive performance enhancement. Some bioethicists speculate that applications of this new "cosmetic neurology" (Chatterjee 2007, 2004) could eventually extend to the spheres of education (Anonymous 2010; Diller 2009), professional performance enhancement (McBeth et al. 2009; Greely et al. 2008; Cheshire 2008c), and penal correction (Farah 2002).

Memory is another target for psychotropic drugs. Certain cholinergic drugs can improve memory in patients with dementia, and drugs currently in development such as ampakines could potentially be used to augment memory in healthy individuals. Some patients are in need of better memory, while others are in need of forgetting. For example, preliminary research has shown that beta-blockers administered shortly after an emotionally traumatic experience can decrease the subsequent risk of developing the recurrent painful memories that occur in post-traumatic stress disorder (Pitman et al. 2002). Interestingly, following reactivation of consolidated, old, traumatic memories, beta-blockers attenuated not only the behavioral expression of the fear component of the memory but also its reconsolidation (Kindt et al. 2009; Debiec and LeDoux 2004).

A noteworthy development is the increasing availability over the Internet of prescription drugs and nonprescription psychotropic supplements claimed to improve brain functions (Schepis et al. 2008).

PSYCHOTROPIC DRUGS AND NEUROETHICS

Whereas the science of neuropharmacology concerns what *can* be done with psychotropic drugs, neuroethics asks how such drugs *should* be used in accordance with ethical principles. The reasons offered for using or not using a drug a certain way may draw from an understanding of human dignity.

Ethical concerns about the appropriate use of psychotropic drugs usually begin with questions of safety. All drugs, whether available by prescription or over the counter, have potential adverse effects, some of which are dose-related, while others depend on genetic and other factors affecting how the patient metabolizes the drug or how the patient's concurrent medications interact. Adverse effects may involve brain or bodily functions within or distant from the targeted system. The history of pharmacology is a story of the discovery of new drugs as well as efforts to modify known drugs and identify alternative agents with fewer and less harmful adverse effects. Following improvements in safety, effective drugs are likely to be more widely pre-scribed, and with greater use, further ethical concerns come into view.

Ethical obligations concerning the use of psychotropic medications include obtaining informed consent, respecting the patient's autonomy in deciding upon treatment or nontreatment, providing care, avoiding psychological and physical harm, minimizing conflicts of interest, separating coercive influences, maintaining truthfulness in claims of drug efficacy, distributing medical resources justly, and adhering to the highest standards of medical care. Within these categories of ethical concern are numerous ways in which the prescribing of psychotropic drugs might, intentionally or unintentionally, impinge on aspects of human dignity. Ethical breaches may result in physi-cal injury, or they may cause emotional distress. They may diminish the patient's autonomy, or they may erode the relationship of trust that ought to exist between patient and caregiver.

The neuroethical questions introduce an additional level of complexity into considerations of human dignity beyond those concerning life and death quandaries. They are also more nuanced, since many aspects of thought and behavior are qualitative and subjective. Whereas appeals to human dignity at the beginning of life evaluate such concerns as the moral status of the human embryo (Cheshire 2005; Cheshire et al. 2003), and at the end of life such concerns as expenditure of health care resources during terminal illness (Foley and Hendin 2002; Kilner 1992), ethical concerns with regard to psy-chotropic drugs raise further questions regarding the meaning of personal identity. Each of the categories of psychotropic drugs outlined above touches on some aspect of personal identity, whether in relation to mood, awareness, suffering, consciousness, intelligence, or memory. Insofar as each of these aspects of personal identity contributes to conceptions of human dignity, the reasons guiding decisions regarding how psychotropic drugs are used can inform discussions about the meaning of human dignity.

THE ELUSIVE CONCEPT OF HUMAN DIGNITY

Because questions of human dignity concern so many aspects of personal life, types of relationships, and categories of decisions, efforts to outline an overarching definition of human dignity have received criticism for seeming too vague (Bostrom 2009: 174; Farah and Heberlein 2007; Macklin 2003). No broad consensus exists on the meaning of human dignity, and many agree that a universal definition of human dignity seems elusive. Francis Fukuyama writes, "Human dignity is one of those concepts that politicians, as well as virtually everyone else in political life, like to throw around, but that almost no one can either define or explain" (Fukuyama 2002: 148). Why is this?

In contrast to the precision for which science strives and the clarity which philosophy seeks, vagueness often typifies political rhetoric. A number of explanations may be offered. For one, specificity invites targeted rebuttals, which might be seen to impede progress toward an objective. Second, reasoned arguments may take a back seat to the emotional plea used to motivate others toward an objective. In these respects, vagueness may have pragmatic utility. Efforts toward a more precise definition of human dignity could be perceived as not useful to some forms of political discourse.

There is a third explanation, in which lies the hope that clarification need not remain completely beyond reach. That is, the current understanding of human dignity may simply be provisional. Just as neuroscience at this time cannot adequately explain how consciousness emerges from a web of neurons, a satisfactory definition of human dignity may lie so far beyond current human language or understanding that to attempt to define it comprehensively would, for the time being, be to describe it incorrectly. In this respect, it is worth noting the astonishing rate at which new information about the brain is becoming available. Much of this information is possible through the study of psychotropic drugs' effects on the brain and, in turn, on personal and interpersonal behavior. The accumulation of more knowledge about the organ of human thought may in time yield new insights into the nature of human dignity.

A complete understanding of human dignity cannot, however, be expected to be possible through the study of the brain in isolation. It is persons, not their brains, that can be said to possess dignity. In considering the implications of psychotropic drugs for an understanding of human dignity, it is important to consider the many stakeholders in society, their diverse perspectives, and the ways in which psychotropic drugs might affect their thoughts and lives.

WHO DECIDES

Who is qualified to assess the meaning of human dignity in health care? The unmedicated patient? The medicated patient? Is there a difference? Or is it the physician, who brings to the bedside his or her own conceptions of what constitutes a good life? The family, whose interests may or may not coincide with those of the patient? The employer, who has an interest in maximizing the employee's productivity and minimizing work absences? The health insurance agency or third-party payer, who balances the medical expenses of the patient against the needs of other customers? Or society, which may function in a more orderly manner if its citizens behaved calmly and predictably? Is more order necessarily better?

A medical perspective might reasonably combine the ethical concerns touching on human dignity under the rubric of medical professionalism. Responsible prescribing of psychotropic drugs in a way that honors human dignity is largely a matter of practicing good medicine, the moral nature of which is determined by the conditions at the bedside. These include the fact of illness, the act of profession in promising to help the patient, and the provision of treatment such as psychotropic drugs (Pellegrino 2006).

In reality, however, gathered around the bedside are a number of other participants who join with the physician in caring for the patient. While the patient's point of contact with biotechnology, at least in regard to prescription drugs and surgical procedures, is traditionally the bedside or medical office, the health care system brings together many ancillary professionals and supporting staff whose decisions and actions may honor or impinge on the dignity of the patient or affect how that dignity is perceived. In the hospital or medical clinic, often the first person to acknowledge or ignore the patient's dignity is the receptionist. Nurses, pharmacists, physician assistants, administrators, phlebotomists, schedulers, clinical technicians, secretaries, and housekeeping personnel all have a role in respecting the patient's dignity. Beyond the walls of the clinic, decisions regarding what type of new drugs to research and develop, how and to whom to market them, what regulatory standards to adopt, and which drugs to cover under health insurance plans all entail ethical choices with potential consequences for patients who, as human beings, are entitled to be treated with dignity.

Deciding what human dignity means, therefore, is not a question just for specialists. It is not exclusively an academic exercise by philosophers but in reality a practical matter in which all people participate in their daily lives and interactions with others. Similarly, considering the ways in which the effects of psychotropic drugs shape conceptions of human dignity is not just a topic for neuroscientists, neurologists, psychiatrists, or pharmacologists but rather is open to all of society. Broad discussion is needed, because there is none untouched directly or indirectly by illness or the influence of psychotropic drugs.

AREAS OF TENSION

Society's continuing exploration of the meaning of human dignity may be an unavoidably untidy process. Consideration of the multifarious effects of psychotropic drugs renders that exploration even more complex. In probing the meaning of human dignity in relation to the development, prescribing, and utilization of psychotropic drugs, a number of key areas of tension are evident.

Philosophical Reductionism

One area of tension is reductionism, for example, of human dignity to a single principle such as autonomy. The principle of autonomy in bioethics respects the right of patients to self-determination, including the rights to choose among medically feasible alternatives and to refuse invasive interventions. Respect for autonomy accommodates the patient's right to choose or refuse psychotropic drugs.

Some writers reject the language of human dignity, claiming that it is merely a placeholder for the established principle of autonomy. Steven Pinker, in an article entitled "The Stupidity of Dignity," alleges that the concept of dignity "springs from a movement to impose a radical political agenda, fed by fervent religious impulses, onto American biomedicine" and is otherwise "just another application of the principle of autonomy" (Pinker 2008: 29). Ruth Macklin argues that "dignity seems to be nothing other than respect for autonomy" and indicts those who appeal to the language of dignity for what she believes to be covert intrusion of religion into the literature of medical ethics (Macklin 2003: 1420). "Dignity," writes Macklin, "is a useless concept in medical ethics and can be eliminated without any loss of content" (Macklin 2003: 1420).

The narrow focus of reductionism, while successful at elucidating small things, is not always the correct conceptual tool to provide explanations for larger realities. Other bioethicists endorse a robust meaning for human dignity. C. Ben Mitchell and colleagues, for example, argue that dignity is not reducible to respect for persons or their autonomy. "Rather," they affirm, "it is the basis for why such respect is warranted" (Mitchell et al. 2007: 63). Gilbert Meilaender adds that autonomy should not preempt all other aspects of human dignity such as embodiment, purpose and self-sacrifice. Pinker's reduction of dignity to autonomy, writes Meilaender, "assumes that freedom is the sole truth about human beings and this reduces the complexity of our humanity" (Meilaender 2009: 29). Likewise, Edmund Pellegrino, in a seminal series of essays on *Human Dignity and Bioethics* by the President's Council on Bioethics, which he chaired, argues emphatically against replacing dignity with autonomy, writing,

> Humans possess autonomy because of their intrinsic dignity; they are not dignified because they are autonomous. Holocaust victims did not lose their dignity or the rights that it entailed because they were despoiled of their autonomy. Nor do infants, the comatose, or the brain-damaged lack dignity because they are not fully autonomous. (Pellegrino 2009: 534)

This distinction is all the more important when considering the use of drugs that may restrict the patient's autonomy. If dignity were ultimately reducible to autonomy, then it would be uncertain whether full respect would be due someone whose decision-making capacity were limited or otherwise altered by a medically necessary psychotropic drug. Nonreductionistic evaluations of dignity, on the other hand, ground more deeply the moral requirement to respect autonomy.

Immanuel Kant, while agreeing that "autonomy is therefore the ground of the dignity of human nature," further elevates dignity, arguing that, "Everything has either a price or a dignity. If it has a price, something else can be put in its place as an equivalent; if it is exalted above all price and so admits of no equivalent, then it has a dignity" (Kant 1785: 102). Extending his point to neuroscience, one might say that everything in the brain has a chemistry or a dignity—or both. A psychotropic drug, by upregulating or downregulating a neurotransmitter, may alter the economy of neurochemistry with the goal of restoring normal function. There may also be thoughts and feelings beyond pharmacologic manipulation. These relate to dignity.

Both secular and religious thinkers have embraced nonreductionistic conceptions of human dignity. The drafters of the United Nations Universal Declaration of Human Rights, who came from diverse cultural, intellectual, and ideological backgrounds, appealed to common language in stating that "recognition of the inherent dignity and of the equal and inalienable rights of all members of the human family is the foundation of freedom, justice and peace in the world" (United Nations General Assembly 1948). In this context, dignity was, as Richard John Neuhaus observes, "largely defined negatively against the background of evils to which the declaration says, in effect, 'Never again!'" (Neuhaus 2009: 218). Neuhaus adds that the concept of human dignity has been sustained by "a form of understanding that is carefully reasoned, frankly moral and, for most people who affirm it, is in fact, if not by theoretical necessity, inseparable from a comprehensive account that is unapologetically acknowledged as religious" (Neuhaus 2009: 226). Likewise, C. Ben Mitchell and colleagues write, "if there is a God who establishes a special relationship with human beings that confers special worth on them, all people may be said to have a dignity that is distinctively human" (Mitchell et al. 2007: 69).

Neurochemical Reductionism

A parallel example of reductionism is of the brain to neurochemicals. To the strict materialist, the human brain is an object composed of three pounds of matter. Ideas consist of informational arrangements of macromolecules. Thoughts are the product of oxygen exchange and other metabolic activity generating electrochemical flux through circuits of synapses. Desire, remorse, yearning, and love are no more than the outpouring of neurotransmitters in response to accidental stimuli and are completely determined by genetic structure and environmental influences, including psychopharmacological interventions. Though complex, the brain is essentially mechanical. The idea of human dignity, according to this view, was a cozy illusion of a prescientific age that has been supplanted by a factual understanding of human nature.

Patricia Churchland, who argues that the mind is fully reducible to the physical brain (Churchland 2002), writes that "human dignity is not a precise concept, in the way that 'electron' or 'hemoglobin' are precise. . . . It does not connote a matter of fact" (Churchland 2009: 100). Her premise is that only what can be empirically measured and quantified counts for what should be accepted as an accurate description of human nature. A similar reductionistic approach appears in the writings of Francis Crick. His "astonishing hypothesis" is "that 'You,' your joys and your sorrows, your memories and your ambitions, your sense of personal identity and free will, are in fact no more than the behaviour of a vast assembly of nerve cells and their associated molecules" (Crick 1994: 3). The words "no more than" delimit a boundary that constrains the meaning of human dignity to what science can directly observe and manipulate.

Materialistic views of human nature are at odds with appeals to human dignity because they omit from consideration the categories of value, meaning, and purpose, which require a nonmaterial frame of reference. They also tend to deconstruct consciousness and agency, which underlie moral responsibility. Crick asserts that personal identity and free will are attributes of consciousness of which one can have only a fallacious sense (Crick 1994). In a similar vein, Daniel Wegner writes, "The fact is, we find it enormously seductive to think of ourselves as having minds, and so we are drawn into an intuitive appreciation of our own conscious will" (Wegner 2002: 26). Wegner argues that this conscious will, on which depend the capacities for making ethical decisions and judgments of value and purpose, is merely an illusion. Raj Persaud writes of "the delusion that we are responsible for all our actions," arguing that the belief in free will is nothing more than a function of the brain (Persaud 1999: 130). If materialism were true, then even the capacity for reason would collapse to automatic consequences of material causes prodding the brain as inexorably as lines of computer code (Cheshire 2010b).

Within a materialistic framework, in principle, psychotropic drugs could have unlimited potential to modify any aspect of human thought and behavior. However, materialism lacks the perspective needed to provide principles by which to guide the appropriate use of psychotropic drugs. Accordingly, advocates of retaining the language of human dignity have resisted confining their understanding of human nature to reductionistic models.

Reductionistic accounts of human nature, moreover, are in tension with the practice of medicine. Whereas medicine owes much of its progress to the contributions of neuroscience, such contributions are necessary, but not sufficient, to explain the attentiveness to human dignity that inspires the care of the sick. A purely scientific description of human nature limited to factual knowledge about the brain is incapable of resolving poignant dilemmas in medical ethics that touch on human dignity. A strictly materialistic appraisal of the indications and effects of psychotropic drugs would impoverish medicine (Cheshire 2008a). A prime example of this is the treatment of pain, which is a paramount obligation in medical practice, whether through analgesic drugs or other palliative means. The related goals of alleviating pain and respecting the dignity of the patient are among the core principles of palliative medicine (Cassell and Foley 1999). Like dignity, pain is often vaguely described, but that does not mean it is not real. The practice of medicine respects the dignity of patients even though a precise definition of human dignity may lie beyond the grasp of reductionistic descriptions. Physicians do not ask for proof to believe that the care of the patient is never futile (Cheshire 2003).

Pellegrino adds that

> Chronic illness, mental illness, dying and death are occasions when the patient's perception of loss of his or her dignity is deep enough and persistent enough to be, itself, an additional source of suffering. This suffering is often more distressing than the pains, discomfort, or disability caused by the disease itself. (Pellegrino 2009: 524–25)

These sources of distress that are so profound that the word *pain* is inadequate to describe them require a descriptor closer to the heart of personal identity. A suitable descriptor in this context is "loss of dignity." Although pain can be addressed with analgesic drugs, the caring response to loss of dignity is more complex. A pill alone is not the way to restore a sense of dignity.

Noting the ever-tightening links between mental phenomena and neuroscientific descriptions, Malcolm Jeeves argues for the ontological primacy of conscious mental events, pointing out that the reductionistic approach of the scientific method affords great benefits methodologically but not metaphysically (Jeeves 2004). Jeeves bases human dignity not in human cognitive capacities themselves but in their inherent potential to relate to the transcendent. "Humans," he writes, "have the critical capacities for personal relatedness, for complex language, for forming a 'theory of mind,' for historical memory, and for contemplating the future. Such highly developed 'soulishness' makes humans unique and, I believe, confers the capacity for personal relationship with God" (Jeeves 2000: 133).

Churchland's point that human dignity is not the same sort of thing as a molecule, a gene, or a neuron is nonetheless helpful. Human dignity cannot be caught in the calipers of quantifiable scientific examination or replaced by a drug effect. Arthur Eddington recognized this distinction, writing in his Gifford Lectures that, "The cleavage between the scientific and the extra-scientific domain of experience is, I believe, not a cleavage between the concrete and the transcendental but between the metrical and the non-metrical" (Eddington 1931). The nonmetrical or nonquantifiable aspect of human thought complements the neurophysiologic representation of human nature and edifies an understanding of dignity. Psychotropic drugs, which have quantifiable effects on the brain, affect the nonmetrical aspect of human thought indirectly. The ways in which psychotropic drugs color the meaning of human dignity in clinical practice may be comprehensible only through the patient's subjective experiences and shareable through empathy.

Privacy

Another area of tension is personal privacy. Stunning brain images now abound in the popular media. They embellish magazine covers and proliferate on the Internet. These images are used in research to study the effects of psychotropic drugs on the brain and to assess clinical indications for prescribing them. Such images also paint pictorial narratives of human nature that offer clues to how people think about the world, others, and themselves—whether, for example, someone is an introvert or extrovert, prefers one brand over another, harbors unconscious bias, is truthful or deceptive, has the capacity for empathy, or has the cravings of a drug addict (Frith 2004). The drugs used as contrast agents for functional MRI scans are in a sense psychotropic without being psychoactive. Without altering brain function, they permit the detection of brain regions that are metabolically active during certain cognitive processes. The information they yield can be quite powerful.

Legal experts now debate whether it is ethical to override the autonomy of a psychiatrically ill defendant and allow the court to order the use of psychotropic medication for the purpose of restoring competency to stand trial. Also under debate is whether the persuasive and potentially misleading potential of brightly colored brain scans should be admissible in courtroom proceedings as evidence of the reliability of witness testimony or of diminished moral culpability on the part of the defendant (Wolpe et al. 2010; Schauer 2010; Greely and Illes 2007; Feigenson 2006). Beyond the courtroom, in medical, employment, and insurance contexts, the ability to peer into brain processes underlying one's intimate thoughts raises concerns about what has been characterized as the dignity of personal privacy (Kennedy 2004). Commenting on potential future applications for brain imaging, Donald Kennedy, the former editor of *Science*, said,

> If my stored brain images said something about my tendency to anger under different kinds of stress, or accounted for the ways in which I make moral choices, or how strangely I perform on certain intelligence tests, then I would be troubled. I don't want anyone to know it, for any purpose whatever, including those offered in my own interest. It's way too close to who I am, and it's my right to keep that most intimate identity to myself. (Sample 2003)

Authenticity

A related area of tension concerns personal authenticity. If a drug changes one's alertness, mood, motivation, memory, or self-perception, how is one to know which is the authentic self—the self under or absent the influence of psychotropic medication?

Peter Kramer's book *Listening to Prozac* (Kramer 1993) initiated a debate over pharmacologic mood enhancement that is ongoing. Prozac is the trade name for fluoxetine, an SSRI used to treat depression, panic disorder, and obsessive-compulsive disorder. Using an antidepressant to restore the mood of a depressed patient usually presents no ethical difficulty. The line distinguishing depression from normal variations in mood, however, is not always clear. Should low self-esteem or a sense of loss always be medicated?

Some patients treated with SSRIs report that the drug makes them feel renewed. Others report that the drug makes them feel like their true selves. Still others report that the drug makes them feel somehow better than themselves (Elliott 2003). If human dignity is to be found in the patient's authentic self, it may seem unclear whether that is the patient on or off medication. In this sense the drug may alter what Gilbert Meilaender calls "human dignity" without in any way changing "personal dignity." In his book *Neither Beast nor God*, Meilaender proposes a conceptual distinction that seeks to resolve some of the confusion that has muddled debates over dignity. He refers to human dignity as that which "has to do with the powers and the limits characteristic of our species—a species marked by the integrated functioning of body and spirit." Human dignity in this sense is grounded in certain cognitive capacities present in some human beings in greater degree than in others as well as attributes of individual character "found in the kind of life that honors and upholds the peculiar nature that is ours." By contrast, he considers personal dignity to be the equal dignity or worth of the individual person, "grounded not in any particular characteristics but in the belief that every person is equidistant from Eternity" and "whose dignity calls for our respect whatever his or her powers and limits may be" (Meilaender 2009: 6–8, 81).

In a similar vein, Renée Mirkes points out that such perceptions of personal identity conflate two distinct realities: "a person's essential identity (that which characterizes his personhood) and a person's accidental identity (that which characterizes his personality)" (Mirkes 2010: 181). "Essential identity" derives from the model of substance dualism (Moreland and Rae 2000) and refers to one's being or substance that endures throughout a lifetime of accidental changes, including psychotropic drug-induced changes, to his or her personality.

Mood enhancement in the healthy also raises more subtle concerns. Leon Kass cautions of the transformation in our humanity that drugs might bring to "our fundamental ways of encountering, enjoying, and acting in and on the world" (Kass 2002: 12). Nick Bostrom writes of the potential for "individuals to clip the wings of their own souls." This, he continues,

> would be the result if we used emotional enhancers in ways that would cause us to become so "well-adjusted" and psychologically adaptable that we lost hold of our ideals, our loves and hates, or of our capacity to respond spontaneously with the full register of human emotions to the exigencies of life. (Bostrom 2009: 191)

While these images intrigue the moral imagination, they fail in the practical yet more difficult task of defining at what point such use would cross the threshold to a soma-steeped dystopia (Huxley 1932).

Memory also contributes to one's sense of personal authenticity. The loss of memory that occurs in dementia and the resulting fragmentation of personal biography assault the patient's dignity (Pacholczyk 2010; Post 1995). Drugs designed to improve memory function are appropriately viewed as means toward preserving that dignity (Dekkers and Rikkert 2007). The efficacy of currently available drugs to restore memory is quite limited. More potent drugs are in development, and, once available, healthy individuals may find such drugs appealing for off-label purposes of memory enhancement (Cheshire 2006b). Whether such use might also distort the self-perception of one's own dignity may be a troublesome question.

Walter Glannon notes that both remembering and forgetting are "distinctly human" phenomena (Glannon 2007). Paradoxically, drug-induced forgetting might also be viewed as an intervention on behalf of dignity. Preventing or reducing the severe emotional burden of recurrent traumatic memories in patients afflicted by post-traumatic stress disorder would be most welcome.

The ability to edit memories by using drugs to suppress their emotional force also raises profound questions in regard to how personal memories shape one's autobiographical identity. The President's Council on Bioethics questioned whether personal identity—"shaped by our own experiences, aspirations, attachments, achievements, disappointments, and feelings . . . that which makes all of us human and that which makes each of us individually who we are"—might be lost in the process of attaining happiness "if the condition for attaining it required that we become someone else" (President's Council on Bioethics 2003: 211). Mirkes takes a different view. Appealing to the untouched integrity of essential personal identity, she concludes that such pharmacologic interventions constitute an accidental change that mimics the memory extinction process of healthy trauma survivors (Mirkes 2010).

Another concern is that memory-blunting drugs could also be misused to sanitize the shameful memories of guilty conduct (Mirkes 2010). Further, if future drugs to dull the emotional impact of painful memories were to work well enough to enter routine medical practice, would that shift society's attitudes toward moral violations? If, for example, the community believed that the availability of a memory-blunting drug could lessen the harm to human dignity, would their response be as urgent? In these and other areas, debates persist over whether thoughts and behaviors represent moral or biological conditions or whether complementary perspectives are still needed (Cochrane 2007).

Intelligence also contributes to one's sense of authenticity. If human dignity were reducible to intelligence, which varies greatly from person to person, it might seem that people possess dignity to unequal degrees. The use of psychotropic drugs to enhance intelligence beyond what is normal could, in principle, further widen that disparity if smarter people had greater access to the means to cognitive enhancements. Biologically enhanced superintellects, perceiving themselves to have attained a higher level of dignity, might be tempted to despise the unenhanced. Meilaender helps to put this problem in perspective by defining cognitive enhancement as augmentation of an aspect of human dignity independent of personal dignity. Retaining the concept of personal dignity is a valuable precaution against negative assessments of human worth solely on a scale of cognitive or any other functional capacity.

A practical consequence of understanding dignity predominantly in terms of autonomy in a meritocratic society would be that cognitive performance-enhancing drugs would find their most ready market in those seeking increasing productivity. In a possible future in which such drugs were to be used routinely, Peter Lawler questions whether people would be free not to choose them and still compete alongside the enhanced. "If there are no natural, relational, and dignified limits to our free choices," he writes, "it will increasingly seem that we have no choice, really, but to maximize our productivity" (Lawler 2009: 49).

From a neurologic perspective, there is a further problem with reducing human dignity to a matter of intelligence, for there is not just one scale of intellect. Intelligence comprises many complementary cognitive domains. These include not only memory but also attention, perception, logical and abstract reasoning, creativity, athletic and musical skills, assessment of consequences, planning, judgment, initiative, restraint, compassion, empathy, and the capacity to develop interpersonal relationships. Even memory may be subdivided into distinct cognitive capacities, which include short-term, long-term, declarative, procedural, semantic, and autobiographical memory. These cognitive capacities are variously subject to the influence of psychotropic drugs. The rapid learning of new information in the hippocampal system and the gradual consolidation of memories in the neocortex are anatomically and functionally distinct brain processes (McClelland et al. 1995). On that basis, a number of writers have speculated that the use of drugs to enhance any one aspect of intelligence, such as the durability of individual memories, might steer cognitive focus away from the tasks of generalizing, contrasting and reflecting deeply that are essential to acquiring wisdom in contrast to accumulating information (Cheshire 2009a; Cheshire 2012).

Autonomy

In addition to the problems resulting from reducing the definition of dignity to autonomy are ethical questions concerning the desirable aims and appropriate boundaries of individual autonomy in choosing how to use psychotropic drugs. Since drugs sometimes affect cognitive functions in opposite ways, dignity would seem to be more closely related to one's ability to choose than the working of the drug itself. Of course, individual choices are not made in isolation. However one defines dignity, personal choices can impinge on the dignity of others.

Wrye Sententia extends the principle of autonomy to embrace the notion of "cognitive liberty," which she defines as "every person's fundamental right to think independently, to use the full spectrum of his or her mind, and to have autonomy over his or her own brain chemistry" (Sententia 2004: 223). She welcomes the voluntary use of psychoactive drugs for mental enhancement purposes and argues that individuals should be free to determine what are acceptable personal risks of such drugs. She also cautions against the involuntary use of psychotropic drugs for social or penal purposes, citing as a hypothetical example court-mandated medication to curb aggression, impulsive violence, drug or alcohol cravings, or sexual drive as a substitute for corrective measures such as anger management classes, which allow the individual to think his or her own thoughts and have his or her own personality. For Sententia, at the core of the principle of autonomy is "the right and freedom to control one's own consciousness and electrochemical thought processes," without or with psychotropic drugs (Sententia 2004: 227).

How far the appeal to autonomy might reach once severed from its conceptual grounding in human dignity is increasingly a biotechnological question. Julian Savulescu argues for pursuing radical human enhancements and abandonment of "the human prejudice" (Savulescu 2009: 216–17). Psychotropic drugs are among the biotechnologies central to the transhumanist vision, which urges a transition to a "posthuman" future with the goal of overcoming fundamental human limitations. The dignity of a given human nature can only be seen as an obstacle to projects that seek to redesign, if not replace, humanity. Savulescu speculates that "Humans may become extinct. . . . We might have reason to save or create such vastly superior lives, rather than continue the human line" (Savulescu 2009: 244).

Human Exceptionalism

Yet another area of tension is the claim of human exceptionalism, which attributes to humankind elevated moral status above that of nonhuman animals. To accept human exceptionalism does not mean to condone the mistreatment of animals. This perspective is implicit in ethical codes governing research which require that newly developed psychotropic and other drugs be tested first on laboratory animals before clinical studies are done in human subjects (Annas and Grodin 1992).

Proponents of human exceptionalism identify an ontological reality that need not root human dignity in genetic fixity (Pollard 2009) or any one cognitive capacity. Of the neurological capacities unique to humans, Steven Rose recognizes language, social existence, and consciousness of self as those that "present a sharp discontinuity with even our closest genetic and evolutionary relatives. Underlying all of them is that special attribute of humans that we call possessing a mind, of being conscious" (Rose 2006: 87).

Disputing the special status of human nature has become fashionable, ironically, in some of academia's most elite quarters. Peter Singer's "new vision," for example, calls for repudiation of the traditional understanding of human exceptionalism, particularly where affirmed through religious teaching (Singer 1994: 180–83).

A more recent departure from "the human prejudice" is the prospect of artificial intelligence advancing to the point of convincingly mimicking human intelligence. Ray Kurzweil predicts a "singularity," at which time computer intelligence will transcend human intelligence, abolishing the distinction between them:

> Human cognition is being ported to machines, and many machines have personalities, skills, and knowledge bases derived from the reverse engineering of human intelligence. Conversely, neural implants based on machine intelligence are providing enhanced perceptual and cognitive functioning to humans. Defining what constitutes a human being is emerging as a significant legal and political issue. (Kurzweil 1999: 222–23)

Whether society would come to think of intelligent machines as having artificial dignity remains to be seen. One may speculate that realization of Kurzweil's vision could lead to powerful computers that themselves claim to possess a superior form of dignity that less intelligent humans need not bother to try to comprehend. Cognitive performance-enhancing pharmaceuticals may not be strong enough to enable human brain function to keep pace with such computers. Whether or not such predictions arrive, it is reasonable to ask whether there is also a dignity in human limitation.

Medicalization of Striving

The final area of tension to be considered is how human finitude relates to what it means to have dignity. This includes the human response to diseases and circumstances beyond control. The question arises in the debate over whether cognitive enhancing drugs developed to treat neurological and psychiatric disorders should also be prescribed to healthy individuals. Answers hinge on how one draws distinctions between therapy and enhancement. The President's Council on Bioethics under Leon Kass offered as a starting point a general framework:

> "Therapy" is the use of biotechnical power to treat individuals with known diseases, disabilities, or impairments, in attempt to restore them to a normal state of health and fitness. "Enhancement," by contrast, is the directed use of biotechnical power to alter, by direct intervention, not disease processes, but the "normal" workings of the human body and psyche, to augment or improve their native capacities and performances. (President's Council on Bioethics 2003: 13)

Requests from parents for cognitive enhancements for their children to secure an academic edge would be particularly problematic. Children are less capable of comprehending medical risks and would be less able to participate in informed consent. Further, the Council raised concern about separating personal gains from the effort required for genuine achievement, which could teach the child "that high performance is to be achieved by artificial, even medical, means" (President's Council on Bioethics 2003: 93).

The Council explored in *Beyond Therapy* the significance of the language of dignity in regard to "excellent human activity." While the members of the Council affirmed the dignity of agency, they also articulated the importance of embodiment, of human striving, of human limitations, and a respect for the given, all of which shape the character of human longing and contribute to an understanding of human dignity. Radical autonomy stripped of these things, cautioned the Council, might "achieve our most desired results at the ultimate cost: getting what we seek or think we seek by no longer being ourselves" (President's Council on Bioethics 2003: 155).

The Council identified two opposing dangers from the growth of medicalization of human achievement. On the one hand is "the risk of viewing everything in human life . . . under the lens of disease and disability" and potentially in need of rescue by medication or other biotechnology. On the other hand is "the risk of attacking human limitation altogether, seeking to produce a more-than-human being, one not only without illnesses, but also without foibles, fatigue, failures, or foolishness" (President's Council on Bioethics 2003: 307). Human dignity, the Council concluded, should not be thought of as something to be gained through medication. If anything, Council members cautioned that excessive trust in biotechnology might leave dignity behind.

It is notable that the language of dignity has prevailed in connection with this cautionary perspective while being generally absent from contrasting views. Underlying the appeal to human dignity is skepticism that psychotropic drugs, even if used benevolently, could smooth away the rough edges of human nature without erasing what is most precious about human freedom and distinctive of human expression and creativity. These elaborations of the nuances of human dignity suggest an open-ended concept ever rich in meaning.

THE RHETORIC OF HUMAN DIGNITY

A memorable 1987 television commercial portrayed a cracked egg sizzling in a frying pan with the narrative, "This is your brain on drugs." That advertisement, which was part of an antinarcotics campaign by the Partnership for a Drug-Free America, was ranked one of the top television ads of all time and helped to change attitudes toward illicit drug use (Alexander 2000). Today, functional brain imaging provides pictures of the neural substrates of drug addiction and craving (Magalhaes 2005; Hommer 1999). The ability to display images of the harmful effects of illicit drugs on the brain provides powerful reinforcement to campaigns against drug abuse in the interest of human dignity. One writer asks, "Are you trading health and dignity for a few minutes of cocaine high?" (Enterline 2010).

Rhetoric has its proper place in discussions about psychotropic drugs. Rhetoric aims to persuade. It may educate, convince, incentivize, or inspire. It may provoke, frighten, enrage, or deceive. Donald Calne points out that, "The essential difference between emotion and reason is that emotion leads to actions while reason leads to conclusions" (Calne 1999: 236). Wisdom needs both, and it is both to which rhetoric appeals.

It is interesting to consider whether a brain-based conceptualization of the rhetoric of human dignity might be possible (Gazzaniga 2005). As it turns out, the rhetoric of neuroscience and the neuroscience of rhetoric are strangely interrelated. Neuroscience is beginning to describe the art of rhetoric in terms of what areas of the brain are activated by emotional appeals, the triggering of memories, or reasoning tasks involving the anticipation of reward or loss. Neurons in the orbitofrontal cortex, for example, encode the value of offered and chosen goods (Padoa-Schioppa and Assad 2006), and another group of neurons in the parietal cortex adjust their firing rate to the magnitude and probability of anticipated reward (Yang and Schladen 2007; Platt and Glimcher 1999). The deciphering of these and other neural mechanisms encoding logic and emotion may in time permit rhetorical scholars to tease out the neural correlates of rhetorical appeals (Jack and Appelbaum 2010).

Brain images might even show interesting differences in how people think about dignity. In other words, disparities in word choice when referring to human dignity, or the ways of thinking about human nature they imply, may have detectable neurocognitive signatures. The cognitive capacities to discern individuality and value in others and to develop interpersonal relationships have discrete neuroanatomical correlates. These include neurons in the fusiform gyrus that encode the recognition of faces (Cheshire 2009b) and mirror neurons, which allow one to infer another's mental state (Rizzolatti and Sinigaglia 2007).

Fukuyama writes,

> Denial of the concept of human dignity—that is, of the idea that there is something unique about the human race that entitles every member of the species to a higher moral status than the rest of the natural world— leads us down a very perilous path. (Fukuyama 2002: 160)

The word *perilous* may stir neurons in the amygdala responsible for emotional arousal, anxiety, and fear, and in this case is united with a rational argument. By contrast, when Macklin grumbles "useless," and Pinker resorts to epithets such as "stupid" and "totalitarian" in rejecting the language of human dignity, such words appeal to less sophisticated neural centers. Expletives are mediated by primitive limbic system structures and serve the social functions of repulsing intruders and expressing anger (Van Lancker and Cummings 1999). By contrast, Bostrom's criticism of human dignity as having "profound vagueness" (Bostrom 2009: 174) is a form of expression that involves higher cerebral centers, as do a number of examples of affirming rhetoric. For example, Kass writes of "excellent human activity" (President's Council on Bioethics 2003), Mitchell of "special worth" (Mitchell et al. 2007: 69), and Jeeves of "highly developed soulishness" and human uniqueness (Jeeves 2000).

The use of rhetoric in commenting on human dignity may be an indication that its full meaning is beyond reasonable description (Cheshire 2006a). Explanations of human dignity also involve the use of metaphor, the complex neural correlates of which are only beginning to be elucidated (Schmidt and Seger 2009; Shibata et al. 2007). Yuval Levin, while advocating for the use of explicit reasoned arguments, argues that there are moral truths that are reasonable but not fully rational, that can be understood but not fully articulated. These, he writes, "are the realms where many ethical limits express themselves not in syllogisms but in shudders" (Levin 2003: 55). Levin's rhetoric bids the reader to reflect on the deep wisdom of moral intuition that, if unheeded, he cautions could lead to "a culture without awe filled with people without souls" (Levin 2003: 65). Similarly, C. S. Lewis employs the haunting metaphor of "men without chests" to depict humans devoid of the dignity of moral sensitivity (Lewis 1947).

Implicit in reductionistic descriptions is a silent rhetoric that undervalues human nature. A case in point is the presentation by Steven Hyman to the President's Council on Bioethics as recounted by Meilaender:

> There is, he said, no fundamental difference between using psychotropic drugs to affect the brain and using our disciplined experience to do so. Both are probably ways of accomplishing the same mechanistic aim: remodeling synapses within the brain. If that is really true, of course, it should make no difference to us which means we choose to reach the desired end when behavior needs to be altered. (Meilaender 2009: 50)

On saying this, Hyman drew back somewhat from the implications of his own assertion, which Meilaender interpreted to be the sign of a humane insight. The cold rhetoric of a reductionistic scientific view of mind and brain stands in tension, writes Meilaender, with "a reluctance to turn at once to medication as a solution to misbehavior and its accompanying discontents" (Meilaender 2009: 50).

DIGNIFYING THE DEBATE

Roger Brownsword writes that the words "human dignity" operate as a "conversation stopper" (Brownsword 2005). This symptom of conversation arrest is in need of a diagnosis. The preceding discussions provide a number of diagnostic clues that lend support to the following assessment.

There are two ways in which the conversation on human dignity tends to get stuck. First is the tendency to rush to judgment and pronounce a more conclusive appraisal of human dignity than the evidence allows. Included in this type of error are the unprovable assertions that human dignity either does not exist or is reducible to a material account of so many collisions of neuropeptides or other molecules. Overconfidence in the sufficiency of neurobiology or of a favorite theory to the exclusion of other sources of knowledge can lead to faulty conclusions.

Second is the substitution of insults and ad hominem attacks for rational argument. Such intimidating, if not undignified, rhetoric arouses primal brain structures, but it fails to engage centers of higher intelligence in the cerebral cortex. Shouting may draw attention, but it cannot establish truth.

The therapeutic response to conversation arrest does not involve psychotropic drugs. The way forward requires an ethical framework for civil discourse. Within that framework, the autonomy of communities and individuals to think for themselves about the meaning of human dignity deserves respect. Contributions from all charitable perspectives should be welcomed for consideration. Rather than give up on the possibility of consensus, interested parties would benefit from pressing forward in mutually respectful dialogue to elaborate the meaning of human dignity in light of developments in neuroscience and psychopharmacology and to consider the implications for all of humanity.

Progress toward enriching the understanding of human dignity also requires an attitude of humility. The diversity of viewpoints on the meaning of human dignity would seem to suggest that any one represents but a partial understanding of something greater. In evaluating the meaning of human dignity, it is necessary to be aware of other arguments and alternative opinions, not at a distance but in community. The meaning of human dignity, moreover, is not the exclusive possession of any one partisan group but is open to discovery by all, provided the conversation is allowed to continue.

A humble attitude entails the willingness to hold in tension competing ideas, any one of which, if elevated to the status of an overriding tenet and taken to the extreme, would violate other valid moral principles. A recurring example is the assumed priority given to ends over means in the choice of the word *useful*. Assertions that the language of human dignity is "useless," whether in regard to psychotropic drugs or in other contexts, imply a utilitarian approach to ethics. Whether the language of human dignity is useful depends very much on the purpose in mind. The phrase "human dignity" does not lend itself well to the formulation of operational criteria to be used for utilitarian goals that emphasize ends without proper regard for means. Regardless of what one believes human dignity to be, acknowledging it restrains undue expediency.

Where biotechnology interfaces with the brain, concepts of human dignity will be tested anew. They will be tested, in particular, by how well they guide decisions about the application of psychotropic drugs in ways that succeed in safeguarding and illuminating what is essential to our common humanity. Glimpses of human dignity will be ever present to those who search for them, oftentimes at moments of greatest human limitation and vulnerability.

Whether the language of human dignity will take hold in discussions about the appropriate use of psychotropic drugs will depend on the lessons of history and the discoveries of the future. Pursuing an understanding of human dignity ultimately can enrich society, not because it is easy, but because it is difficult.

Bibliography

Academy of Medical Sciences (2011) 'Animals Containing Human Material.' Available HTTP: <http://go.nature.com/uy17hx> (accessed11 August 2011).

Agich, G. J. (2007) 'Reflections on the Function of Dignity in the Context of Caring for Old People,' *Journal of Medicine and Philosophy*, 32: 483–94.

Ai, A., and T. R. McCormick (2010) 'Increasing Diversity of Americans' Faiths Alongside Baby Boomers' Aging,' *Journal of Health Care Chaplains*, 16, no. 1–2: 24–41.

Aikman, D. (2003) *Jesus in Beijing: How Christianity Is Transforming China and Changing the Global Balance of Power*, Washington, DC: Regnery.

Alexander, E. (2000) 'Famous Fried Eggs: Students Debate Effectiveness, Accuracy of Well-known Anti-drug Commercial,' CNN. Available HTTP: <http://www.cnn.com/fyi/interactive/news/brain/brain.on.drugs.html#'Good%20persuasive%20information'> (accessed 11 December 2010).

Alexander, T. (1986) *John Dewey's Theory of Art, Experience and Nature: The Horizons of Feeling*, Albany: State University of New York Press.

Alghrani, A. (2009) 'Viability and Abortion: Lessons from Ectogenesis?' *Expert Review of Obstetrics and Gynecology*, 4, no. 6: 625–34.

Almeida-Porada, G., C. Porada, et al. (2007) 'The Human-Sheep Chimeras as a Model for Human Stem Cell Mobilization and Evaluation of Hematopoietic Grafts' Potential,' *Experimental Hematology*, 35, no. 10: 1594–1600.

Almeida-Porada, G., E. D. Zanjani, et al. (2010) 'Bone Marrow Stem Cells and Liver Regeneration,' *Experimental Hematology*, 38, no. 7: 574–80.

Ambady, N., and J. Skowronski, J. (2008) *First Impressions*, New York: Guilford Press.

Andorno, R. (2007) 'Global Bioethics at UNESCO: In Defence of the Universal Declaration on Bioethics and Human Rights,' *Journal of Medical Ethics*, 33: 150–54.

—— (2009) 'Human Dignity and Human Rights as a Common Ground for a Global Bioethics,' *Journal of Medicine and Philosophy*, 34, no. 3: 223–40.

Annas, G. J. (2005) *American Bioethics. Crossing Human Rights and Health Law Boundaries*, New York: Oxford University Press.

Annas, G. J., L. Andrews, and R. Isasi (2002) 'Protecting the Endangered Human: Toward an International Treaty Prohibiting Cloning and Inheritable Alterations,' *American Journal of Law and Medicine*, 28, no. 2–3: 151–78.

Annas, G., and M. A. Grodin (1992) *The Nazi Doctors and the Nuremberg Code: Human Rights in Human Experimentation*, New York: Oxford University Press.

Anonymous (2010) 'Adderall Receives Honorary Degree from Harvard,' *The Onion*, 31 May. Available HTTP: <http://www.theonion.com/articles/adderall-receives-honorary-degree-from-harvard,17527/> (accessed 6 December 2010).

Anversa, R., D. Torella, J. Kajstura, B. Nadal-Ginard, and A. Leri, A. (2002) 'Myocardial Regeneration,' *European Heart Journal Supplements*, 4: G67–G71.

Aquinas, T. (1920) *The Summa Theologica of St. Thomas Aquinas*. 2nd and rev. edn. Literally translated by Fathers of the English Dominican Province. Available HTTP: <http://www.newadvent.org/summa/index.html> (accessed 14 June 2011).

―――― (1997) *Aquinas on Creation: Writings on the 'Sentences' of Peter Lombard*, trans. S. E. Baldner and W. E. Carroll, Toronto: Pontifical Institute of Mediaeval Studies.

Arieli, Y. (2002) 'On the Necessary and Sufficient Conditions for the Emergence of the Doctrine of the Dignity of Man and His Rights,' in D. Kretzmer and E. Klein (eds.), *The Concept of Human Dignity in Human Rights Discourse*, The Hague: Kluwer.

Aristotle (1968) *De Anima, Books II and III*, trans. D. W. Hamlyn, Oxford: Clarendon Press.

―――― (1984) *The Complete Works of Aristotle: The Revised Oxford Translation*, 2 vols., ed. J. Barnes, Princeton, NJ: Princeton University Press.

―――― (1999) *Nicomachean Ethics*, 2nd edn, trans. T. Irwin, Indianapolis, IN: Hackett.

Arnhart, L. (1988) 'Aristotle's Biopolitics: A Defense of Biological Teleology against Biological Nihilism [with Commentaries],' *Politics and the Life Sciences*, 6, no. 2: 173–229. Available HTTP: <http://www.jstor.org/stable/4235573> (accessed 20 August 2010).

―――― (1998) *Darwinian Natural Right: The Biological Ethics of Human Nature*, Albany: State University of New York Press.

Artigas, M., T. F. Glick, and R. A. Martínez (2006) *Negotiating Darwin: The Vatican Confronts Evolution 1877–1902*, Baltimore: Johns Hopkins University Press.

Asai, A., and S. Oe (2005) 'A Valuable Up-to-date Compendium of Bioethical Knowledge,' *Developing World Bioethics* (Special Issue: Reflections on the UNESCO Draft Declaration on Bioethics and Human Rights), 5, no. 3: 216–19.

Ashcroft, R. (2005) 'Making Sense of Dignity,' *Journal of Medical Ethics*, 31, no. 11: 679–82.

Ashworth, W. B., Jr. (2003) 'Christianity and the Mechanistic Universe,' in D. C. Lindberg and R. L. Numbers (eds.), *When Science and Christianity Meet*, Chicago: University of Chicago Press.

Astafiev, S., C. Stanley, G. Shulman, and M. Corbetta, M. (2004) 'Extrastriate Body Area in Human Occipital Cortex Responds to the Performance of Motor Actions,' *Nature Neuroscience*, 7: 542–48.

Atkinson, A., M. Tunstall, and W. Dittrich, W. (2007) 'Evidence for Distinct Contributions of Form and Motion Information to the Recognition of Emotions from Body Gestures,' *Cognition*, 104: 59–72.

Atkinson, A., W. Dittrich, A. Gemmell, and A. Young, A. (2004) 'Emotion Perception from Dynamic and Static Body Expressions in Point-Light and Full-Light Displays,' *Perception*, 33: 717–46.

Atran, S. (1999) 'The Universal Primacy of Generic Species in Folk-Biological Taxonomy: Implications for Human Biological, Cultural, and Scientific Evolution,' *Species: New Interdisciplinary Essays*, 231–61.

Augustine of Hippo [Aurelius Augustinus] (2006) *Confessions*, 2nd edn, trans. F. J. Sheed, ed. M. P. Foley, Indianapolis, IN: Hackett.

Austin, M., T. Riniolo, and S. Porges, S. (2007) 'Borderline Personality Disorder and Emotion Regulation: Insights from the Polyvagal Theory,' *Brain and Cognition*, 65: 69–76.

Aveling, E. B. (1883) *The Religious Views of Charles Darwin*, London: Freethought. Available HTTP: <http://darwin-online.org.uk/converted/Ancillary/1883_Aveling_A234.html> (accessed 15 September 2010).

Ayton-Shenker, D. (1995) *The Challenge of Human Rights and Cultural Diversity: United Nations Background Note*. New York: United Nations Department of

Public Information. Available HTTP: <http://www.un.org/rights/dpi1627e.htm> (accessed 14 June 2011).

Azevedo, F.A.C., L.R.B. Carvalho, L. T. Grinberg, J. M. Farfel, R.E.L. Ferretti, , R.E.P. Leite, W. J. Filho, R. Lent, and S. Herculano-Houzei (2009) 'Equal Numbers of Neuronal and Nonneuronal Cells Make the Human Brain an Isometrically Scaled-up Primate Brain,' *Journal of Comparative Neurology*, 513: 532–41.

Bacon, F. (1964) 'Refutation of Philosophies,' in B. Farrington (ed.), *The Philosophy of Francis Bacon: An Essay on Its Development from 1603 to 1609, with New Translations of Fundamental Texts*, Chicago: University of Chicago Press.

Badcott, D. (2003) 'The Basis and Relevance of Emotional Dignity,' *Medicine, HealthCare and Philosophy*, 6: 123–31.

Bailey, K. (2008) *Jesus through Middle Eastern Eyes*, Downers Grove, IL: InterVarsity Press.

Baker, H. (1947) *The Image of Man: A Study of the Idea of Human Dignity in Classical Antiquity, the Middle Ages, and the Renaissance*, New York: Harper & Brothers.

Baker, R. (2001) 'Bioethics and Human Rights: A Historical Perspective,' *Cambridge Quarterly of Healthcare Ethics*, 10, no. 3: 241–52.

Bal, E., E. Harden, D. Lamb, A. Van Hecke, J. Denver, and S. Porges, S. (2010) 'Emotion Recognition in Children with Autism Spectrum Disorders: Relations to Eye Gaze and Autonomic State,' *Journal of Autism and Developmental Disorders*, 40: 358–70.

Balaban, E. (1997) 'Changes in Multiple Brain Regions Underlie Species Differences in a Complex, Congenital Behavior,' *Proceedings of the National Academy of Science*, 94, no. 5: 2001–6.

—— (2005) 'Brain Switching: Studying Evolutionary Behavioral Changes in the Context of Individual Brain Development,' *International Journal of Developmental Biology*, 49, no. 2–3: 117–24.

Balaban, E., M. A. Teillet, et al. (1988) 'Application of the Quail-Chick Chimera System to the Study of Brain Development and Behavior,' *Science*, 241, no. 4871: 1339–42.

Balaguer-Ballester, E., N. Clark, M. Coath, K. Krumbholz, and S. Denham, S. (2009) 'Understanding Pitch Perception as a Hierarchical Process with Top-Down Modulation,' *PLoS Computational Biology*, 5: e1000301.

Bannerman, R., M. Milders, B. De Gelder, and A. Sahraie, A. (2008) 'Orienting to Threat: Faster Localizations of Fearful Facial Expressions and Body Postures Revealed by Saccadic Eye Movements,' *Proceedings of the Royal Society*, Series B, 276: 1635–41.

Banse, R., and Scherer, K. (1996) 'Acoustic Profiles in Vocal Emotion Expression,' *Journal of Personality and Social Psychology*, 70: 614–36.

Bänziger, T., and K. Scherer (2005) 'The Role of Intonation in Emotional Expressions,' *Speech Communication*, 46: 252–67.

Bar, M., M. Neta, and H. Linz (2006) 'Very First Impressions,' *Emotion*, 6: 269–78.

Barclay, C., J. Cutting, and L. Koslowski (1978) 'Temporal and Spatial Factors in Gait Perception That Influence Gender Recognition,' *Perception and Psychophysics*, 23: 145–52.

Barilan, Y. M. (2009) 'Judaism, Dignity and the Most Vulnerable Women on Earth,' *American Journal of Bioethics*, 9, no. 11: 35–37.

Baron-Cohen, S. (2009) 'Autism: The Empathizing-Systemizing (E-S) Theory,' *Annals of the New York Academy of Sciences*, 1156: 68–80.

Barrett, L., and D. Rendall (2010) 'Out of Our Minds: The Neuroethology of Primate Strategic Behavior,' in M. Platt and A. Ghanzanfar (eds.), *Primate Neuroethology*, New York: Oxford University Press.

Bates, L., P. Lee, N. Njiraini, J. Poole, K. Sayialel, S. Sayialel, et al. (2008) 'Do Elephants Show Empathy?' *Journal of Consciousness Studies*, 15: 204–25.

Batson, C., T. Klein, L. Highberger, and L. Shaw (1995) 'Immorality from Empathy-Induced Altruism: When Compassion and Justice Conflict,' *Journal of Personality and Social Psychology*, 68: 1042–54.

Bauckham, R. (1998) *God Crucified: Monotheism and Christology in the New Testament*, Grand Rapids, MI: Eerdmans.

Bayertz, K. (1996) 'Human Dignity: Philosophical Origin and Scientific Erosion of an Idea,' in K. Bayertz (ed.), *Sanctity of Life and Human Dignity*, Dordrecht: Kluwer.

Baylis, F. (2008) 'Global Norms in Bioethics. Problems and Prospects,' in R. Green, A. Donovan, and S. Jauss (eds.), *Global Bioethics. Issues of Conscience for the Twenty-First Century*, Oxford: Clarendon Press.

Baylis, F., and J. S. Robert (2003) 'Crossing Species Boundaries,' *American Journal of Bioethics*, 3, no. 3: 1–13.

Beabout, G. (2004) 'Personhood as Gift and Task: The Place of the Person in Catholic Social Thought,' *Catholic Social Science Review*, 9. Available HTTP: <http://www.catholicsocialscientists.org/CSSR/Archival/vol_ix.htm> (accessed 14 June 2011).

Beauchamp, T. L., and J. F. Childress (1994) *Principles of Biomedical Ethics*, 4th edn, New York: Oxford University Press.

Beckwith, F. (1998) 'Arguments from Bodily Rights: A Critical Analysis,' in L. Pojman and F. Beckwith (eds.), *The Abortion Controversy: 25 Years after Roe v. Wade, A Reader*, 2nd edn, Belmont, CA: Wadsworth.

—— (2007) *Defending Life: A Moral and Legal Case against Abortion Choice*, New York: Cambridge University Press.

Behe, M. J. (1996) *Darwin's Black Box*, New York: Free Press.

—— (2007) *The Edge of Evolution: The Search for the Limits of Darwinism*, New York: Free Press.

Beilby, J. (ed.) (2002) *Naturalism Defeated? Essays on Plantinga's Evolutionary Argument against Naturalism*, Ithaca, NY: Cornell University Press.

Bekoff, M. (2007) *Encyclopedia of Human-Animal Relationships: A Global Exploration of Our Connections with Animals*, Westport, CT: Greenwood Press.

Bell, R., A. Buchner, and J. Musch (2010) 'Enhanced Old-New Recognition and Source Memory for Faces of Cooperators and Defectors in a Social-Dilemma Game,' *Cognition*, 117: 261–75.

Bell, R., and A. Buchner (2009) 'Enhanced Source Memory for Names of Cheaters,' *Evolutionary Psychology*, 7: 317–30.

—— (2010) 'Justice Sensitivity and Source Memory for Cheaters,' *Journal of Research in Personality*, 44: 677–83.

—— (2011) 'Source Memory for Faces Is Determined by Their Emotional Evaluation,' *Emotion*, 11: 249–61.

Benagiano, G., S. Carrara, and V. Filippi (2011) 'Robert G. Edwards and the Roman Catholic Church,' *Reproductive BioMedicine Online*, 22, no. 7: 665–672.

Benatar, S. (2005) 'Moral Imagination: The Missing Component in Global Health,' *PLoS Medicine*, December. Available HTTP: <http://www.plosmedicine.org/article/info:doi/10.1371/journal.pmed.0020400> (accessed 20 July 2010).

Benedict XVI, Pope (2005) 'Homily: Mass for the Inauguration of the Pontificate of His Holiness Benedict XVI,' 24 April 2005, Vatican City: Libreria Editrice Vaticana. Available HTTP: <http://www.vatican.va/holy_father/benedict_xvi/homilies/2005/documents/hf_ben-xvi_hom_20050424_inizio-pontificato_en.html> (accessed 5 September 2010).

—— (2008) 'Saint Paul (15): The Apostle's Teaching on the Relation between Adam and Christ,' General Audience, 3 December 2008, Vatican City: Libreria Editrice Vaticana. Available HTTP: <http://www.vatican.va/holy_father/

benedict_xvi/audiences/2008/documents/hf_ben-xvi_aud_20081203_en.html> (accessed 5 September 2010).

Benn, P., and A. Chapman (2010) 'Ethical Challenges in Providing Noninvasive Prenatal Diagnosis,' *Current Opinion in Obstetrics and Gynecology*, 22: 128–34.

Berger, P. (1983) 'On the Obsolescence of the Concept of Honor,' in S. Hauerwas and A. MacIntyre (eds.), *Revisions: Changing Perspectives in Moral Philosophy*, Notre Dame, IN: University of Notre Dame Press.

Berridge, K. (2007) 'The Debate over Dopamine's Role in Reward: The Case for Incentive Salience,' *Psychopharmacology*, 191: 391–43.

Berridge, K., and J. Aldridge, J. (2008) 'Decision Utility, Incentive Salience, and Cue-Triggered "Wanting,"' in E. Morsella, J. Bargh, and P. Gollwitzer (eds.), *The Oxford Handbook of Human Action*, New York: Oxford University Press.

Bettenhausen, E. (1998) 'Genes in Society: Whose Body?' in R. Willer (ed.), *Genetic Testing and Screening: Critical Engagement at the Intersection of Faith and Science*, Minneapolis, MN: Kirk House Publishers.

Beyleveld, D., and R. Brownsward (2001) *Human Dignity in Bioethics and Biolaw*, Oxford: Oxford University Press.

Bhuiyan, M. R. (2009) 'Imagining the Consequences of Human Biotechnology,' *Journal of Health Politics, Policy and Law*, 34, no. 5: 829–39.

Bickerton, E. (2008) 'Darwin's Last Word: How Words Changed Cognition,' response to Penn et al. 2008, *Behavioral and Brain Sciences*, 31: 109–78. Available HTTP: <http://proxy.foley.gonzaga.edu:2048/login?url=http://proquest.umi.com/pqdweb?did=1601421561&sid=1&Fmt=2&clientId=10553&RQT=309&VName=PQD> (accessed 17 August 2010).

Birnbacher, D. (1996) 'Ambiguities in the Concept of *Menschenwürde*,' in K. Bayertz (ed.), *Sanctity of Life and Human Dignity*, Dordrecht, Netherlands: Kluwer Academic Publishers.

Black, V. (2000) 'What Dignity Means,' in E. McLean (ed.), *Common Truths: New Perspectives on Natural Law*, Wilmington, DE: ISI Books.

Blair, J., D. Mitchell, and K. Blair (2005) *The Psychopath: Emotion and the Brain*, Malden, MA: Blackwell.

Blake, R., and M. Shiffrar (2007) 'Perception of Human Motion,' *Annual Review of Psychology*, 58: 47–71.

Bobrow, M. (2011) 'Regulate Research at the Animal-Human Interface,' *Nature*, 475, no. 7357: 448.

Boehm, C. (1993) 'Egalitarian Behavior and Reverse Dominance Hierarchy,' *Current Anthropology*, 34: 227–54.

—— (1999) *Hierarchy in the Forest: The Evolution of Egalitarian Behavior*, Cambridge, MA: Harvard University Press.

Bonhoeffer, D. (1955) *Ethics*, trans. N. H. Smith, New York: Macmillan.

—— (2005) *Ethics* (Dietrich Bonhoeffer Works, Vol. 6), trans. I. Tödt, H. E. Tödt, E. Feil, and C. Green, Minneapolis, MN: Fortress Press.

Bontekoe, R. (2008) *The Nature of Dignity*, Lanham, MD: Rowman & Littlefield.

Booth, A., J. Pinto, and B. Bertenthal (2002) 'Perception of Symmetrical Patterning of Human Gait by Infants,' *Developmental Psychology*, 28: 554–63.

Bora, E., S. Gökçen, and B. Veznedaroglu (2008) 'Empathic Abilities in People with Schizophrenia,' *Psychiatry Research*, 160: 23–29.

Bordet, S., Feldman, S. & Knoppers, B.M. (2007) Legal aspects of animal-human combinations in Canada. *McGill Health Law Publication*, 1: 83–99.

Bordet, S., J. Bennett, et al. (2010) 'The Changing Landscape of Human-Animal Chimera Research: A Canadian Regulatory Perspective,' *Stem Cell Research*, 4, no. 1: 10–16.

Bortolotti, L., and J. Harris (2005) 'Stem Cell Research, Personhood and Sentience,' *Reproductive Biomedicine Online*, 10, Supplement 1: 68–75.

Bostanov, V., and B. Kotchoubey (2004) 'Recognition of Affective Prosody: Continuous Wavelet Measures of Event-Related Brain Potentials to Emotional Exclamations,' *Psychophysiology*, 41: 259–68.

Bostrom, N. (2009) 'Dignity and Enhancement,' in E. D. Pellegrino (ed.), *Human Dignity and Bioethics*, Washington, DC: President's Council on Bioethics.

Bouchrika, I., and M. Nixon (1988) 'Gait Recognition by Dynamic Cues,' paper presented at Nineteenth IEEE International Conference on Pattern Recognition, Tampa, Florida.

Boyd, R., H. Gintis, and S. Bowles (2010) 'Coordinated Punishment of Defectors Sustains Cooperation and Can Proliferate When Rare,' *Science*, 328: 617–20.

Boyle, J. (1977) 'The Concept of Health and the Right to Health Care,' *Social Thought*, 3: 5–17.

—— (1996) 'Catholic Social Justice and Health Care Entitlement Packages,' *Christian Bioethics*, 2: 280–92.

—— (2001) 'Fairness in Holdings: A Natural Law Account of Property and Welfare Rights,' *Social Philosophy and Policy*, 18: 206–26.

Braaten, C. (1992) 'Protestants and Natural Law,' *First Things*, January. Available HTTP: http://www.firstthings.com/article/2007/12/002-protestants-and-natural-law-28 (accessed 19 July 2012).

Bradley, G. V. (1998) 'No Intentional Killing Whatsoever: The Case of Capital Punishment,' in R. P. George (ed.), *Natural Law and Moral Inquiry: Ethics, Metaphysics, and Politics in the Work of Germain Grisez*, Washington, DC: Georgetown University Press.

Braine, D. (1988) *The Reality of Time and the Existence of God: The Project of Proving God's Existence*, Oxford: Oxford University Press.

—— (1992) *The Human Person: Animal and Spirit*, Notre Dame, IN: Notre Dame University Press.

Brazelton, T., B. Koslowski, and M. Main (1974) 'The Origins of Reciprocity. (The Early Mother–Infant Interaction),' in M. Lewis and L. Rosenblum (eds.), *The Effect of the Infant on Its Caregiver*, New York: John Wiley.

Brebels, L., D. De Cremer, and C. Sedikides (2008) 'Retaliation as a Response to Procedural Unfairness: A Self-Regulatory Approach,' *Journal of Personality and Social Psychology*, 95: 1511–25.

Brooke, J. H. (2009a) 'Darwin and Victorian Christianity,' in J. Hodge and G. Radick (eds.), *The Cambridge Companion to Darwin*, 2nd edn, Cambridge: Cambridge University Press.

—— (2009b) '"Laws Impressed on Matter by the Creator"? The *Origin* and the Question of Religion,' in M. Ruse and R. J. Richards (eds.), *The Cambridge Companion to the 'Origin of Species,'* Cambridge: Cambridge University Press.

Brown, F. B. (1986) *The Evolution of Darwin's Religious Views*, Macon, GA: Mercer University Press.

Brown, M. T. (2009) 'Moral Complicity in Induced Pluripotent Stem Cell Research,' *Kennedy Institute of Ethics Journal*, 19: 1–22.

Brownback, S. (2004) 'Abortion and the Conscience of the Nation, Revisited,' *Human Life Review*, 30, no. 1: 5–12.

Browne, J. (1995) *Charles Darwin: Voyaging*, New York: Alfred A. Knopf.

Brownsword, R. (2003) 'Bioethics Today, Bioethics Tomorrow: Stem Cell Research and the "Dignitarian Alliance",' *Notre Dame Journal of Law, Ethics, and Public Policy*, 17: 15–51.

Brownsword, R. (2005) 'Stem Cells and Cloning: Where the Regulatory Consensus Fails,' *New England Law Review*, 39: 543.

Brugger, E. C. (2003) *Capital Punishment and Roman Catholic Moral Tradition*, Notre Dame, IN: University of Notre Dame Press.

────── (2008) 'Aquinas on the Immateriality of Intellect: A Non-materialist Reply to Materialist Objections,' *National Catholic Bioethics Quarterly*, 8: 103–19.

────── (2010) 'Parthenotes, iPS Cells, and the Product of ANT-OAR: A Moral Assessment Using the Principles of Hylomorphism,' *National Catholic Bioethics Quarterly*, 10: 123–42.

Buchner, A., R. Bell, B. Mehl, and J. Musch (2009) 'No Enhanced Recognition Memory, But Better Source Memory for Faces of Cheaters,' *Evolution and Human Behavior*, 30: 212–24.

Burt, C. W., L. F. McCaig, and E. A. Rechtsteiner (2007) 'Ambulatory Medical Care Utilization Estimates for 2005,' *Advance Data from Vital and Health Statistics*. Available HTTP: <http://www.cdc.gov/nchs/data/ad/ad388.pdf> (accessed 6 December 2010).

Bush, G. W. (2001) 'On Stem Cell Research,' in M. Ruse, and C. Pynes (eds.), *Stem Cell Controversy: Debating the Issues*, Amherst, NY: Prometheus Books.

────── (2001) 'Stem Cell Science and the Preservation of Life,' *New York Times*, 12 August, WK 13.

Callahan, D. (1995) 'The Puzzle of Profound Respect,' *Hastings Center Report*, 25: 39–40.

Calne, D. B. (1999) *Within Reason: Rationality and Human Behavior*, New York: Pantheon.

Cameron, N. M. de S. (2004) 'Christian Vision for the Biotech Century: Toward a Strategy,' in C. Colson and N. Cameron (eds.), *Human Dignity in the Biotech Century*, Downers Grove, IL: InterVarsity Press.

Cancik, H. (2002) '"Dignity of Man" and "*Persona*" in Stoic Anthropology: Some Remarks on Cicero, *De Officiis I* 105–107,' in D. Kretzmer and E. Klein (eds.), *The Concept of Human Dignity in Human Rights Discourse*, The Hague: Kluwer.

Capron, A. (1991) 'In re Helga Wanglie,' *Hastings Center Report*, 21.

Carlsmith, K. (2006) 'The Roles of Retribution and Utility in Determining Punishment,' *Journal of Experimental Psychology*, 42: 437–51.

Carlsmith, K., and J. Darley (2008) 'Psychological Aspects of Retributive Justice,' *Advances in Experimental Social Psychology*, 40: 193–236.

Carroll, W. E. (2000) 'Creation, Evolution, and Thomas Aquinas,' *Revue des Questions Scientifiques*, 171, no. 4: 319–347.

Casile, A., and M. Giese (2005) 'Critical Features for Recognition of Biological Motion,' *Journal of Vision*, 5: 348–60.

Cassell, C. K., and K. M. Foley (1999) *Principles for Care of Patients at the End of Life: An Emerging Consensus among the Specialties of Medicine*, New York: Milbank Memorial Fund.

Cassidy, J., and P. Shaver (eds.) (2008) *Handbook of Attachment: Theory, Research, and Clinical Applications*, New York: Guilford Press.

Castiglione, J. (2008) 'Human Dignity under the Fourth Amendment,' *Wisconsin Law Review*, 4: 655–711.

Caulfield, T., and A. Chapman (2005) 'Human Dignity as a Criterion for Science Policy,' *PLOS Medicine*, 2, no. 8: 736–38.

Caulfield, T., and R. Brownsword (2006) 'Human Dignity: A Guide to Policy Making in the Biotechnology Era,' *Nature Reviews Genetics*, 7: 72–76.

Chachkin, C. J. (2007) 'What Potent Blood: Noninvasive Prenatal Genetic Diagnosis and the Transformation of Modern Prenatal Care,' *American Journal of Law and Medicine*, 33: 9–53.

Chamberlain, J., T. Yamagami, et al. (2007) 'Efficient Generation of Human Hepatocytes by the Intrahepatic Delivery of Clonal Human Mesenchymal Stem Cells in Fetal Sheep,' *Hepatology*, 46, no. 6: 1935–45.

Chambers, R. (1844) *Vestiges of the Natural History of Creation*, London: John Churchill. Available HTTP: <http://darwin-online.org.uk/content/frameset?view type=text&itemID=A2&pageseq=1> (accessed 1 September 2010).

Chang, D., and N. Troje (2008) 'Perception of Animacy and Direction from Local Biological Motion Signals,' *Journal of Vision*, 8: 1–10.

Chapman, A. R. (1999) *Unprecedented Choices: Religious Ethics at the Frontiers of Genetic Science*, Minneapolis, MN: Fortress Press.

Chapman, A. R., M. S. Frankel, and M. S. Garfinkel (1999) *Stem Cell Research and Applications: Monitoring the Frontiers of Biomedical Research*, Washington, DC: American Association for the Advancement of Science.

Chaskalon, A. (2002) 'Human Dignity as a Constitutional Value,' in D. Kretzmer and E. Klein (eds.), *The Concept of Human Dignity in Human Rights Discourse*, The Hague: Kluwer Law International.

Chatterjee, A. (2004) 'Cosmetic Neurology: The Controversy over Enhancing Movement, Mentation, and Mood,' *Neurology*, 63: 968–74.

——— (2007) 'Cosmetic Neurology and Cosmetic Surgery: Parallels, Predictions, and Challenges,' *Cambridge Quarterly of Healthcare Ethics*, 16: 129–37.

Chayama, K., C. N. Hayes, et al. (2011) 'Animal Model for Study of Human Hepatitis Viruses,' *Journal of Gastroenterol Hepatology*, 26, no. 1: 13–18.

Cheney, D., and R. Seyfarth (2007) *Baboon Metaphysics: The Evolution of a Social Mind*, Chicago: University of Chicago Press.

Cheshire, W. P. (2003) 'The Overlooked Test,' *Parkinsonism and Related Disorders*, 9: 315.

——— (2005) 'Small Things Considered: The Ethical Significance of Human Embryonic Stem Cell Research,' *New England Law Review*, 39: 573–81.

——— (2006a) 'When Eloquence Is Inarticulate,' *Ethics and Medicine*, 22: 135–38.

——— (2006b) 'Drugs for Enhancing Cognition and Their Ethical Implications: A Hot New Cup of Tea,' *Expert Review of Neurotherapeutics*, 6: 263–66.

——— (2007a) 'Glimpsing the Grey Marble,' *Ethics and Medicine*, 23: 119–21.

——— (2007b) 'Can Grey Voxels Resolve Neuroethical Dilemmas?' *Ethics and Medicine*, 23: 135–40.

——— (2008a) 'Till We Have Minds,' *Today's Christian Doctor*, 39: 11–16.

——— (2008b) 'The Synapse and Other Gaps,' *Ethics and Medicine*, 24: 139–43.

——— (2008c) 'The Pharmacologically-Enhanced Physician,' *Virtual Mentor: AMA Journal of Ethics*, 10: 594–98.

——— (2009a) 'Accelerated Thought in the Fast Lane,' *Ethics and Medicine*, 25: 75–78.

——— (2009b) 'Facebook and the Fusiform Gyrus,' *Ethics and Medicine*, 25: 139–44.

——— (2010a) 'Just Enhancement,' *Ethics and Medicine*, 26: 7–10.

——— (2010b) 'Does Alien Hand Syndrome Refute Free Will?' *Ethics and Medicine*, 26: 71–74.

Cheshire, W. P., E. D. Pellegrino, L. K. Bevington, C. B. Mitchell, N. L. Jones, K. T. FitzGerald, C. E. Koop, and J. F. Kilner (2003) 'Stem Cell Research: Why Medicine Should Reject Human Cloning,' *Mayo Clinic Proceedings*, 78: 1010–18.

Churchland, P. S. (2002) *Brain-Wise: Studies in Neurophilosophy*, Cambridge, MA: MIT Press.

——— (2009) 'Human Dignity from a Neurophilosophical Perspective,' in E. D. Pellegrino (ed.), *Human Dignity and Bioethics*, Washington, DC: President's Council on Bioethics.

Cicero, M. T. (1997) *The Nature of the Gods*, trans. P. G. Walsh, Oxford: Clarendon Press.

——— (2000) *On Obligations (De Officiis)*, trans. P. G. Walsh, Oxford: Oxford University Press.

Cimino, A., and A. Delton (2010) 'On the Perception of Newcomers: Toward an Evolved Psychology of Intergenerational Coalitions,' *Human Nature*, 21: 186–202.

Clanton, J. C. (2009) *The Ethics of Citizenship: Liberal Democracy and Religious Convictions*, Waco, TX: Baylor University Press.

Clark, W. P. (2004) 'For Reagan, All Life Was Sacred,' *Human Life Review*, 30, no, 1: 69–70.

Clarke, W. N., S.J. (1992) 'Person, Being, and St. Thomas,' *Communio*, 19: 601–18.

—— (2004) 'Freedom, Equality, Dignity of the Human Person,' *Catholic Social Science Review*, 9. Available HTTP: <http://www.catholicsocialscientists.org/CSSR/Archival/vol_ix.htm> (accessed 14 June 2011).

Cleary, D. (2008) 'Antonio Rosmini,' *The Stanford Encyclopedia of Philosophy* (Winter 2008 edn), E. N. Zalta (ed.). Available HTTP: <http://plato.stanford.edu/archives/win2008/entries/antonio-rosmini/> (accessed 30 December 2010).

Clutton-Brock, T. (2009) 'Cooperation between Non-kin in Animal Societies,' *Nature*, 462: 51–57.

Clutton-Brock, T., and G. Parker (1995) 'Punishment in Animal Societies,' *Nature*, 373: 209–16.

Coady, C. (2009) 'Playing God,' in J. Savulescu and N. Bostrom (eds.), *Human Enhancement*, Oxford: Oxford University Press, 155–80.

Cochrane, T. I. (2007) 'Brain Disease or Moral Condition? Wrong Question,' *American Journal of Bioethics*, 7: 24–25.

Colson, C. W. (2004) 'Preface,' in C. Colson and N. Cameron (eds.), *Human Dignity in the Biotech Century*, Downers Grove, IL: InterVarsity Press.

Committee on Economic, Social and Cultural Rights (2000) 'General Comment 14: The Right to the Highest Attainable Standard of Health' (article 12 of the International Covenant on Economic, Social and Cultural Rights), UN Doc. E/C.12/2000/4.

Condic, M. L. (2008) 'Alternative Sources of Pluripotent Stem Cells: Altered Nuclear Transfer,' *Cell Proliferation*, 41, Supplement 1: 7–19.

Congregation for the Doctrine of the Faith (1974) *Declaration on Procured Abortion*. Available HTTP: <http://www.vatican.va/roman_curia/congregations/cfaith/documents/rc_con_cfaith_doc_19741118_declaration-abortion_en.html> (accessed 14 June 2011).

—— (1987) *Donum Vitae: Respect for Human Life*. Available HTTP: <http://www.vatican.va/roman_curia/congregations/cfaith/documents/rc_con_cfaith_doc_19870222_respect-for-human-life_en.html> (accessed 14 June 2011).

—— (2008) *Dignitas Personae: On Certain Bioethical Questions*. Available HTTP: <http://www.vatican.va/roman_curia/congregations/cfaith/documents/rc_con_cfaith_doc_20081208_dignitas-personae_en.html> (accessed 14 June 2011).

Convention on the Elimination of All Forms of Discrimination against Women (1979) United Nations General Assembly Resolution 34/180. Entered into force 3 September 1981.

Cooper, J. W. (2000) *Body, Soul, and Life Everlasting: Biblical Anthropology and the Monism-Dualism Debate*, updated edn, Grand Rapids, MI: William B. Eerdmans.

Coors, M. E., J. J. Glover, et al. (2010) 'The Ethics of Using Transgenic Non-human Primates to Study What Makes Us Human,' *Nature Reviews Genetics*, 11, no. 9: 658–62.

Copan, P. (2007) *Loving Wisdom: Christian Philosophy of Religion*, St. Louis: Chalice Press.

—— (2008) 'Theism, Naturalism, and the Foundations of Morality,' in R. Stewart (ed.), *The Future of Atheism*, Minneapolis: Fortress Press.

—— (2009) *'True for You, But Not for Me': Overcoming Objections to Christian Faith*, Minneapolis: Bethany House.

Cornetta, K., and E. M. Meslin (2011) 'Ethical and Scientific Issues in Gene Therapy and Stem Cell Research,' in R. Chadwick, H. T. Have, and E. M. Meslin (eds.), *The SAGE Handbook of Health Care Ethics*, Thousand Oaks, CA: Sage.

Cosmides, L., H. Barrett, and J. Tooby, J. (2010) 'Adaptive Specializations, Social Exchange, and the Evolution of Human Intelligence,' *Proceedings of the National Academy of Sciences*, 107, Supplement 2: 9007–14.

Cosmides, L., J. Tooby, L. Fiddick, and G. Bryant (2005) 'Detecting Cheaters,' *Trends in Cognitive Science*, 9: 505–6.

Cottingham, J. (1978) '"A Brute to the Brutes?": Descartes' Treatment of Animals,' *Philosophy*, 53: 551–59.

Cowie, R., and R. Cornelius (2003) 'Describing the Emotional States That Are Expressed in Speech,' *Speech Communication*, 40: 5–32.

Craig, William Lane (2009) 'This Most Gruesome of Guests,' in R. K. Garcia and N. L. King (eds.), *Is Goodness without God Good Enough?* Lanham, MD: Rowman & Littlefield.

Crick, F. (1994) *The Astonishing Hypothesis: The Scientific Search for the Soul*, New York: Simon & Schuster.

Crockett, M. (2009) 'The Neurochemistry of Fairness: Clarifying the Link Between Serotonin and Prosocial Behavior,' *Annals of the New York Academy of Sciences*, 1167: 76–86.

Crockett, M., L. Clark, and T. Robbin (2009) 'Reconciling the Role of Serotonin in Behavioral Inhibition and Aversion: Acute Tryptophan Depletion Abolishes Punishment-Induced Inhibition in Humans,' *Journal of Neuroscience*, 29: 11993–99.

Cunningham, P. C. (2004) 'Learning from Our Mistakes: The Pro-Life Cause and the New Bioethics,' in C. Colson and N. Cameron (eds.), *Human Dignity in the Biotech Century*, Downers Grove, IL: InterVarsity Press.

Cushing, H. (1925) *The Life of Sir William Osler*, Oxford: Clarendon Press.

Cutting, J. (1977) 'Recognizing Friends from Their Walk: Gait Perception without Familiarity Cues,' *Bulletin of the Psychonomic Society*, 9: 353–56.

Dabrock, P. (2009) 'Playing God? Synthetic Biology as a Theological and Ethical Challenge,' *Systems and Synthetic Biology*, 3, no. 1–4: 47–54.

Damasio, H., T. Grabowski, R. Frank, A. M. Galaburda, and A. R. Damasio (1994) 'The Return of Phineas Gage: Clues about the Brain from the Skull of a Famous Patient,' *Science*, 20: 1102–5.

Darwin, C. (1839) *Voyages of the Adventure and Beagle*, vol. 3. London: Henry Colburn. Available HTTP: <http://darwin-online.org.uk/content/frameset?viewty pe=text&itemID=F10.3&pageseq=1> (accessed 23 July 2010).

—— (1859) *On the Origin of Species by Means of Natural Selection, or the Preservation of Favoured Races in the Struggle for Life*, 1st edn, London: John Murray. Available HTTP: <http://darwin-online.org.uk/content/frameset?viewtype=text& itemID=F373&pageseq=1> (accessed 23 July 2010).

—— (1875) *The Variation of Animals and Plants under Domestication*, 2nd edn, 2 vols, London: John Murray. Vol. 2 available HTTP: <http://darwin-online. org.uk/content/frameset?itemID=F880.2&viewtype=text&pageseq=1> (accessed 22 November 2010).

—— (1876) *On the Origin of Species by Means of Natural Selection, or the Preservation of Favoured Races in the Struggle for Life*, 6th edn, London: John Murray. Available HTTP:<http://darwin-online.org.uk/content/frameset?viewtype=te xt&itemID=F401&pageseq=1> (accessed 12 August 2010).

—— (1881) 'Mr. Darwin on Vivisection,' *British Medical Journal* 1 (23 April): 660. Available HTTP: <http://darwin-online.org.uk/content/frameset?itemID=F1 354&viewtype=text&pageseq=1> (accessed 4 September 2010).

—— (1882) *The Descent of Man and Selection in Relation to Sex*, 2nd edn, London: John Murray. Available HTTP: <http://darwin-online.org.uk/content/frame set?itemID=F955&viewtype=text&pageseq=1> (accessed 23 July 2010).

—— (1890) *The Expression of the Emotions in Man and Animals*, New York: Appleton and Company. (Originally published London: John Murray, 1872.)

—— (1958) *The Autobiography of Charles Darwin*, ed N. Barlow, New York: W. W. Norton.

—— (1987) *Charles Darwin's Notebooks, 1836–1844: Geology, Transmutation of Species, Metaphysical Enquiries*, ed. and transcribed P. H. Barrett, P. J. Gautrey, S. Herbert, D. Kohn, and S. Smith, Ithaca, NY: Cornell University Press. Several digital versions of the *Notebooks*, with digital reproductions of the original autographs, available HTTP: <http://darwin-online.org.uk/>.

Darwin, F. (ed.) (1887) *The Life and Letters of Charles Darwin, Including an Autobiographical Chapter*, 3 vols., London: John Murray. Vol. 1 available HTTP: <http://darwin-online.org.uk/content/frameset?itemID=F1452.1&viewtype=text &pageseq=1>. Vol. 2 available HTTP: <http://darwin-online.org.uk/content/fra meset?viewtype=text&itemID=F1452.2&pageseq=1> (accessed 22 July 2010).

Darwin, F., and A. C. Seward (eds.) (1903) *More Letters of Charles Darwin: A Record of His Work in a Series of Hitherto Unpublished Letters*, 2 vols., London: John Murray. Vol. 1 available HTTP: <http://darwin-online.org.uk/content/frame set?itemID=F1548.1&viewtype=text&pageseq=1> (accessed 1 September 2010). Vol. 2 available HTTP: <http://darwin-online.org.uk/content/frameset?itemID=F 1548.2&viewtype=text&pageseq=1> (accessed 18 September 2010).

Davis, D. S. (2010) *Genetic Dilemmas: Reproductive Technology, Parental Choices, and Children's Futures*, New York: Oxford University Press.

Davis, F. D. (2008) 'Human Dignity and Respect for Persons: A Historical Perspective on Public Bioethics,' in *Human Dignity and Bioethics: Essays Commissioned by the President's Council on Bioethics*, Washington, DC: President's Council on Bioethics. Available HTTP: <http://bioethics.georgetown.edu/pcbe/ reports/human_dignity/human_dignity_and_bioethics.pdf> (accessed 21 July 2010).

Dawkins, R. (1986) *The Blind Watchmaker*, New York: W. W. Norton.

—— (1989) *The Selfish Gene*, new edn, Oxford: Oxford University Press.

—— (1998) 'When Religion Steps on Science's Turf: The Alleged Separation between the Two Is Not So Tidy,' *Free Inquiry* 18, no. 2. Available HTTP: <http://www. secularhumanism.org/library/fi/dawkins_18_2.html> (accessed 25 April 2007).

De Dreu, C., L. Greer, M. Handgraaf, S. Shalvi, G. Van Kleef, M. Baas, et al. (2010) 'The Neuropeptide Oxytocin Regulates Parochial Altruism in Intergroup Conflict among Humans,' *Science*, 328: 1408–11.

de Duve, C. (2008) 'Summary,' in Address of His Holiness Benedict XVI to the Members of the Pontifical Academy of Sciences on the Occasion of Their Plenary Assembly, Extra Series 33, Vatican City: Pontifical Academy of Sciences. Available HTTP: <http://www.vatican.va/roman_curia/pontifical_academies/acdscien/ documents/newpdf/es33.pdf> (accessed 5 September 2010).

De Gelder, B. (2006) 'Towards the Neurobiology of Emotional Body Language,' *Nature Reviews Neuroscience*, 7: 242–49.

—— (2009) 'Why Bodies? Twelve Reasons for Including Bodily Expressions in Affective Neuroscience,' *Philosophical Transactions of the Royal Society, Series B*, 364: 3475–84.

De Gelder, B., J. Van den Stock, H. Meeren, C. Sinke, M. Kret, and M. Tamietto (2010) 'Standing Up for the Body: Recent Progress in Uncovering the Networks Involved in the Perception of Bodies and Bodily Expressions,' *Neuroscience and Biobehavioral Reviews*, 34: 513–27.

de Melo-Martin, I. (2008) 'Chimeras and Human Dignity,' *Kennedy Institute of Ethics Journal*, 18, no. 4: 331–46.

De Waal, F. (1996) *Good Natured: The Origins of Right and Wrong in Humans and Other Animals*, Cambridge, MA: Harvard University Press.

—— (2006) *Primates and Philosophers: How Morality Evolved*, ed. S. Macedo and J. Ober, Princeton, NJ: Princeton University Press.

—— (2008) 'Putting the Altruism Back into Altruism: The Evolution of Empathy,' *Annual Review of Psychology*, 59: 279–300.

—— (2009) *The Age of Empathy: Nature's Lessons for a Kinder Society*, New York: Harmony Books.

Death with Dignity Act of Oregon. Available HTTP: http://www.oregon.gov/DHS/ph/pas/docs/year11.pdf (accessed 19 November 2009).

Death with Dignity Act of Washington. Available HTTP: <http://www.doh.wa.gov/dwda/> (accessed 19 November 2009).

Debiec, J., and J. E. LeDoux (2004) 'Noradrenergic Signaling in the Amygdala Contributes to the Reconsolidation of Fear Memory: Treatment Implications for PTSD,' *Annals of the New York Academy of Science*, 1071: 521–24.

Decety, J., and P. Jackson (2004) 'The Functional Architecture of Human Empathy,' *Behavioral and Cognitive Neuroscience Reviews*, 3: 71–100.

—— (2006) 'A Social-Neuroscience Perspective on Empathy,' *Current Directions in Psychological Science*, 15: 54–58.

Decety, J., and Y, Moriguchi (2007) 'The Empathic Brain and Its Dysfunction in Psychiatric Populations: Implications for Intervention across Different Clinical Conditions,' *BioPsychoSocial Medicine*, 1: 22–43.

Decety, J., P. Jackson, J. Sommerville, T. Chaminade, and A. Meltzoff (2004) 'The Neural Basis of Cooperation and Competition: An fMRI Investigation,' *NeuroImage*, 23: 744–51.

DeGrazia, D. (2007) 'Human-Animal Chimeras: Human Dignity, Moral Status, and Species Prejudice,' *Metaphilosophy*, 38, nos. 2–3: 309–29.

Dekkers, W., and M. O. Rikkert (2007) 'Memory Enhancing Drugs and Alzheimer's Disease: Enhancing the Self or Preventing the Loss of It?' *Medicine, Health Care, and Philosophy*, 10: 141–51.

Delgado, M., R. Frank, and E. Phelps (2005) 'Perceptions of Moral Character Modulate the Neural Systems of Reward during the Trust Game,' *Nature Neuroscience*, 8: 1611–18.

Dembski, W. A. (1994) 'On the Very Possibility of Intelligent Design,' in J. P. Moreland (ed.), *The Creation Hypothesis*, Downers Grove, IL: InterVarsity Press.

Dennett, D. (1995) *Darwin's Dangerous Idea: Evolution and the Meanings of Life*, New York: Simon & Schuster.

—— (2003) 'How Has Darwin's Theory of Natural Selection Transformed Our View of Humanity's Place in the Universe?,' in W. K. Purves, D. E. Sadava, G. H. Orians, and H. C. Heller (eds.), *Life: The Science of Biology*, 7th ed, Sinauer Associates. Available HTTP: <http://ase.tufts.edu/cogstud/papers/Lifeessay.htm> (accessed 18 August 2010).

—— (2008) 'How to Protect Human Dignity from Science,' in *Human Dignity and Bioethics: Essays Commissioned by the President's Council on Bioethics*, Washington, DC: President's Council on Bioethics. Available HTTP: <http://bioethics.georgetown.edu/pcbe/reports/human_dignity/human_dignity_and_bioethics.pdf> (accessed 21 July 2010).

Denver, J., S. Reed, and S. Porges (2007) 'Methodological Issues in the Quantification of Respiratory Sinus Arrhythmia,' *Biological Psychology*, 74: 286–94.

Department of Health, Education, and Welfare (1979) 'Protection of Human Subjects in Biomedical and Behaviorial Research,' Ethics Advisory Board, Final Report. *Federal Register*.

Descartes, R. (1970) *Philosophical Letters*, trans. and ed. A. Kenny, Oxford: Clarendon Press.

—— (1984) *The Philosophical Writings of Descartes*, vol. 2, trans. J. Cottingham, R. Stoothoff, and D. Murdoch, Cambridge: Cambridge University Press.

—— (1993) *Discourse on Method* and *Meditations on First Philosophy*, trans. Donald A. Cress, Indianapolis, IN: Hackett.

Desmond, A., and J. Moore (1991) *Darwin*, New York: W. W. Norton.

—— (2009) *Darwin's Sacred Cause: How a Hatred of Slavery Shaped Darwin's Views on Human Evolution*, New York: Houghton Mifflin Harcourt.

Dewey, J. (1922) *Human Nature and Conduct*, New York: Modern Library.

—— (1938) *Logic: The Theory of Inquiry*, ed. Jo Ann Boydston. Carbondale: Southern Illinois University Press.

—— (1960) *The Quest for Certainty*, New York: G. P. Putnam's Sons.

—— (1997) *How We Think*, New York: Dover.

Dicke, K. (2002) 'The Founding Function of Human Dignity in the Universal Declaration of Human Rights,' in D. Kretzmer and E. Klein (eds.), *The Concept of Human Dignity in Human Rights Discourse*, The Hague: Kluwer.

Diller, L. H. (2009) 'A Misuser's Guide to Adderall,' *The Harvard Crimson*. Available HTTP <http://www.thecrimson.harvard.edu/article/2009/4/28/a-misusers-guide-to-adderall-the/> (accessed 6 December 2010).

Dilley, S. C. (2008) 'Enlightenment Science and Globalization,' *Journal of Interdisciplinary Studies* 20, no. 1/2: 135–54.

—— (2011) 'Charles Darwin's Use of Theology in the Origin of Species,' *British Journal for the History of Science*, doi: 10.1017/S000708741100032X, Published online by Cambridge University Press, 4 May.

Diogenes Laertius (1972) *Lives of Eminent Philosophers*, 2 vols., trans. R. D. Hicks, Cambridge, MA: Harvard University Press. Available HTTP: <http://www.perseus.tufts.edu/hopper/text?doc=Perseus%3atext%3a1999.01.0258> (accessed 3 September 2010).

Dobzhansky, T. (1973) 'Nothing in Biology Makes Sense Except in Light of Evolution,' *American Biology Teacher*, 35 (March): 125–29.

Dolgin, J., and L. Shepherd (2005) *Bioethics and the Law*, New York: Aspen.

Donchin, A. (1996) 'Feminist Critiques of New Fertility Technologies: Implications for Social Policy,' *Journal of Medicine and Philosophy*, 21: 475–98.

Donnelly, J. (1989) *Universal Human Rights in Theory and Practice*, Ithaca, NY: Cornell University Press.

Doussard-Roosevelt, J., L. Montgomery, and S. Porges (2003) 'Short-Term Stability of Physiological Measures in Kindergarten Children: Respiratory Sinus Arrhythmia, Heart Period, and Cortisol,' *Developmental Psychobiology*, 43: 230–42.

Downing, P., Y. Jiang, M. Shuman, and N. Kanwisher (2001) 'A Cortical Area Selective for Visual Processing of the Human Body,' *Science*, 293: 2470–73.

Draper, J. W. (1874) *History of the Conflict between Religion and Science*, New York: D. Appleton. Available HTTP: <http://etext.virginia.edu/toc/modeng/public/DraHist.html> (accessed 29 April 2010).

Drummond, H. (1895) *The Ascent of Man*, 5th edn, New York: James Pott.

Dubreuil, B. (2008) 'Strong Reciprocity and the Emergence of Large-Scale Societies,' *Philosophy of the Social Sciences*, 38: 192–210.

—— (2010) 'Punitive Emotions and Norm Violations,' *Philosophical Explorations*, 13: 35–50.

Duclos, S., J. Laird, E. Schneider, M. Sexter, L. Stern, and O. Van Lighten (1989) 'Emotion-Specific Effects of Facial Expressions and Postures on Emotional Experience,' *Journal of Personality and Social Psychology*, 57: 100–8.

Duff, R. (2005) 'Punishment, Dignity, and Degradation,' *Oxford Journal of Legal Studies*, 25: 141–55.

Dupre, J. (2001) *Human Nature and the Limits of Science*, Oxford: Clarendon Press.

Dworkin, R. (1994) *Life's Dominion. An Argument about Abortion, Euthanasia and Individual Freedom*, New York: Vintage.

Earls, F. (2010) 'Darwin and Lincoln: Their Legacy of Human Dignity,' *Perspectives in Biology and Medicine* 53, no. 1: 3–13. Available HTTP: <http://muse.jhu.edu/journals/pbm/summary/v053/53.1.earls.html> (accessed 16 July 2010).

Eberl, J. T., and R. A. Ballard (2009) 'Metaphysical and Ethical Perspectives on Creating Animal-Human Chimeras,' *Journal of Medicine and Philosophy*, 34, no. 5: 470–86.

Eddington, A. S. (1931) *The Nature of the Physical World*, New York: Macmillan, 1928.

Edwards, R. G., P. C. Steptoe, and J. M. Purdy (1970) 'Fertilization and Cleavage In Vitro of Preovulator Human Oocytes,' *Nature*, 227: 1307–9.

Egas, M., and A. Riedl (2008) 'The Economics of Altruistic Punishment and the Maintenance of Cooperation,' *Proceedings of the Royal Society, Series B*, 275: 871–78.

Eisenstadt v. Baird (1972) 405 U.S. 438.

Ekert, J. (2002) 'Legal Roots of Human Dignity in German Law,' in D. Kretzmer and E. Klein (eds.), *The Concept of Human Dignity in Human Rights Discourse*, The Hague: Kluwer.

Elfenbein, H., and N. Ambady (2002) 'On the Universality and Cultural Specificity of Emotion Recognition: A Meta-analysis,' *Psychological Bulletin*, 128: 203–35.

—— (2003) 'Cultural Similarity's Consequences: A Distance Perspective on Cross-Cultural Differences in Emotion Recognition,' *Journal of Cross Cultural Psychology*, 34: 92–109.

Elliott, C. (2003) *Better than Well: American Medicine Meets the American Dream*, New York: W. W. Norton.

Elmer, K. R., and A. Meyer (2010) 'Sympatric Speciation without Borders?' *Molecular Ecology*, 19, no 10: 1991–93.

Enard, W., M. Przeworski, et al. (2002) 'Molecular Evolution of FOXP2, a Gene Involved in Speech and Language,' *Nature*, 418, no. 6900: 869–72.

Enard, W., S. Gehre, et al. (2009) 'A Humanized Version of FOXP2 Affects Cortico-basal Ganglia Circuits in Mice,' *Cell*, 137, no. 5: 961–71.

Engel, P., and R. Rorty (2007) *What's the Use of Truth?* New York: Columbia University Press.

Enterline, J. (2010) 'Are You Trading Health and Dignity for a Few Minutes of Cocaine High? Available HTTP: <http://ezinearticles.com/?Are-You-Trading-Health-and-Dignity-for-a-Few-Minutes-of-Cocaine-High?&id = 5022149> (accessed 11 December 2010).

Epictetus (2008) *Discourses and Selected Writings*, trans. and ed. R. Dobbin, London: Penguin.

Ersek, A., J. S. Pixley, et al. (2010) 'Persistent Circulating Human Insulin in Sheep Transplanted In Utero with Human Mesenchymal Stem Cells,' *Experimental Hematology*, 38, no. 4: 311–20.

Espmark, Y., T. Amundsen, and G. Rosenqvist (eds.) (2000) *Animal Signals*, Trondheim: Tapir Forlag.

Ethofer, T., S. Anders, M. Erb, C. Droll, L. Royen, R. Saur, et al. (2006) 'Impact of Voice on Emotional Judgment of Faces: An Event-Related fMRI study,' *Human Brain Mapping*, 27: 707–14.

Etkin, A., T. Egner, and R. Kalisch (2011) 'Emotional Processing in Anterior Cingulate and Medial Prefrontal Cortex,' *Trends in Cognitive Science*, 15: 85–93.

Evans, M. J., and M. H. Kaufman (1981) 'Establishment in Culture of Pluripotential Cells from Mouse Embryos,' *Nature*, 292: 154–56.

Falk, A., E. Fehr, and U. Fischbacher (2005) 'Driving Forces behind Informal Sanctions,' *Econometrica*, 73: 2017–30.

Farah, M. J. (2002) 'Emerging Ethical Issues in Neuroscience,' *Nature Neuroscience*, 5: 1123–29.

Farah, M. J., and A. S. Heberlein (2007) 'Personhood and Neuroscience: Naturalizing or Nihilating?' *American Journal of Bioethics*, 7: 37–48.

Farrelly, D., and N. Turnbull (2008) 'The Role of Reasoning Domain on Face Recognition: Detecting Violations of Social Contract and Hazard Management Rules,' *Evolutionary Psychology*, 6: 523–37.

Fehr, E. (2004) 'Don't Lose Your Reputation,' *Nature*, 432: 449–50.

—— (2005) 'Reply to Fowler, Johnson and Smirnov, 2005,' *Nature*, 433: E1–E2.

Fehr, E., and S. Gächter (2002) 'Altruistic Punishment in Humans,' *Nature*, 415: 137–40.

Fehr, E., and U. Fischbacher (2004) 'Third-Party Punishment and Social Norms,' *Evolution and Human Behavior*, 25: 63–87.

Feigenson, N. (2006) 'Brain Imaging and Courtroom Evidence: On the Admissibility and Persuasiveness of fMRI,' *International Journal of Law in Context*, 2: 233–55.

Feldman, R. (2007a) 'Mother-Infant Synchrony and the Development of Moral Orientation in Childhood and Adolescence: Direct and Indirect Mechanisms of Developmental Continuity,' *American Journal of Orthopsychiatry*, 77: 582–97.

—— (2007b) 'Parent-Infant Synchrony and the Construction of Shared Timing: Physiological Precursors, Developmental Outcomes, and Risk Conditions,' *Journal of Child Psychology and Psychiatry*, 48: 329–54.

Fiddick, L. (2008) 'Which Norms Are Strong Reciprocators Supposed to Enforce? Not All Norms Are Psychologically the Same,' *International Review of Economics*, 55: 77–89.

Field, T., and M. Diego (2008) 'Vagal Activity, Early Growth, and Emotional Development,' *Infant Behavior and Development*, 31: 361–73.

Finnis, J. (1980) *Natural Law and Natural Rights*, Oxford: Clarendon Press.

—— (1995) 'The Fragile Case for Euthanasia: A Reply to John Harris,' in J. Keown (ed.), *Euthanasia Examined*, Cambridge: Cambridge University Press.

—— (1998) *Aquinas: Moral, Political, and Legal Theory*, Oxford: Oxford University Press.

Fischbach, R. L., and J. D. Loike (2008) 'Postmortem Fatherhood: Life after Life,' *The Lancet*, 371, no. 9631: 2166–67.

Fisher, S. E., and C. Scharff (2009) 'FOXP2 as a Molecular Window into Speech and Language,' *Trends in Genetics*, 25, no. 4: 166–77.

Fiske, A. (1991) *Structures of Social Life: The Four Elementary Forms of Human Relations*, New York: Free Press.

—— (1992) 'The Four Elementary Forms of Sociality: Framework for a Unified Theory of Social Relations,' *Psychological Review*, 99: 689–723.

Foley, K., and H. Hendin (2002) *The Case against Assisted Suicide: For the Right to End-of-Life Care*, Baltimore: Johns Hopkins University Press.

Foot, P. (1978) *Virtues and Vices and Other Essays in Moral Philosophy*, Berkeley: University of California Press.

Foucault, Michel (1994) *The Order of Things: An Archaeology of the Human Sciences*, New York: Random House/Vintage.

Frede, M. (1992) 'On Aristotle's Conception of the Soul,' in M. C. Nussbaum and A. O. Rorty (eds.), *Essays on Aristotle's De Anima*, Oxford: Clarendon Press.

Freud, S. (1920) *A General Introduction to Psychoanalysis*, New York: Horace Liveright. Available HTTP: <http://books.google.com/books?pg=PA251&dq=Freud+general+introduction+to+psychoanalysis&id=4cgtdgIOCzAC#v=onepage&q&f=false> (accessed 21 August 2010).

—— (1964) 'New Introductory Lectures on Psycho-Analysis,' in J. Strachey (ed.), *The Standard Edition of the Complete Psychological Works of Sigmund Freud*, Vol. 22, London: Hogarth Press.

—— (1974) 'A Difficulty in the Path of Psycho-Analysis,' in J. Strachey (ed.), *The Standard Edition of the Complete Psychological Works of Sigmund Freud*, Vol. 17, *An Infantile Neurosis and Other Works*, London: Hogarth Press.

Frith, S. (2004) 'Who's Minding the Brain?' *Pennsylvania Gazette*. Available HTTP: <http://www.upenn.edu/gazette/0104/frith1.html> (accessed 11 December 2010).

Frowein, J. (2002) 'Human Dignity in International Law,' in D. Kretzmer and E. Klein (eds.), *The Concept of Human Dignity in Human Rights Discourse*, The Hague: Kluwer Law International.

Fukuda, K., and S. Yuasa (2006) 'Stem Cells as a Source of Regenerative Cardiomyocytes,' *Circulation Research*, 98: 1002–13.

Fukuyama, F. (2002) *Our Posthuman Future: Consequences of the Biotechnology Revolution*, New York: Farrar, Straus and Giroux.

Gächter, S., and B. Hermann (2009) 'Reciprocity, Culture and Human Cooperation: Previous Insights and a New Cross-Cultural Experiment,' *Philosophical Transactions of the Royal Society, Series B*, 364: 791–806.

Gage, L. P. (2010) 'Can a Thomist Be a Darwinist?' in J. Richards (ed.), *God and Evolution*, Seattle: Discovery Institute Press.

Galilei, G. (2008) *The Essential Galileo*, Indianapolis, IN: Hackett.

Gallese, V. (2001) 'The 'Shared Manifold' Hypothesis: From Mirror Neurons to Empathy,' *Journal of Consciousness Studies*, 8: 33–50.

Gazzaniga, M. S. (2005) *The Ethical Brain*, New York: Dana Press.

—— (2009) 'Humans: The Party Animal,' *Daedalus*, 138, no. 3: 21–34.

Gazzola, V., L. Aziz-Zadeh, and C. Keysers (2006) 'Empathy and the Somatotopic Auditory Mirror System in Humans,' *Current Biology*, 16: 1824–29.

Gelernter, D. (2008) 'The Irreducibly Religious Character of Human Dignity,' in *Human Dignity and Bioethics: Essays Commissioned by the President's Council on Bioethics*, Washington, DC: President's Council on Bioethics. Available HTTP: <http://bioethics.georgetown.edu/pcbe/reports/human_dignity/human_dignity_and_bioethics.pdf> (accessed 21 July 2010).

George, R. P., and C. Tollefsen (2008). *Embryo: A Defense of Human Life*, New York: Doubleday.

Gewirth, A. (1996) *The Community of Rights*, Chicago: University of Chicago Press.

Giese, M., I. Thornton, and S. Edelman (2008) 'Metrics of the Perception of Body Movement,' *Journal of Vision*, 8: 1–18.

Gintis, H. (2000) 'Strong Reciprocity and Human Sociality,' *Journal of Theoretical Biology*, 206: 169–79.

Giudice, M., B. Ellis, and E. Shirtcliff (2011) 'The Adaptive Calibration Model of Stress Responsivity,' *Neuroscience and Biobehavioral Reviews*, 35: 1562–92.

Glannon, W. (2007) *Bioethics and the Brain*, New York: Oxford University Press.

Glendon, M. A. (2002) *A World Made New. Eleanor Roosevelt and the Universal Declaration of Human Rights*, New York: Random House.

—— (2001) *The World Made New: Eleanor Roosevelt and the Universal Declaration of Human Rights*, New York: Random House.

Gobl, C., and A. Chasaide (2003) 'The Role of Voice Quality in Communicating Emotion, Mood, and Attitude,' *Speech Communication*, 40: 189–212.

Goble, D., J. M. Scott, et al. (2006) *The Endangered Species Act at Thirty*, Washington, DC: Island Press.

Gomez, C. (1991) *Regulating Death: Euthanasia and the Case of the Netherlands*, New York: Free Press.

Gonzales v. Carhart (2007) 127 S. Ct. 1610.

Goodenough, U. (1998) *The Sacred Depths of Nature*, Oxford: Oxford University Press.

Goodfield, J. (1977) *Playing God: Genetic Engineering and the Manipulation of Life*, London: Hutchinson.

Gopnik, A., A. Meltzoff, and P. Kuhl (1999) *The Scientist in the Crib: Minds, Brains, and How Children Learn*, New York: William Morrow.

Gormally, L. (2004) 'Pope John Paul II's Teaching on Human Dignity and Its Implications for Bioethics," in C. Tollefsen (ed.), *John Paul II's Contribution to Catholic Bioethics*, Dordrecht, Netherlands: Springer.

Gould, J. L., and C. G. Gould (1994) *The Animal Mind*, New York: Scientific American Library.

Gould, S. J. (1977) *Ever Since Darwin: Reflections in Natural History*, New York: W. W. Norton.

—— (1981) *The Mismeasure of Man*, New York: W. W. Norton.

—— (1989) *Wonderful Life: The Burgess Shale and the Nature of History*, New York: W. W. Norton.

—— (1997) 'Nonoverlapping Magisteria,' *Natural History*, 106: 16–22.

Gourbin, C. (2005) 'Foetal Mortality, Infant Mortality, and Age of Parents. An Overview,' *Revue d'épidémiologie et de santé publique*, 53: 81–86.

Graham, G. (2001) *Evil and Christian Ethics*, Cambridge: Cambridge University Press.

Grandjean, D., B. Sander, G. Pourtois, S. Schwartz, M. Seghier, et al. (2005) 'The Voices of Wrath: Brain Responses to Angry Prosody in Meaningless Speech,' *Nature Neuroscience*, 8: 145–46.

Grandjean, D., T. Bänziger, and K. Scherer (2006) 'Intonation as an Interface between Language and Affect,' *Progress in Brain Research*, 156: 235–47.

Gray, A. (1861) *Natural Selection Not Inconsistent with Natural Theology: A Free Examination of Darwin's Treatise on the Origin of Species, and of Its American Reviewers* (reprinted from the *Atlantic Monthly* for July, August, and October, 1860), London: Trübner; Boston: Ticknor and Fields. Available HTTP: <http://darwin-online.org.uk/content/frameset?viewtype=text&itemID=A567&pageseq=1> (accessed 25 August 2010).

Gray, P. (2009) 'Play as a Foundation for Hunter-Gatherer Social Existence,' *American Journal of Play*, 1: 476–522.

Greely H. T., and J. Illes (2007) 'Neuroscience-based Lie Detection: The Urgent Need for Regulation,' *American Journal of Law and Medicine*, 33: 377–431.

Greely, H., B. Sahakian, J. Harris, R. C. Kessler, M. Gazzaniga, P. Campbell, and M. J. Farah (2008) 'Towards Responsible Use of Cognitive-Enhancing Drugs by the Healthy,' *Nature*, 456: 702–5.

Green, J. (2008) *Body, Soul, and Human Life: The Nature of Humanity in the Bible*, Grand Rapids, MI: Baker Academic.

Green, R. M. (1974) 'Conferred Rights and the Fetus,' *Journal of Religious Ethics*, 2, no. 1: 55–75.

—— (2001) *The Human Embryo Research Debates: Bioethics in the Vortex of Controversy*, New York: Oxford University Press.

—— (2002) 'Benefiting from "Evil": An Incipient Moral Problem in Human Stem Cell Research,' *Bioethics*, 16: 544–56.

Grene, M. (1961) 'Statistics and Selection,' *British Journal for the Philosophy of Science*, 12, no. 45: 25–42. Available HTTP: <http://www.jstor.org/stable/685536> (accessed 20 August 2010).

Grézes, J., S. Pichon, and B. De Gelder (2007) 'Perceiving Fear in Dynamic Body Expressions,' *NeuroImage*, 35: 959–67.

Griffin, D. R. (2001) *Animal Minds: Beyond Cognition to Consciousness*, rev. edn, Chicago: University of Chicago Press.

Grisez, G. (1993) *The Way of The Lord Jesus Christ*, Vol. 2: *Living a Christian Life*, Quincy, IL: Franciscan Press.

—— (2004) *God? Philosophical Preface to Faith*, Notre Dame, IN: St. Augustine's Press.

—— (2008) 'The True Ultimate End of Human Beings: The Kingdom, not God Alone,' *Theological Studies*, 69: 38–61.

Grisez, G., J. Boyle, and J. Finnis (1987) 'Practical Principles, Moral Truth, and Ultimate Ends,' *American Journal of Jurisprudence*, 32: 99–151.

Gros Espiell, H. (1999) 'Introduction,' in UNESCO's Division of Ethics of Science and Technologies (ed.), *Genèse de la Déclaration universelle sur le génome humain et les droits de l'homme*, Paris: UNESCO.

Gruber, H. E. (1974) *Darwin on Man: A Psychological Study of Scientific Creativity*, New York: E. P. Dutton.

Guenin, L. (2006) 'The Nonindividuation Argument against Zygotic Personhood,' *Philosophy*, 81: 463.

—— (2004) 'A Failed Noncomplicity Scheme,' *Stem Cells and Development*, 13: 456–59.

Güroğlu, B., W. van den Bos, S. Rombouts, and E. Crone (2010) 'Unfair? It Depends: Neural Correlates of Fairness in Social Context,' *Social Cognitive and Affective Neuroscience*, 5: 414–23.

Gustafson, J. (1974) *Theology and Christian Ethics*, Philadelphia: Pilgrim Press.

Habermas, J. (2003) *The Future of Human Nature*, Cambridge, UK, and Malden, MA: Polity Press.

—— (2006) *Time of Transitions*, trans. C. Cronin and M. Pensky, Cambridge: Polity.

Hagen, E., and G. Bryant (2003) 'Music and Dance as a Coalition Signaling System,' *Human Nature*, 14: 21–51.

Hahn, R. (1986) 'Laplace and the Mechanistic Universe,' in D. C. Lindberg and R. L. Numbers (eds.), *God and Nature: Historical Essays on the Encounter between Christianity and Science*, Berkeley: University of California Press.

Haidt, J. (2001) 'The Emotional Dog and Its Rational Tail: A Social Intuitionist Approach to Moral Judgment,' *Psychological Review*, 108: 818–34.

—— (2009) 'Moral Psychology and the Misunderstanding of Religion,' in J. Schloss and M. Murray (eds.), *The Believing Primate: Scientific, Philosophical, and Theological Reflections on the Origin of Religion*, New York: Oxford University Press.

Haidt, J., and C. Joseph (2004) 'Intuitive Ethics: How Innately Prepared Intuitions Generate Culturally Variable Virtues,' *Daedalus*, 133: 55–66.

Haidt, J., and J. Graham (2007) 'When Morality Opposes Justice: Conservatives Have Moral Intuition That Liberals May Not Recognize,' *Social Justice Research*, 20: 98–116.

—— (2009) 'Planet of the Durkheimians, Where Community, Authority, and Sacredness Are Foundations of Morality,' in J. Jost, A. Kay, and H. Thorisdottir (eds.), *Social and Psychological Bases of Ideology and System Justification*, New York: Oxford University Press.

Handyside, A. H., E. H. Kontogianni, K. Hardy, and R. M. Winston (1990) 'Pregnancies from Biopsied Human Preimplantation Embryos Sexed by Y-specific DNA Amplification,' *Nature*, 344: 768–70.

Hare, John M. (2007) *God and Morality: A Philosophical History*, Oxford: Blackwell.

Harris, J. (2005) 'Cloning,' in R. G. Frey and C. W. Heath (eds.), *A Companion to Applied Ethics*, Malden, MA: Blackwell.

—— (2008) 'Global Norms, Informed Consent, and Hypocrisy in Bioethics,' in R. Green, A. Donovan, and S. Jauss (eds.), *Global Bioethics. Issues of Conscience for the Twenty-First Century*, Oxford: Clarendon Press.

Harrison, P. (1992) 'Descartes on Animals,' *Philosophical Quarterly*, 42, no. 167: 219–27.

Hasker, W. (1999) *The Emergent Self*, Ithaca, NY: Cornell University Press.

Hauerwas, S. (1977) *Truthfulness and Tragedy*, Notre Dame, IN: University of Notre Dame Press.

Haught, J. F. (2003) *Deeper than Darwin: The Prospect for Religion in the Age of Evolution*, Boulder, CO: Westview Press.

—— (2008) *God after Darwin: A Theology of Evolution*, 2nd edn, Boulder, CO: Westview.

—— (2010) *Making Sense of Evolution: Darwin, God, and the Drama of Life*, Louisville, KY: Westminster John Knox Press.

Hayutin, A. M., M. Dietz, and L. Mitchell (2010) 'New Realities of an Older America: Challenges, Changes and Questions,' *Stanford Center on Longevity*, 5.

Hebets, E., and D. Papaj (2005) 'Complex Signal Functions: Developing a Framework of Testable Hypotheses,' *Behavioral Ecology and Sociobiology*, 57: 197–214.

Heffernan, J. (2004) 'Making the Christian Case for Democracy,' *Catholic Social Science Review*, 9. Available HTTP: <http://www.catholicsocialscientists.org/CSSR/Archival/vol_ix.htm> (accessed 14 June 2011).

Heidegger, M. (1959) *An Introduction to Metaphysics*, trans. R. Manheim, New Haven, CT: Yale University Press.

—— (1962) *Being and Time*, trans. J. Macquarrie and E. Robinson, New York: Harper & Row.

—— (1977) *The Question Concerning Technology and Other Essays*, trans. W. Lovitt, New York: Garland.

Heilman, K., E. Bal, O. Bazhenova, and S. Porges (2007) 'Respirational Sinus Arrhythmia and Tympanic Membrane Compliance Predict Spontaneous Eye Gaze Behaviors in Young Children: A Pilot Study,' *Developmental Psychobiology*, 49: 531–42.

Henrich, J., R. Boyd, S. Bowles, C. Camerer, E. Fehr, H. Gintis, et al. (2005) '"Economic Man" in Cross-Cultural Perspective: Behavioral Experiments in 15 Small-Scale Societies,' *Behavioral and Brain Sciences*, 28: 795–855.

Hermann, B., C. Thöni, and S. Gächter (2008) 'Antisocial Punishment across Societies,' *Science*, 319: 1362–67.

Hess, N., and E. Hagen (2006) 'Psychological Adaptations for Assessing Gossip Veracity,' *Human Nature*, 17: 337–54.

Hickman, L. (2002) 'Pragmatic Resources for Biotechnology,' in J. Keulartz et al. (eds.), *Pragmatist Ethics for a Technological Culture*, Boston: Kluwer Academic Publishers.

—— (2007) *Pragmatism as Post-Postmodernism*, New York: Fordham University Press.

Himmelfarb, G. (1968) *Darwin and the Darwinian Revolution*, New York: W. W. Norton.

Hittinger, R. (2006) 'Human Nature and States of Nature in John Paul II's Theological Anthropology,' in D. Robinson, G. M. Sweeney, and R. Gill, L.C. (eds.), *Human Nature in Its Wholeness: A Roman Catholic Perspective*, Washington, DC: Catholic University of America Press.

Hobbes, T. (1886) *Leviathan, or, The Matter, Form, and Power of a Commonwealth, Ecclesiastical and Civil*, 2nd edn, London: George Routledge and Sons. Available HTTP: <http://books.google.com/books?id=8-QtAAAAIAAJ&printsec=frontcover&source=gbs_ge_summary_r&cad=0#v=onepage&q&f=false> (accessed 21 December 2010).

Hobson, J., R. Harris, R. Garcia-Perez, and P. Hobson (2009) 'Anticipatory Concern: A Study in Autism,' *Developmental Science*, 12: 249–63.

Hobson, P. (2004) *The Cradle of Thought: Exploring the Origins of Thinking*, New York: Oxford University Press.

Hodge, J. and G. Radick (2003) 'Introduction,' in J. Hodge and G. Radick (eds.), *The Cambridge Companion to Darwin*, New York: Cambridge University Press.

Hodges, S., and K. Klein (2001) 'Regulating the Costs of Empathy: The Price of Being Human,' *Journal of Socio-Economics*, 30: 437–52.

Hoffman, D. et al. (2003) 'Cryptopreserved Embryos in the United States and Their Availability for Research,' *Fertility and Sterility*, 79: 1063–69.

Holland, S. (2007) 'Market Transactions in Reprogenetics: A Case for Regulation,' in L. K.a.G. Kaebnicht (ed.), *Reprogenetics: Law, Policy and Ethical Issues*, Baltimore: Johns Hopkins University Press.

Holland, S. (ed.) (2001) *Beyond the Embryo: A Feminist Appraisal of the Embryonic Stem Cell Debate*, London: MIT Press.

Holy Bible, Revised Standard Version (1952) New York: Thomas Nelson.

Hommer, D. W. (1999) 'Functional Imaging of Craving,' *Alcohol Research and Health*, 23: 187–96.

Horton, R. (2004) 'Rediscovering Human Dignity,' *Lancet*, 364: 1081–85.

House of Commons Science and Technology Committee (2005) *Human Reproductive Technologies and the Law*, vol. 1, *Fifth Report of Session 2004–2005*, London: House of Commons, Stationery Office Limited.

Hug, K. (2009) 'Research on Human-Animal Entities: Ethical and Regulatory Aspects in Europe,' *Stem Cell Reviews*, 5, no. 3: 181–94.

Hull, D. L. (1973) *Darwin and His Critics: The Reception of Darwin's Theory of Evolution by the Scientific Community*, Cambridge, MA: Harvard University Press.

—— (1991) 'The God of the Galapagos,' *Nature*, 352: 485–86.

Hume, D. (1988) *Dialogues Concerning Natural Religion and the Posthumous Essays of the Immortality of the Soul and of Suicide*, 2nd edn, R. Popkin (ed.), Indianapolis, IN: Hackett.

Hunt, A., and F. Halper (2008) 'Disorganizing Biological Motion,' *Journal of Vision*, 8: 1–5.

Huntley, W. B. (1972) 'David Hume and Charles Darwin,' *Journal of the History of Ideas*, 33, no. 3: 457–70.

Huther, C. (2009). 'Chimeras: The Ethics of Creating Human–Animal Interspecifics.' PhD diss. Ludwig-Maximilians-Universität Munich, Germany.

Huxley, A. (1932) *Brave New World*, New York: Harper & Brothers.

Huxley, T. H. (1901) *Life and Letters of Thomas Henry Huxley*, vol. 1, New York: D. Appleton. Available HTTP: <http://books.google.com/books?id=QVh87yn5cs EC&printsec=frontcover&dq=huxley+life+and+letters&hl=en&ei=fmp1TIDdJo WKlwer18nOAg&sa=X&oi=book_result&ct=result&resnum=2&ved=0CC8Q 6AEwAQ#v=onepage&q&f=false> (accessed 25 August 2010).

—— (1887) 'On the Reception of the *Origin of Species*,' in F. Darwin (ed.), *The Life and Letters of Charles Darwin, Including an Autobiographical Chapter*, vol. 2, London: John Murray. Available HTTP: <http://darwin-online.org.uk/content/frameset?itemID=F1452.2&viewtype=text&pageseq=1> (accessed 22 July 2010).

—— (1959) *[Evidence as to] Man's Place in Nature*, Ann Arbor: University of Michigan.

Iglesias, T. (2001) 'Bedrock Truths and the Dignity of the Individual,' *Logos*, 4: 114–34.

Ignatieff, M. (2001) *Human Rights as Politics and Idolatry*, Princeton, NJ: Princeton University Press.

In re Quinlan (1976), 70 N. J. 10, 355 A.2d 647, 79 ALR3d 205.

International Covenant on Civil and Political Rights (1966) United Nations General Assembly Resolution 2200 A (XXI). Entered into force 23 March 1976.

Jack, J., and L. G. Appelbaum (2010) 'This Is Your Brain on Rhetoric: Research Directions for Neurorhetorics,' *Rhetoric Society Quarterly*, 40: 411–437.

James, W. (1949) *Essays on Faith and Morals*, New York: Longmans, Green.

—— (1956) *The Will to Believe and Other Essays in Popular Philosophy*, New York: Dover.

—— (1985) *The Varieties of Religious Experience: A Study in Human Nature* (*The Works of William James*, vol. 15), Cambridge, MA: Harvard University Press.

Jameson, F. (1991) *Postmodernism or, The Cultural Logic of Late Capitalism*, Durham, NC: Duke University Press.

Järvinen-Pasley, A., S. Peppé, G. King-Smith, and P. Heaton (2008) 'The Relationship between Form and Function Level Receptive Prosodic Abilities in Autism,' *Journal of Autism and Developmental Disorders*, 38: 1328–40.

Jeeves, M. (2000) 'Whatever Became of the Soul?' in R. Stannard (ed.), *God for the 21st Century*, Philadelphia: Templeton Foundation Press.

—— (2004) 'Toward a Composite Portrait of Human Nature,' in M. Jeeves (ed.), *From Cells to Souls and Beyond: Changing Portraits of Human Nature*, Grand Rapids, MI: Eerdmans.

Jenkins, P. (2007) *The Next Christendom: The Coming of Global Christianity*, New York: Oxford University Press.

—— (2008) *The New Faces of Christianity: Believing the Bible in the Global South*, New York: Oxford University Press.

Jing-Bao N. (2005) 'Cultural Values Embodying Universal Norms: A Critique of a Popular Assumption about Cultures and Human Rights,' *Developing World Bioethics* (Special Issue: Reflections on the UNESCO Draft Declaration on Bioethics and Human Rights), 5, no. 3: 251–57.

John Paul II, Pope (1981) *Familiaris Consortio*. Available HTTP: <http://www.vatican.va/holy_father/john_paul_ii/apost_exhortations/documents/hf_jp-ii_exh_19811122_familiaris-consortio_en.html> (accessed 14 June 2011).

—— (1985) *Dilecti Amice*. Available HTTP: <http://www.vatican.va/holy_father/john_paul_ii/apost_letters/documents/hf_jp-ii_apl_31031985_dilecti-amici_en.html> (accessed 14 June 2011).

—— (1988) *Mulieris Dignitatem*. Available HTTP: <http://www.vatican.va/holy_father/john_paul_ii/apost_letters/documents/hf_jp-ii_apl_19880815_mulieris-dignitatem_lt.html> (accessed 14 June 2011).

—— (1991) *Centesimus Annus*, Encyclical issued 1 May 1991, Vatican City: Libreria Editrice Vaticana. Available HTTP: <http://www.vatican.va/holy_father/john_paul_ii/encyclicals/documents/hf_jp-ii_enc_01051991_centesimus-annus_en.html> (accessed 30 December 2010).

—— (1993) *Veritatis splendor: The Splendor of Truth*. Available HTTP: <http://www.vatican.va/holy_father/john_paul_ii/encyclicals/documents/hf_jp-ii_enc_06081993_veritatis-splendor_en.html> (accessed 14 June 2011).

—— (1995) *Evangelium vitae: The Gospel of Life*. Available HTTP: <http://www.vatican.va/holy_father/john_paul_ii/encyclicals/documents/hf_jp-ii_enc_25031995_evangelium-vitae_en.html> (accessed 14 June 2011).

—— (1997) *The Theology of the Body: Human Love in the Divine Plan*, Boston: Pauline Books.

—— (2003) 'Address to the Plenary Session on the Subject "The Origins and Early Evolution of Life," ' 22 October 1996, in *Papal Addresses to the Pontifical Academy of Sciences 1917–2002 and to the Pontifical Academy of Social Sciences 1994–2002, Scripta Varia* 100, Vatican City: Pontifical Academy of Sciences. Available HTTP: <http://www.vatican.va/roman_curia/pontifical_academies/acd-scien/documents/newpdf/sv100.pdf> (accessed 4 September 2010).

John XXIII, Pope (1963) *Pacem in Terris: On Establishing Universal Peace in Truth, Justice, Charity, and Liberty*, Encyclical issued 11 April 1963, Vatican City: Libreria Editrice Vaticana. Available HTTP: <http://www.vatican.va/holy_father/john_xxiii/encyclicals/documents/hf_j-xxiii_enc_11041963_pacem_en.html> (accessed 30 December 2010).

Johnson, B., and J. Pierce (2008) 'Morality in Animals? Yes, No, Maybe,'. May 15. Online. Available HTTP: , <http://scholar.googleusercontent.com/scholar?q=cache:mNr_bALXuOsJ:scholar.google.com/+animal+morality+%2b+Johnson+Pierce&hl=en&as_sdt=0,48> (accessed 24 July 2012).

Johnson, D., M. Price, and M. Takezawa (2008) 'Renaissance of the Individual: Reciprocity, Positive Assortment, and the Puzzle of Human Cooperation,' in C.

Crawford and D. Krebs (eds.), *Foundations of Evolutionary Psychology*, New York: Lawrence Erlbaum.

Johnson, P. (1976) *A History of Christianity*, New York: Simon & Schuster.

Johnson, P. E. (1991) *Darwin on Trial*, Washington, DC: Regnery Gateway.

Johnson, P.R.S. (1998) 'An Analysis of "Dignity,' *Theoretical Medicine and Bioethics*, 19: 337–52.

Jonas, H. (1985) *Technik, Medizin und Genetik. Zur Praxis des Prinzips Verantwortung*, Frankfurt: Insel.

Jones, D. G. (2010) 'Peering into People's Brains: Neuroscience's Intrusion into Our Inner Sanctum,' *Perspectives on Science and Christian Faith*, 62: 122–32.

Jones, D. G., and M. I. Whitaker (2009) 'Religious Traditions and Embryo Science,' *American Journal of Bioethics*, 9: 41–43.

Jonsen, A. (1998) *The Birth of Bioethics*, New York: Oxford University Press.

Jonsen, A., M. Siegler, and W. Winslade (2006) *Clinical Ethics*, 6th edn, New York: McGraw-Hill Medical.

Joubert, O., G. Rousselet, D. Fize, and M. Fabre-Thorpe (2007) 'Processing Scene Context: Fast Categorization and Object Interference,' *Vision Research*, 47: 3286–97.

Kant, I. (1785) *Groundwork of the Metaphysics of Morals*, trans. H. J. Paton (1964), New York: Harper & Row.

Kant, Immanuel (1998) *Groundwork on the Metaphysics of Morals*, Cambridge Texts in the History of Philosophy, ed. Mary McGregor, Cambridge: Cambridge University Press.

Karen, R. (1998) *Becoming Attached: First Relationships and How They Shape Our Capacity for Love*, New York: Oxford University Press.

Karpowicz, P., C. B. Cohen, et al. (2005) 'Developing Human-Nonhuman Chimeras in Human Stem Cell Research: Ethical Issues and Boundaries,' *Kennedy Institute of Ethics Journal*, 15, no. 2: 107–34.

Kass, L. (1974) 'Averting One's Eyes, or Facing the Music?—On Dignity in Death,' *Hastings Center Studies*, 2, no. 2: 67–80.

—— (1985) *Toward a More Natural Science: Biology and Human Affairs*, New York: Free Press.

—— (2004) *Life, Liberty and the Defense of Dignity: The Challenge for Bioethics*, San Francisco: Encounter Books.

—— (2007) 'Science, Religion, and the Human Future,' *Commentary*, 123, no. 4: 36–48.

—— (2008) 'Defending Human Dignity,' in E. D. Pellegrino (ed.), *Human Dignity and Bioethics*, Washington, DC: President's Council on Bioethics.

—— (2009) 'Forbidding Science: Some Beginning Reflections,' *Science and Engineering Ethics*, 15, no. 3: 271–82.

—— (1997b) Testimony to National Bioethics Advisory Commission. 14 March. Washington DC: Eberlin Reporting Service.

—— (1998) 'The Wisdom of Repugnance: Why We Should Ban the Cloning of Humans,' *New Republic*, 2 June: 20–21.

Katz, J. (1992) 'The Consent Principle of the Nuremberg Code: Its Significance Then and Now,' in G. J. Annas and M. A. Grodin (eds.), *The Nazi Doctors and the Nuremberg Code*, New York: Oxford University Press.

Kennedy, D. (2004) 'Neuroscience and Neuroethics,' *Science*, 306: 373.

Kerstein, S. (2009) 'Kantian Condemnation of Commerce in Organs,' *Kennedy Institute of Ethics Journal*, 19, no. 2: 147–69.

Keynes, R. (2002) *Darwin, His Daughter, and Human Evolution*, New York: Riverhead/Penguin Putnam.

Kierkegaard, S. (1980) *The Sickness unto Death: A Christianity Psychological Exposition for Upbuilding and Awakening* (Kierkegaard's Writings, 19), trans. H. V. Hong and E. H. Hong, Princeton, NJ: Princeton University Press.

Killmister, S. (2010) 'Dignity: Not Such a Useless Concept,' *Journal of Medical Ethics*, 36: 160–64.

Kilner, J. F. (1992) *Life on the Line: Ethics, Aging, Ending Patients' Lives, and Allocating Vital Resources*, Grand Rapids, MI: Eerdmans.

Kindt, M., M. Soeter, and B. Vervliet (2009) 'Beyond Extinction: Erasing Human Fear Responses and Preventing the Return of Fear,' *Nature Neuroscience*, 12: 256–58.

King-Casas, B., C. Sharp, L. Lomax-Bream, T. Lohrenz, P. Fonagy, and P. Montague (2008) 'The Rupture and Repair of Cooperation in Borderline Personality Disorder,' *Science*, 321: 806–10.

King-Casas, B., D. Tomlin, C. Anen, C. Camerer, S. Quartz, and P. Montague (2005) 'Getting to Know You: Reputation and Trust in a Two-Person Economic Exchange,' *Science*, 308: 78–83.

Kirksey, S., and S. Helmreich (2010) 'The Emergence of Multispecies Ethnography,' *Cultural Anthropology*, 25, no. 4: 545–76.

Kits Nieuwenkamp, J. (2000) 'The Convention on Human Rights and Biomedicine,' in J. Dahl Rentdorff and P. Kemp (eds.), *Basic Ethical Principles in European Bioethics and Biolaw*, vol. 2, Guissona (Catalonia): Barnola.

Koons, R. C. (2000) 'The Incompatibility of Naturalism and Scientific Realism,' in W. L. Craig and J. P. Moreland (eds.), *Naturalism: A Critical Analysis*, New York, Routledge.

Korsgaard, C. M. (2006) 'Morality and the Distinctiveness of Human Action,' in S. Macedo and J. Ober (eds.), Frans de Waal, *Primates and Philosophers: How Morality Evolved*, Princeton, NJ: Princeton University Press.

Kramer, P. D. (1993) *Listening to Prozac*, New York: Penguin Books.

Kraynak, R. P. (2008) 'The Mystery of the Human Soul,' in *Human Dignity and Bioethics: Essays Commissioned by the President's Council on Bioethics*, Washington, DC: President's Council on Bioethics.

——— (2001) *Christian Faith and Modern Democracy: God and Politics in the Fallen World*, Notre Dame, IN: University of Notre Dame Press.

Kreifelts, B., T. Ethofer, W. Grodd, M. Erb, and D. Wildgruber (2007) 'Audiovisual Integration of Emotional Signals in Voice and Face: An Event-Related fMRI Study,' *NeuroImage*, 37: 1445–56.

Kret, M., and B. De Gelder (2010) 'Social Context Influences Recognition of Bodily Expressions,' *Experimental Brain Research*, 203: 169–80.

Kretzmer, D., and E. Klein (2002) *The Concept of Human Dignity in Human Rights Discourse*, The Hague: Kluwer.

Kriegeskorte, N., M. Mur, D. Ruff, R. Kiani, J. Bodurka, H. Esteky, et al. (2008) 'Matching Categorical Object Representations in Inferior Temporal Cortex of Man and Monkey,' *Neuron*, 60: 1126–41.

Kujala, T., E. Aho, E. Jansson-Verkamlo, T. Nieminen-von Wendt, L. Von Wendt, and R. Näätänen (2007) 'Atypical Pattern of Discriminating Sound Features in Adults with Asperger Syndrome as Reflected by the Mismatch Negativity,' *Biological Psychology*, 75: 109–14.

Kujala, T., T. Lepistö, T. Nieminen-von Wendt, P. Näätänen, and R. Näätänen (2005) 'Neuro-Physiological Evidence for Cortical Discrimination Impairment of Prosody in Asperger Syndrome,' *Neuroscience Letters*, 383: 260–65.

Kure, J. (2009) 'Interspecies Gestation: Ethical Considerations,' in J. Taupitz and M. Weschka (eds.), *CHIMBRIDS: Chimeras and Hybrids in Comparative European and International Research. Scientific, Ethical, Philosophical and Legal Aspects*, New York: Springer.

Kurt, S., M. Groszer, et al. (2009) 'Modified Sound-Evoked Brainstem Potentials in FOXP2 Mutant Mice,' *Brain Research*, 1289: 30–36.

Kurzban, R., and M. Leary (2001) 'Evolutionary Origins of Stigmatization: The Functions of Social Exclusion,' *Psychological Bulletin*, 127: 187–208.

Kurzweil, R. (1999) *The Age of Spiritual Machines: When Computers Exceed Human Intelligence*, New York: Penguin Books.

Kutukdjian, G. (1999) 'Institutional Framework and Elaboration of the Revised Preliminary Draft of a Universal Declaration on the Human Genome and Human Rights,' in M.G.K. Menon, P. N. Tandon, S. S. Agarwal, and V. P. Sharma (eds.), *Human Genome Research: Emerging Ethical, Legal, Social, and Economic Issues*, New Delhi: Allied Publishers.

Laflamme, M. A., and C. E. Murry (2005) 'Regenerating the Heart,' *Nature Biotechnology*, 23: 845–56.

Laflamme, M. A., J. Gold, C. Xu, M. Hassanipour, E. Rosler, S. Police, V. Muskheli, and C. E. Murry (2005) 'Formation of Human Myocardium in the Rat Heart from Human Embryonic Stem Cells,' *American Journal of Pathology*, 167: 663–71.

Laflamme, M. A., K. Y. Chen, A. V. Naumova, V. Muskheli, J. A. Fugate, S. K. Dupras, H. Reinecke, C. Xu, M. Hassanipour, S. Police, et al. (2007) 'Cardiomyocytes Derived from Human Embryonic Stem Cells in Pro-survival Factors Enhance Function of Infarcted Rat Heart,' *Nature Biotechnology*, 25: 1015–24.

Landry, D. W., and H. A. Zucker (2004) 'Embryonic Death and the Creation of Human Embryonic Stem Cells,' *Journal of Clinical Investigation*, 114: 1184–86.

Langford, D., S. Crager, Z. Shehzad, S. Smith, S. Sotocinal, J. Levenstadt, et al. (2006) 'Social Modulation of Pain as Evidence for Empathy in Mice,' *Science*, 312: 1967–70.

Lanza, R., and N. Rosenthal (2004) 'The Stem Cell Challenge,' *Scientific American*, June: 92–99.

LaPlace, P. S., Marquis de (1902) *A Philosophical Essay on Probabilities*, trans. F. W. Truscott and F. L. Emory, London: John Wiley. Available HTTP: <http://books.google.com/books?printsec=frontcover&pg=PA4&id=WxoPAAAAIAAJ#v=onepage&q&f=false> (accessed 19 September 2010).

Larson, D. G., and D. R. Tobin (2000) 'End-of-Life Conversations,' *Journal of the American Medical Association*, 284, no. 12 (27 September): 1573.

Latour, Bruno. (2004) *Politics of Nature: How to Bring the Sciences into Democracy*, Cambridge, MA: Harvard University Press.

Lauritzen, P. (2004) 'Report on the Ethics of Stem Cell Research,' in L. Kass (ed.), *Monitoring Stem Cell Research*, Washington, DC: President's Council on Bioethics.

Lawler, P. A. (2009) 'The Human Dignity Conspiracy,' *The Intercollegiate Review*, 44: 40–50.

Layman, C. S. (1991) *The Shape of the Good: Christian Reflections on the Foundations of Ethics*, Notre Dame, IN: University of Notre Dame.

Leake, J., and S. Templeton (2008) 'Mice Produce Human Sperm to Raise Hope for Infertile Men,' *The Sunday Times* (London), 6 July.

Lee, P. (1997) *Abortion and Unborn Human Life*, Washington, DC: Catholic University of America Press.

—— (2001) 'Germain Grisez's Christian Humanism,' *American Journal of Jurisprudence*, 46: 137–51.

—— (2004) 'The Pro-Life Argument from Substantial Identity,' *Bioethics*, 18, no. 3: 249–63.

Lee, P., and R. P. George (2008) 'The Nature and Basis of Human Dignity,' in *Human Dignity and Bioethics: Essays Commissioned by the President's Council on Bioethics*, Washington, DC: President's Council on Bioethics. Available HTTP: <http://bioethics.georgetown.edu/pcbe/reports/human_dignity/index.html> (accessed 14 June 2011).

—— (2009) *Body-Self Dualism in Contemporary Ethics and Politics*, Cambridge: Cambridge University Press.

Lenox-Conyngham, A. (2005) [Review of] Ambrose: De officiis/Introduction, Text, and Translation/Commentary, *Journal of Theological Studies*, 56, no. 2: 677–81.

Leo XIII, Pope (1891) *Rerum Novarum: On Capital and Labor*. Available HTTP: <http://www.vatican.va/holy_father/leo_xiii/encyclicals/documents/hf_l-xiii_enc_15051891_rerum-novarum_en.html> (accessed 14 June 2011).

—— (1891) *Rerum Novarum: On Labor and Capital*, Encyclical issued 15 May 1891, Vatican City: Libreria Editrice Vaticana. Available HTTP: <http://www.vatican.va/holy_father/leo_xiii/encyclicals/documents/hf_l-xiii_enc_15051891_rerum-novarum_en.html> (accessed 30 December 2010).

Leopold, A. (1989) *A Sand County Almanac and Sketches Here and There*, Special Commemorative Edition, New York: Oxford University Press.

Levin, Y. (2003) 'The Paradox of Conservative Bioethics,' *The New Atlantis*, 1: 53–65.

Levinas, E. (1969) *Totality and Infinity: An Essay on Exteriority*, trans. Alphonso Lingis, Pittsburgh, PA: Duquesne University Press.

Lewis, C. S. (1947a) 'The Self-Refutation of the Naturalist,' in *Miracles: A Preliminary Study*, New York: Macmillan. Revised as (1978) *The Cardinal Difficulty of Naturalism*, New York: Macmillan.

—— (2001) *Mere Christianity*, New York: HarperCollins.

—— (1947b) *The Abolition of Man*, New York: Macmillan.

—— (2001) *The Weight of Glory*, New York: HarperCollins.

Lewis, C. T., and C. Short (1879) *A Latin Dictionary*, Oxford: Clarendon Press. Available HTTP: <http://www.perseus.tufts.edu/hopper/text?doc=Perseus%3atext%3a1999.04.0059> (accessed 21 July 2010).

Lewis, V. B. (2009) 'Theory and Practice of Human Rights: Ancient and Modern,' *Journal of Law, Philosophy and Culture*, 3: 277–96.

Lewy, G. (1996) *Why America Needs Religion*, Grand Rapids, MI: Eerdmans.

Liddell, H., and R. Scott (1940) *A Greek-English Lexicon*, rev. H. S. Jones and R. McKenzie, Oxford: Clarendon Press. Available HTTP: <http://www.perseus.tufts.edu/hopper/text?doc=Perseus%3atext%3a1999.04.0057> (accessed 21 July 2010).

Lindberg, D. C. (1986) 'Science and the Early Church,' in D. C. Lindberg and R. L. Numbers (eds.), *God and Nature: Historical Essays on the Encounter between Christianity and Science*, Berkeley: University of California Press.

—— (2007) *The Beginnings of Western Science: The European Scientific Tradition in Philosophical, Religious, and Institutional Context, Prehistory to A.D. 1450*, 2nd edn, Chicago: University of Chicago Press.

Lindberg, D. C., and R. L. Numbers (eds.) (1986) *God and Nature: Historical Essays on the Encounter between Christianity and Science*, Berkeley: University of California Press.

Lindemann, H. (2009) '" . . . But *I* Could Never Have One": The Abortion Intuition and Moral Luck,' *Hypatia*, 24, no. 1: 41–55.

Lindsay, R. (2008) *Future Bioethics: Overcoming Taboos, Myths, and Dogmas*, Amherst, NY: Prometheus Books.

Linville, M. (2009) 'The Moral Argument,' in W. L. Craig and J. P. Moreland (eds.), *The Blackwell Companion to Natural Theology*, Oxford: Blackwell.

Liu, Z., and S. Sarkar (2007) 'Outdoor Recognition at a Distance by Fusing Gait and Face,' *Image and Vision Computing*, 25: 817–32.

Livingstone, D. N. (2003) 'Re-Placing Darwinism and Christianity,' in D. C. Lindberg and R. L. Numbers (eds.), *When Science and Christianity Meet*, Chicago: University of Chicago Press.

Lo, J.-Y. (2007) 'Universalism Challenged: Human Rights and Asian Values,' *Online Opinion*, February 1. Available HTTP: <http://www.onlineopinion.com.au/view.asp?article=5418> (accessed 19 July 2012).

Locke, J. (1690) *Two Treatises of Government*, CreateSpace [2010].

Loike, J., and M. D. Tendler (2002) 'Revisiting the Definition of Homo Sapiens,' *Kennedy Institute of Ethics Journal*, 12, no. 4: 343–50.

—— (2003) '*Ma adam va-teda-ehu*: Halakhic Criteria for Defining Human Beings,' *Tradition*, 37, no. 2: 1–19.

—— (2008) 'Reconstituting a Human Brain in Animals: A Jewish Perspective on Human Sanctity,' *Kennedy Institute of Ethics Journal*, 18, no. 4: 347–67.

—— (2011) 'Halachic Bioethics Guidelines,' *Journal of Halacha in Contemporary Society*, 16: 92–118.

Lombardo, M., J. Barnes, S. Wheelwright, and S. Baron-Cohen (2007) 'Self-referential Cognition and Empathy in Eutism,' *PLoS ONE*, Issue 9: e883.

Long, A. A. (1986) *Hellenistic Philosophy: Stoics, Epicureans, Sceptics*, 2nd edn, Berkeley: University of California Press.

—— (2008) 'The Concept of the Cosmopolitan in Greek and Roman Thought,' *Daedalus*, 137, no. 3: 50–58. Available HTTP: <http://proxy.foley.gonzaga.edu:2048/login?url=http://proquest.umi.com/pqdweb?did=1533329371&sid=1&Fmt=2&clientId=10553&RQT=309&VName=PQD> (accessed 3 September 2010).

Lorberbaum, Y. (2002) 'Blood and the Image of God: On the Sanctity of Life in Biblical and Early Rabbinic Law, Myth, and Ritual,' in D. Kretzmer and E. Klein (eds.), *The Concept of Human Dignity in Human Rights Discourse*, The Hague: Kluwer.

Luik, J. (1998) 'Humanism,' in E. Craig (ed.), *Routledge Encyclopedia of Philosophy*. London: Routledge. Available HPPT: <http://www.rep.routledge.com/article/N025SECT3> (accessed 22 August 2010).

Lupton, M. L. (1997) 'Artificial Wombs: Medical Miracle, Legal Nightmare,' *Medicine and Law*, 16, no. 3: 621–33.

Lyell, C. (1830) *Principles of Geology, Being an Attempt to Explain the Former Changes of the Earth's Surface, by Reference to Causes Now in Operation*, vol. 1, London: John Murray. Available HTTP: <http://darwin-online.org.uk/content/frameset?viewtype=text&itemID=A505.1&pageseq=1> (accessed 25 July 2010).

—— (1863) *The Geological Evidences of the Antiquity of Man with Remarks on the Origin of Species by Variation*, 3rd edn, rev., London: John Murray. Available HTTP: <http://darwin-online.org.uk/content/frameset?viewtype=text&itemID=A282&pageseq=1> (accessed 20 September 2010).

Lyotard, J. (1979) *The Postmodern Condition: A Report on Knowledge*, trans. Geoff Bennington and Brian Massumi. Minneapolis: University of Minnesota Press.

—— (1993) 'The Other's Rights,' in S. Shute and S. Hurley (eds.), *On Human Rights*, New York: Basic Books.

MacIntyre, A. (1999) *Dependent Rational Animals: Why Human Beings Need the Virtues*, Chicago: Open Court.

—— (2007) *After Virtue: A Study in Moral Theory*, 3rd edn, Notre Dame, IN: University of Notre Dame Press.

Macklin, R. (1997a) 'Human Cloning? Don't Just Say No,' *U. S. News and World Report*, 10 March: 64.

—— (2003) 'Dignity Is a Useless Concept,' *British Medical Journal*, 327: 1419–20.

Magalhaes, A. C. (2005) 'Functional Magnetic Resonance and Spectroscopy in Drug and Substance Abuse,' *Topics in Magnetic Resonance Imaging*, 16: 247–51.

Malby, S. (2002) 'Human Dignity and Human Reproductive Cloning,' *Health and Human Rights*, 6, no. 1: 103–36.

Mann, J. (1998) 'Dignity and Health: The UDHR's Revolutionary First Article,' *Health and Human Rights*, 3, no. 2: 30–38.

Manninen, B. A. (2008) 'Are Human Embryos Kantian Persons? Kantian Considerations in Favor of Embryonic Stem Cell Research,' *Philosophy, Ethics, and Humanities in Medicine*, 3: 4.

Marcel, G. (1963) *The Existential Background of Human Dignity*, Cambridge, MA: Harvard University Press.

Margalit, A. (1998) *The Decent Society*, Cambridge, MA: Harvard University Press.
—— (2002) *The Ethics of Memory*, Cambridge, MA: Harvard University Press.
Maritain, J. (1951) *Man and the State*, Chicago: University of Chicago Press.
—— (1964) *Moral Philosophy*, New York: Charles Scribner's Sons.
Marks, J. (2002) *What It Means To Be 98% Chimpanzee: Apes, People, and Their Genes*, Berkeley: University of California Press.
Marks, S. P. (2002) 'Human Rights Assumptions of Restrictive and Permissive Approaches to Human Reproductive Cloning,' *Health and Human Rights*, 6, no. 1: 81–100.
Marshall, J. R., G. M. Abroms, and M. H. Miller (1967) 'The Doctor, the Dying Patient, and the Bereaved,' *Annals of Internal Medicine*, 70, no. 3: 615–20.
Martin, G. R. (1981) 'Isolation of a Pluripotent Cell Line from Early Mouse Embryos Cultured in Medium Conditioned by Teratocarcinoma Stem Cells,' *Proceedings of the National Academies of Science of the United States of America*, 78: 7634–38.
Mathieu, B. (2000) *Génome humain et droits fondamentaux*, Paris: Economica.
Mayden, R. L. (1997) 'A Hierarchy of Species Concepts: The Denouement in the Saga of the Species Problem,' *Systematics Association*, Special Volume, 54: 381–424.
—— (1999) 'Consilience and a Hierarchy of Species Concepts: Advances Toward Closure on the Species Puzzle,' *Journal of Nematology*, 31, no. 2: 95.
McBeth, B. D., R. M. McNamara, F. K. Ankel, E. J. Mason, L. J. Ling, T. J. Flottemesch, and B. R. Asplin (2009) 'Modafinil and Zolpidem Use by Emergency Medicine Residents,' *Academic Emergency Medicine*, 16: 1311–17.
McClelland, J. L., B. L. McNaughton, and R. C. O'Reilly (1995) 'Why There Are Complementary Learning Systems in the Hippocampus and Neocortex: Insights from the Successes and Failures of Connectionist Models of Learning and Memory,' *Psychology Review*, 102: 419–57.
McClelland, R. (2010) 'Normal Narcissism and Its Pleasures,' *Journal of Mind and Behavior*, 31: 85–125.
—— (2011) 'A Naturalistic View of Human Dignity,' *Journal of Mind and Behavior*, 32: 5–48.
McCormick, T. R. (1991) 'The Terminal Phase of Illness,' in M. Patrick et al. (eds.), *Medical-Surgical Nursing*, 2nd edn, Philadelphia: J. B. Lippincott.
McCrudden, C. (2008) 'Human Dignity and Judicial Interpretation of Human Rights,' *European Journal of International Law*, 19, no. 4: 655–724.
McCune, J., H. Kaneshima, et al. (1991) "The SCID-hu Mouse: A Small Animal Model for HIV Infection and Pathogenesis,' *Annual Review of Immunology*, 9: 399–429.
McKirahan, R. D., Jr. (1994) *Philosophy before Socrates: An Introduction with Texts and Commentary*, Indianapolis, IN: Hackett.
McMahan, J. (2002) 'Our Fellow Creatures,' *Journal of Ethics*, 9: 353–80.
Meeren, H., C. Van Heijnsbergen, and B. De Gelder (2005) 'Rapid Perceptual Integration of Facial Expression and Emotional Body Language,' *Proceedings of the National Academy of Sciences*, 102: 16518–23.
Meilaender, G. (1995) *Body, Soul, and Bioethics*, Notre Dame, IN: University of Notre Dame Press.
—— (2002) 'Spare Embryos,' *Weekly Standard*, August–September: 25.
—— (2007) 'Human Dignity and Public Bioethics,' *The New Atlantis*, 17: 33–52.
—— (2008) 'Human Dignity: Exploring and Explicating the Council's Vision,' in E. D. Pellegrino (ed.), *Human Dignity and Bioethics*, Washington, DC: President's Council on Bioethics.
—— (2009) *Neither Beast Nor God: The Dignity of the Human Person*, New York: New Atlantis Books.
Meuleman, P., L. Libbrecht, et al. (2005) 'Morphological and Biochemical Characterization of a Human Liver in a uPA-SCID Mouse Chimera,' *Hepatology*, 41, no. 4: 847–56.

Meyer, S. C. (2000) 'DNA and Other Designs,' *First Things*, 102: 30–38.

Michel, J. B., Y. K. Shen, A. P. Aiden, A. Veres, M. K. Gray, J. P. Pickett, D. Hoiberg, D. Clancy, P. Norvig, J. Orwant, S. Pinker, M. A. Nowak, and E. L. Aiden (2010) 'Quantitative Analysis of Culture Using Millions of Digitized Books," *Science*. Published online 16 December 2010. doi:10.1126/science.1199644. Available HTTP: <http://www.ncbi.nlm.nih.gov/pmc/articles/PMC3279742/?tool=pubmed>

Midgley, M. (1999) 'Towards an Ethic of Global Responsibility,' in T. Dunne and N. J. Wheeler (eds.), *Human Rights in Global Politics*, Cambridge: Cambridge University Press.

Mikels, J., B. Fredrickson, G. Larkin, C. Lindberg, S. Maglio, and P. Reuter-Lorenz (2005) 'Emotional Category Data on Images from the IAPS,' *Behavior Research Methods*, 37: 626–30.

Mill, J. S. (1869) *On Liberty*, London: Longman, Roberts and Green.

—— (1963–1991) *Collected Works of John Stuart Mill*, 33 vols., ed. J. M. Robson, Toronto: University of Toronto Press; London: Routledge and Kegan Paul. Includes vol. 10 (1985), available HTTP: <http://files.libertyfund.org/files/241/0223.10.pdf> (accessed 15 December 2010); vol. 18 (1977), available HTTP: <http://files.libertyfund.org/files/233/0223.18.pdf> (accessed 15 December 2010).

Miller, K. R. (1999) *Finding Darwin's God: A Scientist's Search for Common Ground between God and Evolution*, New York: HarperCollins.

Milton, J. (1674) *Paradise Lost: A Poem in Twelve Books*, 2nd edn, London: S. Simmons. Available at the Milton Reading Room, ed. T. H. Luxon, HTTP: <http://www.dartmouth.edu/~milton/reading_room/pl/intro/index.shtml> (accessed 23 December 2010).

Mirkes, R. (2010) 'Does Pharmacologically-Altered Memory Change Personal Identity?' *Ethics and Medicine*, 26: 175–87.

Mitchell, C. B., E. D. Pellegrino, J. B. Elshtain, J. F. Kilner, and S. B. Rae (2007) *Biotechnology and the Human Good*, Washington, DC: Georgetown University Press.

Moll, J., and J. Schulkin (2009) 'Social Attachment and Aversion in Human Moral Cognition,' *Neuroscience and Biobehavioral Reviews*, 33: 456–65.

Montag, C., A. Heinz, D. Kunz, and J. Gallinat (2007) 'Self-reported Empathic Abilities in Schizophrenia,' *Schizophrenia Research*, 92: 85–89.

Montepare, J., and H. Dobish, H. (2003) 'The Contribution of Emotion Perceptions and Their Over-generalizations to Trait Impressions,' *Journal of Nonverbal Behavior*, 27: 237–54.

Montepare, J., S. Goldstein, and A. Clausen (1987) 'The Identification of Emotions from Gait Information,' *Journal of Nonverbal Behavior*, 11: 33–42.

Moore, J. R. (1979) *The Post-Darwinian Controversies: A Study of the Protestant Struggle to Come to Terms with Darwin in Great Britain and America 1870–1900*, Cambridge: Cambridge University Press.

Moraczewski, A. (1997) Testimony to the National Bioethics Advisory Commission. 13 March. Washington, DC: Eberlin Reporting Service.

—— (2003) 'The Moral Traditions of the Catholic Church,' in A. R. Chapman and M. S. Frankel (eds.), *Designing Our Descendants: The Promises and Perils of Genetic Modifications*, Baltimore: Johns Hopkins University Press.

Moreland, J. P., and S. B. Rae (2000) *Body and Soul: Human Nature and the Crisis in Ethics*, Downers Grove, IL: InterVarsity Press.

Morris, R. (ed.) (2010) *Can the Subaltern Speak? Reflections on the History of an Idea*, New York: Columbia University Press.

Morsink, J. (1984) 'The Philosophy of the Universal Declaration,' *Human Rights Quarterly*, 6, no. 3: 309–34.

—— (1993) 'World War Two and the Universal Declaration,' *Human Rights Quarterly*, 15: 357–405. Available HTTP: <http://www.jstor.org/stable/762543> (accessed 29 August 2010).

Movius, H., and J. Allen (2005) 'Cardiac Vagal Tone, Defensiveness and Motivational Style,' *Biological Psychology*, 68: 147–62.

Mu, Y., and D. Tao (2010) 'Biologically Inspired Feature Manifold for Gait Recognition,' *Neurocomputing*, 73: 895–902.

Murphy, T. F. (2010) Using Embryos to Produce Saviour Cell Lines,' *Reproductive Biomedicine Online*, 20, no. 5: 569–71.

Murray, T. H. (1996) *The Worth of a Child*, Berkeley: University of California Press.

Nagan, W., and L. Atkins (2001) 'The International Law on Torture: From Universal Proscription to Effective Application and Enforcement,' *Harvard Human Rights Journal*, 14: 87–121.

Nagel, T. (1993) 'Moral Luck,' in D. Statman (ed.), *Moral Luck*, Albany: State University of New York Press, 51–71.

—— (1997) *The Last Word*, Oxford: Oxford University Press.

Nagy, A., and J. Rossant (2001) 'Chimaeras and Mosaics for Dissecting Complex Mutant Phenotypes,' *International Journal of Developmental Biology*, 45, no. 3: 577–82.

Narayan, A. D., J. L. Chase, et al. (2006) 'Human Embryonic Stem Cell-Derived Hematopoietic Cells Are Capable of Engrafting Primary As Well As Secondary Fetal Sheep Recipients,' *Blood*, 107, no. 5: 2180–83.

National Academy of Sciences (2005) *Guidelines for Human Embryonic Stem Cell Research*, Washington, DC: National Academy of Sciences.

National Bioethics Advisory Commission (1997) *Cloning Human Beings: Report and Recommendations*, Rockville, MD: Author.

—— (1999) *Ethical Issues in Human Stem Cell Research*, Rockville, MD: National Bioethics Advisory Commission.

National Commission for the Protection of Human Subjects of Biomedical and Biobehavioral Research (1979) *The Belmont Report: Ethical Principles and Guidelines for the Protection of Human Subjects of Research*, National Commission for the Protection of Human Subjects of Biomedical and Biobehavioral Research, 18 April. Washington, DC: U.S. Government Printing Office. 20402.

National Institutes of Health (1994) *Report of the Human Embryo Research Panel*, Washington, DC: National Institutes of Health.

Natural Death Act (RCW 70.122) (1979).

Neuhaus, R. (2008) 'Human Dignity and Public Discourse,' in E. D. Pellegrino (ed.), *Human Dignity and Bioethics*, Washington, DC: President's Council on Bioethics.

—— (1984) *The Naked Public Square*, Grand Rapids, MI: Eerdmans.

Newbury, D. F., S. E. Fisher, et al. (2010) 'Recent Advances in the Genetics of Language Impairment,' *Genome Medicine*, 2, no. 1: 6.

Nickel, J. (1987) *Making Sense of Human Rights: Philosophical Reflections on the Universal Declaration of Human Rights*, Berkeley: University of California Press.

Nietzsche, F. (1954) *The Portable Nietzsche*, trans. W. Kaufmann, New York: Penguin Books.

—— (1961) *Thus Spoke Zarathustra: A Book for Everyone and No One*, trans. R. J. Hollingdale, NY: Penguin Books.

—— (2006) *The Nietzsche Reader*, ed. K. A. Pearson and D. Large, Malden, MA: Blackwell.

Noonan, J. P. (2010) 'Neanderthal Genomics and the Evolution of Modern Humans,' *Genome Research*, 20, no. 5: 547.

Noonan, J. P., G. Coop, et al. (2006) 'Sequencing and Analysis of Neanderthal Genomic DNA,' *Science*, 314, no. 5802: 1113.

Nordenfelt, L. (2004) 'The Varieties of Dignity,' *Health Care Analysis*, 12, no. 2 (June): 69.

Nowak, M. (2006) 'Five Rules for the Evolution of Cooperation,' *Science*, 314: 1560–63.

Numbers, R. (1988) 'George Frederick Wright: From Christian Darwinist to Fundamentalist,' *Isis*, 79: 624–45.

—— (ed.) (2009) *Galileo Goes to Jail and Other Myths about Science and Religion*, Cambridge, MA: Harvard University Press.

Nussbaum, M. C. (1998) 'Political Animals: Love, Luck, and Dignity,' *Metaphilosophy*, 29, no. 4: 273–87.

—— (2000) *Women and Human Development: The Capabilities Approach*, Cambridge: Cambridge University Press.

—— (2001a) 'Animal Rights: The Need for a Theoretical Basis' (review of S. M. Wise, *Rattling the Cage*), *Harvard Law Review*, 114, no. 5: 1506–49.

—— (2001b) *The Fragility of Goodness: Luck and Ethics in Greek Tragedy and Philosophy*, rev. edn, Cambridge: Cambridge University Press.

—— (2006) *Frontiers of Justice: Disability, Nationality, Species Membership*, Cambridge, MA: Harvard University Press.

—— (2008) 'Human Dignity and Political Entitlements,' in *Human Dignity and Bioethics: Essays Commissioned by the President's Council on Bioethics*, Washington, DC: President's Council on Bioethics.

Nussbaum, M. C., and H. Putnam (1992) 'Changing Aristotle's Mind,' in M. C. Nussbaum and A. O. Rorty, *Essays on Aristotle's De Anima*, Oxford: Clarendon Press.

Oduncu, F. S. (2003) 'Stem Cell Research in Germany: Ethics of Healing vs. Human Dignity,' *Medicine, Healthcare and Philosophy*, 6: 5–16.

Olfson, M., and S. C. Marcus (2009) 'National Patterns in Antidepressant Medication Treatment,' *Archives of General Psychiatry*, 66: 848–56.

Oosterhof, N., and A. Todorov (2009) 'Shared Perceptual Basis of Emotional Expressions and Trustworthiness Impressions from Faces,' *Emotion*, 9: 128–33.

Orr, H. A., J. P. Masly, et al. (2004) 'Speciation Genes,' *Current Opinion in Genetics and Development*, 14, no. 6: 675–79.

Ospovat, D. (1981) *The Development of Darwin's Theory: Natural History, Natural Theology, and Natural Selection, 1838–1859*, Cambridge: Cambridge University Press.

Oxnard, C. E. (2004) 'Brain Evolution: Mammals, Primates, Chimpanzees, and Humans,' *International Journal of Primatology*, 25, no. 5: 1127–58.

Pacholczyk, T. (2010) 'Defending the Dignity of Those with Dementia,' *The Bulletin*, 10 October. Available HTTP: <http://thebulletin.us/articles/2010/10/10/commentary/op-eds/doc4cb2225774c26650775374.txt> (accessed 11 December 2010).

Padoa-Schioppa, C., and J. A. Assad (2006) 'Neurons in the Orbitofrontal Cortex Encode Economic Value,' *Nature*, 441: 223–26.

Page, K., and M. Nowak (2002) 'Empathy Leads to Fairness,' *Bulletin of Mathematical Biology*, 64: 1101–16.

Pakkenberg, B., and H.J.G. Gundersen (1997) 'Neocortical Neuron Number in Humans: Effect of Sex and Age,' *Journal of Comparative Neurology*, 384: 312–20.

Paley, W. (1809) *Natural Theology: or, Evidences of the Existence and Attributes of the Deity*, 12th edn, London: J. Faulder. Available HTTP: <http://darwin-online.org.uk/content/frameset?itemID=A142&viewtype=text&pageseq=1> (accessed 24 July 2010).

Panchanathan, K., and R. Boyd (2004) 'Indirect Reciprocity Can Stabilize Cooperation without the Second-Order Free Rider Problem,' *Nature*, 432: 499–502.

Paris, P. J. (2001) 'Moral Exemplars in the Global Community,' in M. L. Stackhouse and D. S. Browning (eds.), *God and Globalization*: Vol. 2: *The Spirit and the Modern Authorities*, Harrisburg, PA: Trinity Press International.

Patient Self-Determination Act (1991).

Paton, H. J. (1971) *Categorical Imperative: A Study in Kant's Moral Philosophy*, Philadelphia: University of Pennsylvania Press.

Patrick, C. (ed.) (2006) *Handbook of Psychopathy*, New York: Guilford Press.

Pavlischek, K. (1998) 'Abortion Logic and Paternal Responsibilities,' in L. Pojman and F. Beckwith (eds.), *The Abortion Controversy: 25 Years after Roe v. Wade, A Reader*, Belmont, CA: Wadsworth, 176–99.

Pavlova, M., N. Birbaumer, and A. Sokolov (2006) 'Attentional Moderation of Cortical Neuromagnetic Gamma Response to Biological Motion,' *Cerebral Cortex*, 16: 321–27.

Pavlova, M., W. Lutzenberger, A. Sokolov, and N. Birbaumer (2004) 'Dissociable Cortical Processing of Recognizable and Non-recognizable Biological Motion: Analyzing Gamma MEG Activity,' *Cerebral Cortex*, 14: 181–88.

Payne, C. D. (2010) 'Stem Cell Research and Cloning for Human Reproduction: An Analysis of the Laws, The Direction in Which They May Be Heading in Light of Recent Developments, and Potential Constitutional Issues,' *Mercer Law Review*, 61: 943–1283.

Peelen, M., A. Wiggett, and P. Downing (2006) 'Patterns of fMRI Activity Dissociate Overlapping Functional Brain Areas That Respond to Biological Motion,' *Neuron*, 49: 815–22.

Peirce, C. S. (1949) *Chance, Love, and Logic: Philosophical Essays*, New York: Peter Smith.

—— (1955) *Philosophical Writings of Peirce*, New York: Dover.

Pellegrino, E. D. (2006) 'Toward a Reconstruction of Medical Morality,' *American Journal of Bioethics*, 6: 65–71.

—— (2009) 'The Lived Experience of Human Dignity,' in E. D. Pellegrino (ed.), *Human Dignity and Bioethics*, Washington, DC: President's Council on Bioethics.

Penn, D. C., K. J. Holyoak, and D. J. Povinelli (2008) 'Darwin's Mistake: Explaining the Discontinuity between Human and Animal Minds,' *Behavioral and Brain Sciences*, 31: 109–78. Available HTTP: <http://proxy.foley.gonzaga.edu:2048/login?url=http://proquest.umi.com/pqdweb?did=1601421561&sid=1&Fmt=2&clientId=10553&RQT=309&VName=PQD> (accessed 17 August 2010).

Pennock, R. T. (2000) *Tower of Babel: The Evidence against the New Creationism*, Cambridge, MA: MIT Press.

Percy, W. (1991) 'The Fateful Rift: The San Andreas Fault in the Modern Mind,' in P. Samway (ed.), *Signposts in a Strange Land*. New York: Farrar, Straus and Giroux.

Persaud, R. (1999) 'Sensory Alien Hand Syndrome,' *Journal of Neurology, Neurosurgery and Psychiatry*, 67: 130–31.

Peters, T. (2001) 'Embryonic Stem Cells and the Theology of Dignity,' in S. Holland, K. Lebacqz, and L. Zoloth (eds.), *The Human Embryonic Stem Cell Debate: Science, Ethics, and Public Policy*, London: MIT Press.

Pichon, S., B. De Gelder, and J. Grèzes (2008) 'Emotional Modulation of Visual and Motor Areas by Dynamic Body Expressions of Anger,' *Social Neuroscience*, 3: 199–212.

Pico della Mirandola, G. (1956) *Oration on the Dignity of Man*, trans. A. R. Caponigri, Washington, DC: Regnery Gateway.

Pierce, J. (2008) 'Mice in the Sink: On the Expression of Empathy,' *Environmental Philosophy*, 5: 75–96.

Pihan, H. (2006) 'Affective and Linguistic Processing of Speech Prosody: DC Potential Studies,' *Progress in Brain Research*, 156: 269–84.

Pihan, H., M. Tabert, S. Assuras, and J. Borod (2008) 'Unattended Emotional Intonations Modulate Linguistic Prosody Processing,' *Brain and Language*, 105: 141–47.

Pinker, S. (2008) 'The Stupidity of Dignity,' *The New Republic*, 238, no. 9: 28–31.

Pinsk, M., K. DeSimone, T. Moore, C. Cross, and S. Kastner (2005) 'Representations of Faces with Body Parts in Macaque Temporal Cortex: A Functional MRI Study,' *Proceedings of the National Academy of Sciences*, 102: 6996–7001.

Pinto, J. (2006) 'Developing Body Representations: A Review of Infants' Response to Biological Motion Displays,' in G. Knoblich, I. Thornton, M. Grosjean, and

M. Shiffrar (eds.), *Human Body Perception from the Inside Out*, New York: Plenum Press.

Pirraglia, P. A., R. S. Stafford, and D. E. Singer (2003) 'Trends in Prescribing of Selective Serotonin Reuptake Inhibitors and Other Newer Antidepressant Agents in Adult Primary Care,' *Primary Care Companion to the Journal of Clinical Psychiatry*, 5: 153–57.

Pitman, R. K., K. M. Sanders, R. M. Zusman, A. R. Healy, F. Cheema, N. B. Lasko, L. Cahill, and S. P. Orr (2002) 'Pilot Study of Secondary Prevention of Posttraumatic Stress Disorder with Propranolol,' *Biological Psychiatry*, 51: 189–92.

Pius XI, Pope (1937) *Divini Redemptoris*, Encyclical issued 19 March 1937, Vatican City: Libreria Editrice Vaticana. Available HTTP: <http://www.vatican.va/holy_father/pius_xi/encyclicals/documents/hf_p-xi_enc_19031937_divini-redemptoris_en.html> (accessed 30 December 2010).

Pius XII, Pope (1950) *Humani Generis*, Encyclical issued 12 August 1950, Rome. Available HTTP: <http://www.vatican.va/holy_father/pius_xii/encyclicals/documents/hf_p-xii_enc_12081950_humani-generis_en.html> (accessed 5 September 2010).

Planned Parenthood v. Casey (1992) 505 U.S. 833.

Plantinga, A. (1993) *Warrant and Proper Function*. Oxford: Oxford University Press.

—— (1997a) 'Methodological Naturalism, Part 1,' *Origins and Design*, 18, no. 2. Available HTTP: <http://www.arn.org/docs/odesign/od181/methnat181.htm> (accessed 12 December 2005).

—— (1997b) 'Methodological Naturalism, Part 2,' *Origins and Design*, 18, no. 2. Available HTTP: <http://www.arn.org/docs/odesign/od182/methnat182.htm> (accessed 12 December 2005).

Plato (1997) *Collected Works*, ed. J. M. Cooper, Indianapolis, IN: Hackett.

Platt, M. L., and P. W. Glimcher (1999) 'Neural Correlates of Decision Variables in Parietal Cortex,' *Nature*, 400: 233–38.

Pollard, K. S. (2009) 'What Makes Us Human?' *Scientific American*, May: 44–49.

Pontifical Academy for Life (2004) 'The Dignity of Human Procreation and Reproductive Technologies: Anthropological and Ethical Aspects,' Tenth General Assembly, Final Communique. Available HTTP: http://www.vatican.va/roman_curia/pontifical_academies/acdlife/documents/rc_pont-acd_1 . . . (accessed July 25, 2012).

Porges, S. (2007) 'The Polyvagal Perspective,' *Biological Psychology*, 74: 116–43.

—— (2011) *The Polyvagal Theory: Neurophysiological Foundations of Emotions, Attachment, Communication, and Self-regulation*, New York: Norton.

Porter, J. (1999) *Natural and Divine Law: Reclaiming the Tradition for Christian Ethics*, Grand Rapids, MI: Eerdmans.

Post, S. G. (1995) *The Moral Challenge of Alzheimer Disease*, Baltimore: Johns Hopkins University Press.

Pourtois, G., M. Peelen, L. Spinelli, M. Seeck, and P. Vuilleumier (2007) 'Direct Intracranial Recording of Body-selective Responses to Human Extrastriate Visual Cortex,' *Neuropsychologia*, 45: 2621–25.

President's Council on Bioethics (2002) *Human Cloning and Human Dignity: An Ethical Inquiry*, Washington, DC: President's Council on Bioethics.

—— (2003) *Beyond Therapy: Biotechnology and the Pursuit of Happiness*. Washington, DC: President's Council on Bioethics.

—— (2004a) *Monitoring Stem Cell Research*, Washington, DC: President's Council on Bioethics.

—— (2004b) *Reproduction and Responsibility: The Regulation of New Biotechnologies*, Washington, DC: President's Council on Bioethics.

—— (2005) *Alternative Sources of Pluripotent Cells: A White Paper*, Washington, DC: President's Council on Bioethics.

——— (2008) *Human Dignity and Bioethics*, Washington, DC: President's Council on Bioethics. Available HTTP: <http://bioethics.georgetown.edu/pcbe/reports/human_dignity/human_dignity_and_bioethics.pdf> (accessed 21 July 2010).

Preston, S., and F. De Waal (2002) 'Empathy: Its Ultimate and Proximate Bases,' *Behavioral and Brain Sciences*, 25: 1–20.

Price, M. (2011) 'Cooperation as a Classic Problem in Behavioral Biology,' in V. Swami (ed.), *Evolutionary Psychology: A Critical Introduction*, Hoboken, NJ: Wiley-Blackwell.

Provine, W. B. (1988) 'Progress in Evolution and Meaning in Life,' in M. H. Nitecki (ed.), *Evolutionary Progress*, Chicago: University of Chicago Press.

Pufendorf, S. (1964) *De Jure Naturae et Gentium Libri Octo*, 2 vols., trans. C. H. Oldfather and W. A. Oldfather, New York: Oceana; London: Wildy & Sons.

——— (1994) *The Political Writings of Samuel Pufendorf*, trans. M. J. Siedler, ed. C. L. Carr, New York: Oxford University Press.

Quadflieg, S., H. Mentzel, W. Miltner, and T. Straube (2008) 'Modulation of the Neural Network Involved in the Processing of Angry Prosody: The Role of Task-Relevance and Social Phobia,' *Biological Psychology*, 27: 445–56.

Rachels, J. (1990) *Created from Animals: The Moral Implications of Darwinism*, Oxford: Oxford University Press, 1990.

——— (1993) 'The Challenge of Cultural Relativism,' in J. Rachels (ed.), *The Elements of Moral Philosophy*, New York: McGraw-Hill.

Rakoczy, H., F. Warneken, and M. Tomasello (2009) 'Young Children's Selective Learning of Rule Games from Reliable and Unreliable Models,' *Cognitive Development*, 24: 61–69.

Rakoczy, H., N. Brosche, F. Warneken, and M. Tomasello (2009) 'Young Children's Understanding of the Context-Relativity of Normative Rules in Conventional Games,' *British Journal of Developmental Psychology*, 27: 445–56.

Rand, D., H. Ohtsuki, and M. Nowak (2009) 'Direct Reciprocity with Costly Punishment: Generous Tit-for-Tat Prevails,' *Journal of Theoretical Biology*, 256: 45–57.

Rawls, J. (1973) *A Theory of Justice*, Oxford: Oxford University Press.

——— (1999), *Theory of Justice*, rev. edn, Cambridge, MA: Harvard/Belknap Press.

Raymo, C. (1999) 'Celebrating Creation: A Scientific Skeptic's Encounter with the Supernatural,' *Skeptical Inquirer*, July–August 1999.

Reagan, R. (1983) 'Abortion and the Conscience of the Nation,' *Human Life Review*, 9, no. 2: 1–7.

Reed, E. D. (2007) *The Ethics of Human Rights: Contested Doctrinal and Moral Issues*, Waco, TX: Baylor University Press.

Rees, M. (2011) 'Are We All Doomed?" *New Statesman*, 9 June.

Reid, V., S. Hoehl, J. Landt, and J. Striano. (2008) 'Human Infants Dissociate Structural and Dynamic Information in Biological Motion: Evidence from Neural Systems,' *Social Cognitive and Affective Neuroscience*, 3: 161–67.

Reiner, A., N. Del Mar, et al. (2007) 'R6/2 Neurons with Intranuclear Inclusions Survive for Prolonged Periods in the Brains of Chimeric Mice,' *Journal of Comparative Neurology*, 505, no. 6: 603–29.

Reppert, V. (2003) *C. S. Lewis's Dangerous Idea: A Philosophical Defense of Lewis's Argument from Reason*, Downers Grove, IL: InterVarsity Press.

Richards, R. A. (2009) 'Classification in Darwin's *Origin*,' in M. Ruse and R. J. Richards (eds.), *The Cambridge Companion to the 'Origin of Species,'* Cambridge: Cambridge University Press.

Richards, R. J. (2009) 'Darwin on Mind, Morals and Emotions,' in J. Hodge and G. Radick (eds.), *The Cambridge Companion to Darwin*, 2nd edn, Cambridge: Cambridge University Press.

—— (2009) 'Darwin's Theory of Natural Selection and Its Moral Purpose,' in M. Ruse and R. J. Richards (eds.), *The Cambridge Companion to the 'Origin of Species,'* Cambridge: Cambridge University Press.

Rilling, J., D. Goldsmith, A. Glenn, M. Jairam, H. Elfenbein, J. Dagenais, et al. (2008) 'The Neural Correlates of the Affective Response to Unreciprocated Cooperation,' *Neuropsychologia*, 46: 1256–66.

Riskind, J., and C. Gotay (1982) 'Physical Posture: Could It Have Regulatory or Feedback Effects on Motivation and Emotion?' *Motivation and Emotion*, 6: 273–98.

Rizzolatti, G., and C. Sinigaglia (2007) 'Mirror Neurons and Motor Intentionality,' *Functional Neurology*, 22: 205–10.

Robert, J. S. (2006) 'The Science and Ethics of Making Part-human Animals in Stem Cell Biology,' *The FASEB Journal*, 20, no. 7: 838–45.

Robert, J., and F. Baylis (2003) 'Crossing Species Boundaries,' *American Journal of Bioethics*, 3, no. 3: 1–13.

Roberts, J. H. (2009) 'Myth 18: That Darwin Destroyed Natural Theology,' in R. L. Numbers (ed.), *Galileo Goes to Jail and Other Myths about Science and Religion*, Cambridge, MA: Harvard University Press.

Robertson, J. A. (1994) *Children of Choice: Freedom and the New Reproductive Technologies*, Princeton, NJ: Princeton University Press.

Rockenbach, B., and M. Milinski (2006) 'The Efficient Interaction of Indirect Reciprocity and Costly Punishment,' *Nature*, 444: 718–23.

Rolston, H., III (1988) *Environmental Ethics: Duties to and Values in the Natural World*, Philadelphia: Temple University Press.

—— (1999) *Genes, Genesis and God: Values and Their Origins in Natural and Human History*, Cambridge: Cambridge University Press.

Rorty, R. (1982) *Consequences of Pragmatism (Essays: 1972–1980)*, Minneapolis: University of Minnesota Press.

—— (1993) *Human Rights, Rationality, and Sentimentality*, ed. Stephen Shute and Susan Hurley. On Human Rights: The Oxford Amnesty Lectures 1993. New York: Basic Books.

—— (2007) 'Main Statement,' *What's the Use of Truth?* ed. Patrick Savidan. New York: Columbia University Press.

Rose, S. (2006) *The Future of the Brain: The Promise and Perils of Tomorrow's Neuroscience*, New York: Oxford University Press.

Rosen, C. (2003) 'Why Not Artificial Wombs?' *The New Atlantis. A Journal of Technology and Society*, 3: 67–76.

Rowe, C. (1999) 'Receiver Psychology and the Evolution of Multicomponent Signals,' *Animal Behaviour*, 58: 921–31.

Rue, L. (1994) *By the Grace of Guile: The Role of Deception in Natural History and Human Affairs*, New York: Oxford University Press.

Rusbult, C., and C. Agnew (2010) 'Prosocial Motivation and Behavior in Close Relationships,' in M. Mikulincer and P. Shaver (eds.), *Prosocial Motives, Emotions and Behavior: The Better Angels of Our Nature*, Washington, DC: American Psychological Association.

Ruse, M. (2001) *Can a Darwinian Be a Christian? The Relationship between Science and Religion*, Cambridge: Cambridge University Press.

—— (2005) 'Methodological Naturalism under Attack,' *South African Journal of Philosophy*, 24, no. 1: 44–60.

Russell, Bertrand (1963). 'A Free Man's Worship,' in *Mysticism and Logic and Other Essays*, London: Allen & Unwin.

Russo, M. B. (2007) 'Recommendations for the Ethical Use of Pharmacologic Fatigue Countermeasures in the U.S. Military,' *Aviation Space and Environmental Medicine*, 78, Supplement 4: B119–B127.

Ryle, G. (1949) *The Concept of Mind*, rep., New York: Barnes & Noble.

Saastamoinen, K. (2010) 'Pufendorf on Natural Equality, Human Dignity, and Self-Esteem,' *Journal of the History of Ideas*, 71, no. 1: 39–62.

Said, C., S. Baron, and A. Todorov (2008) 'Nonlinear Amygdala Response to Face Trustworthiness: Contributions of High and Low Spatial Frequency Information,' *Journal of Cognitive Neuroscience*, 21: 519–28.

Sample, I. (2003) 'Secrets of the Mind Must Remain Private Property, Says Scientist,' *The Guardian*, 20 November. Available HTTP: <http://www.guardian.co.uk/uk/2003/nov/20/health.businessofresearch> (accessed 11 December 2010).

Sandel, M. J. (2007) *The Case against Perfection: Ethics in the Age of Genetic Engineering*, Cambridge, MA: Belknap Press of Harvard University Press.

Sartre, J.-P. (1966) *Being and Nothingness: An Essay on Phenomenological Ontology*, trans. H. E. Barnes, New York: Washington Square Press.

—— (1985) *Existentialism and Human Emotions*, trans. B. Frechtman and H. E. Barnes, New York: Philosophical Library.

Sato, T., K. Katagiri, et al. (2011) 'In Vitro Production of Functional Sperm in Cultured Neonatal Mouse Testes,' *Nature*, 471, no. 7339: 504–7.

Savulescu, J. (2002) 'The Embryonic Stem Cell Lottery and the Cannibalization of Human Beings,' *Bioethics*, 16: 508–29.

—— (2009) 'The Human Prejudice and the Moral Status of Enhanced Beings: What Do We Owe the Gods?' in J. Savulescu and N. Bostrom (eds.), *Human Enhancement*, Oxford: Oxford University Press.

Schauer, F. (2010) 'Neuroscience, Lie-Detection, and the Law: Contrary to the Prevailing View, the Suitability of Brain-Based Lie-Detection for Courtroom or Forensic Use Should Be Determined According to Legal and Not Scientific Standards,' *Trends in Cognitive Science*, 14: 101–3.

Schepis, T. S., D. B. Marlowe, and R. F. Forman (2008) 'The Availability and Portrayal of Stimulants over the Internet,' *Journal of Adolescent Health*, 42: 458–65.

Scherer, K. (2003) 'Vocal Communication of Emotion: A Review of Research Paradigms,' *Speech Communication*, 40: 227–56.

Scherer, K., R. Banse, and H. Wallbott (2001) 'Emotion Inferences from Vocal Expression Correlate across Languages and Cultures,' *Journal of Cross Cultural Psychology*, 32: 76–92.

Schilhab, T.S.S. (2004) 'What Mirror Self-Recognition in Nonhumans Can Tell Us about Aspects of Self,' *Biology and Philosophy*, 19: 111–26.

Schmidt, G. L., and C. A. Seger (2009) 'Neural Correlates of Metaphor Processing: The Roles of Figurativeness, Familiarity and Difficulty,' *Brain and Cognition*, 71: 375–86.

Schnall, S., and J. Laird (2003) 'Keep Smiling: Enduring Effects of Facial Expressions and Postures on Emotional Experience and Memory,' *Cognition and Emotion*, 17: 787–97.

Schopenhauer, A. (1901) *The Wisdom of Life and Other Essays*, trans B. Saunders and E. B. Bax, Washington, D.C.: M. Walter Dunne. Available HTTP: <http://books.google.com/books?id=AvAZAAAAYAAJ&dq=schopenhauer%20wisdom%20of%20life%20and%20other%20essays&pg=PP1#v=onepage&q&f=false> (accessed 22 July 2010)

—— (1965) *On the Basis of Morality*, trans. E.F.J. Payne, Indianapolis, IN: Bobbs-Merrill.

Schore, A. (1994) *Affect Regulation and the Origin of the Self: The Neurobiology of Emotional Development*, Hillsdale, NJ: Lawrence Erlbaum.

Schore, J., and A. Schore (2008) 'Modern Attachment Theory: The Central Role of Affect Regulation in Development and Treatment,' *Clinical Social Work Journal*, 36: 9–20.

Schroeder, D. (2008) 'Dignity: Two Riddles and Four Concepts,' *Cambridge Quarterly of Healthcare Ethics*, 17: 230–38.

———— (2010) 'Dignity—One, Two, Three, Four, Five, Still Counting,' *Cambridge Quarterly of Healthcare Ethics*, 19, no. 1: 118–25.

Schuklenk, U., and W. Landman (2005) 'From the Editors,' in *Developing World Bioethics*. Special Issue: Reflections on the UNESCO Draft Declaration on Bioethics and Human Rights, 5, no. 3: iii–vi.

Schulkin, J. (1999) *The Neuroendocrine Regulation of Behavior*, New York: Cambridge University Press.

———— (2004) *Allostasis, Homeostasis, and the Costs of Physiological Adaptation*, New York: Cambridge University Press.

Schulman, A. (2008) 'Bioethics and the Question of Human Dignity,' in President's Council on Bioethics, *Human Dignity in Bioethics*, Washington, DC: President's Council on Bioethics, 3–18.

Searcy, W., and S. Nowicki (eds.) (2005) *The Evolution of Animal Communication: Reliability and Deception in Signaling Systems*, Princeton, NJ: Princeton University Press.

Searle, J. (1984) *Minds, Brains and Science*, Cambridge, MA: Harvard University Press.

Second Vatican Council (1965a) *Dignitatis Humanae: Declaration on Religious Freedom*. Available HTTP: <http://www.vatican.va/archive/hist_councils/ii_vatican_council/documents/vat-ii_decl_19651207_dignitatis-humanae_en.html> (accessed 14 June 2011).

———— (1965b) *Gaudium et Spes: Pastoral Constitution on the Church in the Modern World*. Available HTTP: <http://www.vatican.va/archive/hist_councils/ii_vatican_council/documents/vat-ii_cons_19651207_gaudium-et-spes_en.html> (accessed 14 June 2011).

Secord, J. A. (2000) *Victorian Sensation: The Extraordinary Publication, Reception, and Secret Authorship of* 'Vestiges of the Natural History of Creation,' Chicago: University of Chicago Press.

Select Committee on Assisted Dying for the Terminally Ill (2005) *Minutes of Evidence*, 3 February, United Kingdom Parliament.

Sen, A. (1998) 'Universal Truths: Human Rights and the Westernizing Illusion,' *Harvard International Review*, 20, no. 3: 40–43.

Sententia, W. (2004) 'Cognitive Liberty and Converging Technologies for Improving Human Cognition,' *Annals of the New York Academy of Science*, 1013: 221–8.

Shapin, S. (1996) *The Scientific Revolution*, Chicago: University of Chicago Press.

Shea, J. B. (2003) 'The Moral Status of In Vitro Fertilization (IVF) Biology and Method,' *Catholic Insight*, January/February.

Shell, S. (2008) 'Kant's Concept of Human Dignity,' in E. D. Pellegrino (ed.), *Human Dignity and Bioethics*, Washington, DC: President's Council on Bioethics.

Shibata, M., J. Abe, A. Terao, and T. Miyamoto (2007) 'Neural Mechanisms Involved in the Comprehension of Metaphoric and Literal Sentences: An fMRI Study,' *Brain Research*, 1166: 92–102.

Shizuru, J. A., R. S. Negrin, et al. (2005) 'Hematopoietic Stem and Progenitor Cells: Clinical and Preclinical Regeneration of the Hematolymphoid System,' *Annual Review of Medicine*, 56: 509–38.

Shultziner, D. (2003) 'Human Dignity—Functions and Meanings,' *Global Jurist Topics*, 3: 1–21.

Shuster, E. (2003) 'Human Cloning: Category, Dignity, and the Role of Bioethics,' *Bioethics*, 17, no. 5–6: 517–25.

Siegel, R. (2007) 'Sex Equality Arguments for Reproductive Rights: Their Critical Basis and Evolving Constitutional Expression,' *Emory Law Journal*, 56: 815–42.

———— (2008) 'Dignity and the Politics of Protection: Abortion Restrictions under *Casey/Carhart*,' *Yale Law Journal*, 117: 1694–1800.

Silbersweig, D., J. Clarkin, M. Goldstein, O. Kernberg, K. Levy, and E. Stern (2007) 'Failure of Frontolimbic Inhibitory Function in the Context of Negative Emotion in Borderline Personality Disorder,' *American Journal of Psychiatry*, 164: 1832–41.

Silveira, M. J., Y. H. Kim, and K. M. Langa (2010) 'Advance Directives and Out-comes of Surrogate Decision Making before Death,' *New England Journal of Medicine*, 362, no. 13 (1 April): 1211–18.

Simpson, G. G. (1966) 'The Biological Nature of Man,' *Science*, 152: 472–78.

Singer, P. (1981) *The Expanding Circle: Ethics and Sociobiology*, New York: Farrar, Straus & Giroux.

—— (1986) 'All Animals Are Equal,' in P. Singer (ed.), *Applied Ethics*, Oxford: Oxford University Press.

—— (1994) *Rethinking Life and Death: The Collapse of Our Traditional Ethics*, New York: St. Martin's Griffin.

Singer, T., and C. Lamm (2009) 'The Social Neuroscience of Empathy,' *Annals of the New York Academy of Sciences*, 1156: 81–96.

Singer, T., B. Seymour, J. O'Doherty, K. Stephan, R. Dolan, and C. Frith (2006) 'Empathic Neural Responses Are Modulated by the Perceived Fairness of Others,' *Nature*, 439: 466–69.

Sinke, C., M. Kret, and B. De Gelder (2011) 'Body Language: Embodied Perception of Emotion,' in B. Berglund, G. Rossi, J. Townsend, and L. Pendrill (eds.), *Measurement with Persons: Theory, Methods and Implementation Areas*, New York: Psychology Press.

Sinke, T., B. Sorger, R. Goebel, and B. De Gelder (2010) 'Tease or Threat? Judging Social Interactions from Bodily Expressions,' *NeuroImage*, 49: 1717–27.

Sinnott-Armstrong, W. (2009) *Morality without God*, Oxford: Oxford University Press.

Skene, C., G. Testa, I. Hyun, K. W. Jung, A. McNab, J. Robertson, C. T. Scott, J. H. Solbakk, P. Taylor, and L. Zoloth (2009) 'Ethics Report on Interspecies Somatic Cell Nuclear Transfer Research,' *Cell Stem Cell*, 5, no. 1: 27–30.

Skinner, B. F. (1971) *Beyond Freedom and Dignity*, New York: Bantam.

Smith, A. (1984) *The Theory of Moral Sentiments*, ed. D. D. Raphael and A. L. Macfie, Indianapolis, IN: Liberty Fund. Available HTTP: <http://files.libertyfund.org/files/192/0141–01_Bk.pdf> (accessed 21 December 2010).

Snead, O. C. (2005a), 'Dynamic Complementarity: Terri's Law and Separation of Powers Principles in the End-of-Life Context,' *Florida Law Review*, 57: 53–89.

—— (2005b), 'Preparing the Groundwork for a Responsible Debate on Stem Cell Research and Human Cloning,' *New England Law Review*, 39: 479–88.

—— (2005c), 'The (Surprising) Truth About Schiavo: A Defeat for the Cause of Autonomy,' *Constitutional Commentary*, 22: 383–404.

—— (2009) 'Public Bioethics and the Bush Presidency,' *Harvard Journal of Law and Public Policy*, 32: 867.

—— (2010) 'Science, Public Bioethics, and the Problem of Integration,' *University of California, Davis Law Review*, 43: 1529–1604.

Soloveitchik, J. (1983) *Halakhic man*, Philadelphia: Jewish Publication Society of America.

—— (1993) *Yemei Hashivah u-Kedushah*, Jerusalem: WZO.

Solzhenitsyn, A. I. (1974) *The Gulag Archipelago 1918–1956: An Experiment in Literary Investigation*, vol. 1, trans. T. P. Whitney, New York: Harper & Row.

Somerville, M. (2009) *The Ethical Imagination: Journeys of the Human Spirit*, Montreal: McGill-Queen's University Press.

—— (2010) 'Is the Concept of Human Dignity Useful, Useless or Dangerous?' Human Dignity and the Future of Health Care: Baylor Symposium on Faith and Culture, 28 October 2010, Baylor University, Waco, TX. Available HTTP: <http://www.baylor.edu/player/index.php?id=126003&gallery_id=5664','770','550')> (accessed 12 November 2010).

Spaemann, R. (1991) *Moralische Grundbegriffe*, Munich: Beck.

Spivak, G. (2010) 'Can the Subaltern Speak?' in R. Morris (ed.) *Can the Subaltern Speak? Reflections on the History of an Idea*, New York: Columbia University Press.

Sroufe, L. (1995) *Emotional Development: The Organization of Emotional Life in the Early Years*, New York: Cambridge University Press.

Stackhouse, M. L. (1986) 'Theology and Human Rights,' *Perkins Journal*, October: 11–18.

—— (1996) 'Public Theology and Christological Politics: A Response to Mark A. Noll,' in L. Lugo (ed.), *Adding Cross to Crown: The Political Significance of Christ's Passion*, Washington, DC: Center for Public Justice.

—— (1999–2000) 'Universal Absolutes,' *Journal of Law and Religion*, 14, no. 1: 97–112.

—— (2002) 'Covenantal Justice in a Global Era,' Institute for Reformed Theology, January. Available HTTP: <http://reformedtheology.org/SiteFiles/PublicLectures/StackhousePL.html> (accessed 19 July 2012).

—— (2004) 'A Christian Perspective on Human Rights,' *Society*, 41, no. 2: 23–28.

—— (2005). 'Why Human Rights Needs God: A Christian Perspective,' in E. M. Bucar and Barbra Barnett (eds.), *Does Human Rights Need God?* Grand Rapids, MI: Eerdmans.

Stackhouse, M. L., and S. E. Healey (1996) 'Religion and Human Rights: A Theological Apologetic,' in J. Witte Jr. and J. D. van der Vyer (eds.), *Religious Human Rights in Global Perspective*, Dordrecht, Netherlands: Kluwer.

Starck, C. (2002) 'The Religious and Philosophical Background of Human Dignity and Its Place in Modern Constitutions,' in D. Kretzmer and E. Klein (eds.), *The Concept of Human Dignity in Human Rights Discourse*, The Hague: Kluwer.

Stark, R. (2003) *For the Glory of God: How Monotheism Led to Reformations, Science, Witch-Hunts, and the End of Slavery*, Princeton, NJ: Princeton University Press.

Steinbock, B. (1992) *Life Before Birth: The Moral and Legal Status of Fetuses and Embryos*, New York: Oxford University Press.

—— (2001) 'Respect for Human Embryos,' in P. Lauritzen (ed.), *Cloning and the Future of Human Embryo Research*, Oxford: Oxford University Press.

Steiner, G. (2005) *Anthropocentrism and Its Discontents: The Moral Status of Animals in the History of Western Philosophy*, Pittsburgh, PA: University of Pittsburgh Press.

Stepper, S., and F. Strack (1993) 'Proprioceptive Determinants of Emotional and Nonemotional Feelings,' *Journal of Personality and Social Psychology*, 64: 211–20.

Stevenson, R., and T. James (2008) 'Affective Auditory Stimuli: Characterization of the International Affective Digitized Sounds (IADS) by Discrete Emotional Categories,' *Behavioral Research Methods*, 40: 315–21.

Stith, R. (2007) 'Why Pro-Life Arguments Sound Absurd,' *Human Life Review*, 33, no. 1: 23–26.

Stolberg, S. (2001) 'Scientists Create Scores of Embryos to Harvest Cells,' *New York Times*, 11 July: A1.

Streiffer, R. (2005) 'At the Edge of Humanity: Human Stem Cells, Chimeras, and Moral Status,' *Kennedy Institute of Ethics Journal*, 15, no. 4: 347–70.

Stuber, G., D. Sparta, A. Stamatakis, W. Van Leeuwen, J. Hardjoprajitno, S. Cho, et al. (2011) 'Excitatory Transmission from the Amygdala to Nucleus Accumbens Facilitates Reward Seeking,' *Nature*, 475: 377–80.

Stump, E. (1995) 'Non-Cartesian Substance Dualism and Materialism without Reductionism,' *Faith and Philosophy*, 12, no. 4: 505–23.

Sugiyama, L., J. Tooby, and L. Cosmides (2002) 'Cross-Cultural Evidence of Cognitive Adaptations for Social Exchange among the Shiwiar of Ecuadorian Amazonia,' *Proceedings of the National Academy of Sciences*, 99: 11537–42.

Sulmasy, D. (2008) 'Dignity and Bioethics: History, Theory, and Selected Applications,' in President's Council on Bioethics, *Human Dignity and Bioethics*, Washington, DC: President's Council on Bioethics, 469–501.

Sulmasy, D. P. (2006) 'Dignity and the Human as a Natural Kind,' in C. Taylor and R. Dell'Oro (eds.), *Health and Human Flourishing: Religion, Medicine, and Moral Anthropology*, Washington, DC: Georgetown University Press.

—— (2006a) 'The Concept of Dignity,' presentation to President's Council on Bioethics, 2 February 2006. Available HTTP: <http://bioethics.georgetown.edu/pcbe/transcripts/feb06/session2.html> (accessed 14 December 2010).

—— (2007) 'Human Dignity and Human Worth,' in J. Malpas and N. Lickiss (eds.), *Perspectives on Human Dignity*, Dordrecht, Netherlands: Springer.

—— (2008) 'Dignity, Rights, Health Care, and Human Flourishing,' in D. N. Weisstub and G. D. Pintos (eds.), *Autonomy and Human Rights in Health Care*, Dordrecht, Netherlands: Springer.

Swinburne, R. (1997) *The Evolution of the Soul*, rev. edn, Oxford: Clarendon Press.

Tabibnia, G., A. Satpute, and M. Lieberman (2008) 'The Sunny Side of Fairness: Preference for Fairness Activates Reward Circuitry (And Disregarding Unfairness Activates Self-control Circuitry),' *Psychological Science*, 19: 339–47.

Tabibnia, G., and M. Lieberman (2007) 'Fairness and Cooperation Are Rewarding: Evidence from Social Cognitive Neuroscience,' *Annals of the New York Academy of Sciences*, 1118: 90–101.

Takagishi, H., T. Takahashi, A. Toyomura, N. Takashino, M. Koisumi, and T. Yamagishi (2009) 'Neural Correlates of the Rejection of Unfair Offers in the Impunity Game,' *Neuroendocrinology Letters*, 30: 496–500.

Takahashi, K. et al. (2007) 'Induction of Pluripotent Stem Cells from Adult Human Fibroblasts by Defined Factors,' *Cell*, 131: 861–72.

Takahashi, K., and S. Yamanaka (2006) 'Induction of Pluripotent Stem Cells from Mouse Embryonic and Adult Fibroblast Cultures by Defined Factors,' *Cell*, 126: 663–76.

Tang, Y., J. R. Nyengaard, D. M. De Groot, and H. J. Gundersen (2001) 'Total Regional and Global Number of Synapses in the Human Brain Neocortex,' *Synapse*, 41: 258–73.

Tartaglia, N., S. Howell, et al. (2010) 'A Review of Trisomy X (47, XXX),' *Orphanet Journal of Rare Diseases*, 5, no. 1: 8.

Tauer, C. (2001) 'Responsibility and Regulation: Reproductive Technologies, Cloning, and Embryo Research,' in P. Lauritzen (ed.), *Cloning and the Future of Human Embryo Research*, New York: Oxford University Press.

Taylor, C. (1992) *The Ethics of Authenticity*, Cambridge, MA: Harvard University Press.

—— (1999) 'Conditions of an Unforced Consensus on Human Rights,' in J. R. Bauer and D. A. Bell (eds.), *The East Asian Challenge for Human Rights*, Cambridge: Cambridge University Press.

—— (2007) *A Secular Age*, Cambridge, MA: Harvard University Press.

Taylor, J., A. Wiggett, and P. Downing (2007) 'Functional MRI Analysis of Body and Body Part Representations in the Extrastriate and Fusiform Body Areas,' *Journal of Neurophysiology*, 98: 1626–33.

Taylor, P. W. (1986) *Respect for Nature: A Theory of Environmental Ethics*, Princeton, NJ: Princeton University Press.

Taylor, P., O. Hasson, and D. Clark (2000) 'Body Postures and Patterns as Amplifiers of Physical Condition,' *Proceedings of the Royal Society, Biological Sciences*, 267: 917–22.

Teno, J. M., J. Lynn, R. S. Phillips, et al. (1994) 'Do Formal Advance Directives Affect Resuscitation Decisions and the Use of Resources for Seriously Ill Patients?' *Journal of Clinical Ethics*, 5: 23–30.

Thompson, P. (1999) 'Evolutionary Ethics: Its Origins and Contemporary Face,' *Zygon*, 34, no. 4: 473–84.

Thomson, J. (1971) 'A Defense of Abortion,' *Philosophy and Public Affairs*, 1, no. 1: 47–66.

Thomson, J. A., J. Itskovitz-Eldor, S. S. Shapiro, M. A. Waknitz, J. J. Swiergiel, V. S. Marshall, and J. M. Jones (1998) 'Embryonic Stem Cell Lines Derived from Human Blastocysts,' *Science*, 282: 1145–47.

Tkacz, M. (2008) 'Aquinas vs. Intelligent Design,' *This Rock*, 19, no. 9. Available HTTP: <http://www.catholic.com/thisrock/2008/0811fea4.asp> (accessed 27 July 2010).

Todorov, N., and A. Engell (2008) 'The Role of the Amygdala in Implicit Evaluation of Emotionally Neutral Faces,' *Social and Cognitive Affective Neuroscience*, 3: 303–12.

Tollefsen, C. (2008) *Artificial Nutrition and Hydration: The New Catholic Debate*, Dordrecht, Netherlands: Springer.

Tomasello, M. (2009) *Why We Cooperate*, Cambridge, MA: MIT Press.

Tomlin, D., A. Kayali, B. King-Casas, C. Anen, C. Camerer, S. Quartz, et al. (2006) 'Agent-Specific Responses in the Cingulate Cortex during Economic Exchanges,' *Science*, 312: 1047–50.

Tooby, J., L. Cosmides, and M. Price (2006) 'Cognitive Adaptations for N-person Exchange: The Evolutionary Roots of Organizational Behavior,' *Managerial and Decision Economics* 27: 103–29.

Trinkaus, C. (1973) 'Renaissance Idea of the Dignity of Man,' in P. P. Weiner (ed.), *Dictionary of the History of Ideas*, vol. 4, New York: Charles Scribner's Sons.

Trivers, R. (1971) 'The Evolution of Reciprocal Altruism,' *Quarterly Review of Biology*, 46: 35–57.

Tronick, E. (2007) *The Neurobehavioral and Social-Emotional Development of Infants and Children*, New York: Norton.

Turing, A. (1981) 'Computing Machinery and Intelligence,' in D. Hofstadter and D. C. Dennett (eds.), *The Mind's I: Fantasies and Reflections on Self and Soul*, New York: Basic Books.

Turrini, P., R. Sasso, et al. (2006) 'Development of Humanized Mice for the Study of Hepatitis C Virus Infection,' *Transplantation Proceedings*, 38, no. 4: 1181–84.

U.S. National Center for Health Statistics, National Vital Statistics Reports (2009) *Deaths: Final Data for 2006*, 57, no. 14 (17 April).

UNESCO (1997) *Universal Declaration on the Human Genome and Human Rights*, 29th Sess. 29 C/Resolution 19. Available HTTP: <www.unesco.org/new/ . . . human . . . /human-genome-and-human-rights> (accessed July 25, 2012).

UNHCR's Expert Group on Human Rights and Biotechnology (2002) 'Conclusions on Human Reproductive Cloning,' *Health and Human Rights*, 6, no. 1: 153–59. Available HTTP: <http://www.hhrjournal.org/archives/vol6-no1.php> (accessed July 25, 2012).

United Nations (1948) Universal Declaration on Human Rights, adopted and proclaimed by United Nations General Assembly Resolution 217A (III) 10 December.

United Nations General Assembly (1948) 'Universal Declaration of Human Rights,' General Assembly Resolution 217 A (III). Paris and New York: United Nations. Available HTTP: <http://daccess-dds-ny.un.org/doc/RESOLUTION/GEN/NR0/043/88/IMG/NR004388.pdf> (accessed 21 July 2010).

United States Catholic Conference (1997) *Catechism of the Catholic Church*, 2nd edn, trans. from 1993 Latin text, rev. 1997, Vatican City: Libreria Editrice Vaticana. Available HTTP: <http://www.usccb.org/catechism/text/> (accessed 27 July 2010).

United States Conference of Catholic Bishops (1996) *Critical Decisions: Genetic Testing and Its Implications*. Available HTTP: <http://.www.old.usccb.org/shv/testing.html> (accessed July 25, 2012).

—— (2001) *Interventions upon Human Procreation*. Available HTTP: <http://www.old.usccb.org/prolife/tdocs/part2.shtml> (accessed July 25, 2012).

Urgesi, C., B. Calvo-Merino, P. Haggard, and S. Aglioti (2007) 'Transcranial Magnetic Stimulation Reveals Two Cortical Pathways for Visual Body Processing,' *Journal of Neuroscience*, 27: 8023–30.

Van den Stock, J., J. Grèzes, and B. De Gelder (2008) 'Human and Animal Sounds Influence Recognition of Body Language,' *Brain Research*, 1242: 185–90.

Van den Stock, J., R. Righart, and B. De Gelder (2007) 'Body Expressions Influence Recognition of Emotions in the Face and Voice,' *Emotion*, 7: 487–94.

Van Hecke, A., J. Lebow, E. Bal, D. Lamb, F. Harden, A. Kramer, et al. (2009) 'EEG and Heart Rate Regulation to Familiar and Unfamiliar People in Children with Autism Spectrum Disorders,' *Child Development*, 80: 1118–33.

Van Lancker, D., and J. L. Cummings (1999) 'Expletives: Neurolinguistic and Neurobehavioral Perspectives on Swearing,' *Brain Research Reviews*, 31: 83–104.

Van Lange, P. (2008) 'Does Empathy Trigger Only Altruistic Motivation? How About Selflessness or Justice?' *Emotion*, 8: 766–74.

Van Rijn, S., A. Aleman, E. Van Diessen, C. Berkmoes, and G. Vingerhoets, G. (2005) 'What Is Said or How It Is Said Makes a Difference: Role of the Right Fronto-Parietal Operculum in Emotional Prosody as Revealed by Repetitive TMS,' *European Journal of Neuroscience*, 21: 195–200.

Van Vugt, M. (2008) 'The Origins of Leadership: Understanding Why and How Leadership Evolved Helps Us Understand Our Ambivalent Relationship with Those in Power Today,' *New Scientist*, 198: 42–44.

Van Vugt, M., R. Hogan, and R. Kaiser, R. (2008) 'Leadership, Followership, and Evolution,' *American Psychologist*, 63: 182–96.

Vatican (2011). 'Vatican Health Experts "Dismayed" by Nobel Prize for IVF Co-developer,' Catholic News Agency, 5 October 2010.

Vatican Congregation for the Doctrine of the Faith (1968) *Humanae Vitae*. Available HTTP: <http://www.vatican.va/holy_father/paul_vi/encyclicals/documents/hf_p-vi_enc_25071968_humanae-vitae_en.html> (accessed April 4, 2010).

—— (1987) *Donum Vitae (Gift of Life)*. Available HTTP: <http://www.vatican.va/roman_curia/congregations/cfaith/documents/rc_con_cfaith_doc_19870222_respect-for-human-life_en.html> (accessed July 25, 2012).

—— (2008) *Dignitas Personae*. Available HTTP: <http://www.catholic-ew.org.uk/Catholic-Church/What-does-the-Catholic-Church-teach/Vatican-Documents/Dignitas-Personae> (accessed April 4, 2010).

Vickers, H. (2010) 'Involuntary Childlessness Has a Significant Impact, Study Shows,' *Bionews*, 23 August. Available HTTP: <http://www.bionews.org.uk/page_68973.asp?print=1> (accessed July 25, 2012).

Vieillard, S., and M. Guidetti, M. (2009) 'Children's Perception and Understanding of (Dis)similarities among Dynamic Bodily/Facial Expressions of Happiness, Pleasure, Anger, and Irritation,' *Journal of Experimental Child Psychology*, 102: 78–95.

Vlamings, P., L. Jonkman, and C. Kemner (2010) 'An Eye for Detail: An Event-Related Potential Study of the Rapid Processing of Fearful Facial Expressions in Children,' *Child Development*, 81: 1304–19.

Volodin, V. (ed.) (2009) *Human Rights. Major International Instruments. Status as at 31 May 2009*. Paris: UNESCO. Available HTTP: <http://unesdoc.unesco.org/images/0018/001834/183452m.pdf> (accessed 14 June 2011).

Wallace, A. R. (1871) '[Review of] The Descent of Man and Selection in Relation to Sex, by Charles Darwin,' *The Academy*, 2, no. 20 (15 March): 177–83. Available HTTP: <http://darwin-online.org.uk/content/frameset?viewtype=text&itemID=A270&pageseq=1> (accessed 24 July 2010).

—— (1895) *Natural Selection and Tropical Nature: Essays on Descriptive and Theoretical Biology*. London: Macmillan. Available HTTP: <http://darwin-online.org.uk/content/frameset?viewtype=text&itemID=A238&pageseq=1> (accessed 25 July 2010).

Warren, M. A. (1997) *Moral Status: Obligations to Persons and Other Living Things*, New York: Oxford University Press.

Washington Natural Death Act (1992) Chapter 70.122, RCW 1992 [1992 c 98 § 2; 1979 c 112 § 3.].

Washington v. Glucksberg (1997) 521 U.S. 702.

Watkins, K. (2011) 'Developmental Disorders of Speech and Language: From Genes to Brain Structure and Function,' *Progress in Brain Research*, 189: 225–38.

Webster's New Collegiate Dictionary (1973) Springfield, MA: G & C Merriam.

Wedgwood Museum (2010) 'Slavery.' Available HTTP: <http://www.wedgwoodmuseum.org.uk/learning/discovery_packs/2179/pack/2184/chapter/2351> (accessed 31 August 2010).

Wegner, W. M. (2002) *The Illusion of Conscious Will*, Cambridge, MA: MIT Press.

Weikart, R. (2004) *From Darwin to Hitler: Evolutionary Ethics, Eugenics, and Racism in Germany*, New York: Palgrave Macmillan.

Weissmann, G. (2010) 'A Nobel Prize Is Out of Order: "American Idol" vs. Hypatia of Alexandria,' *The FASEB Journal*, 24, no. 12: 4627.

White, A. D. (1896) *A History of the Warfare of Science with Theology in Christendom*, 2 vols., New York: D. Appleton. Available HTTP: <http://cscs.umich.edu/~crshalizi/White/> (accessed 29 April 2010).

White, R. J. (2007) 'Chimeric Monkeys with Human Brains: A New Bioethical Black Hole,' *Artificial Organs*, 31, no. 6: 423–24.

Wiethoff, S., D. Wildgruber, B. Kreifelts, H. Becker, C. Herbert, W. Grodd, et al. (2008) 'Cerebral Processing of Emotional Prosody—Influence of Acoustic Parameters and Arousal,' *NeuroImage*, 39: 885–93.

Wilberforce, S. (1860) 'On the Origin of Species,' *Quarterly Review*, 102: 225–64. Available HTTP: <http://www.victorianweb.org/science/science_texts/wilberforce.htm> (accessed 17 April 2006).

Wildgruber, D., H. Ackermann, B. Kreifelts, and T. Ethofer, T. (2006) 'Cerebral Processing of Linguistic and Emotional Prosody: fMRI studies,' *Progress in Brain Research*, 156: 249–68.

Wildgruber, D., T. Ethofer, D. Grandjean, and B. Kreiflets, B. (2009) 'A Cerebral Network Model of Speech Prosody Comprehension,' *International Journal of Speech-Language Pathology*, 11: 277–81.

Wilkinson, R., and K. Pickett (2009) *The Spirit Level: Why More Equal Societies Almost Always Do Better*, London: Allen Lane.

Williams, R. W., and K. Herrup (1988) 'The Control of Neuron Number,' *Annual Review of Neuroscience*, 11: 423–53.

Willis, J., and A. Todorov (2006) 'First Impressions: Making Up Your Mind after a 100-ms Exposure to a Face,' *Psychological Science*, 17: 592–98.

Wilson, D. (2006) 'The Evolutionary Neuroscience of Human Reciprocal Sociality: A Basic Outline for Economists,' *Journal of Socio-Economics*, 35: 626–33.

Wilson, E. O. (1975) *Sociobiology: The New Synthesis*, Cambridge, MA: Harvard University Press.

—— (1978) *On Human Nature*, Cambridge, MA: Harvard University Press.

—— (1998) *Consilience: The Unity of Knowledge*, New York: Alfred A. Knopf.

Wiredu, K. (2001) 'An Akan Perspective on Human Rights,' in Patrick Hayden (ed.), *The Philosophy of Human Rights*, St. Paul, MN: Paragon House.

Witte, J. (2009) 'A Dickensian Era of Religious Rights,' in Frederick M. Shepherd (ed.), *Christianity and Human Rights: Christians and the Struggle for Global Justice*, Lanham, MD: Rowman & Littlefield.

Wolf, M., G. Van Doorn, and F. Weissing (2011) 'On the Coevolution of Social Responsiveness and Behavioral Consistency,' *Proceedings of the Royal Society*, Series B, 278: 440–48.

Wolf, N. (1995) 'Our Bodies, Our Souls: Rethinking Pro-Choice Rhetoric,' *The New Republic*, October 16: 26–35.

Wollstonecraft, M. (1891) *A Vindication of the Rights of Woman*, new edn, London: T. Fisher Unwin. Available HTTP: <http://books.google.com/books?id=hxwEAA AAYAAJ&printsec=frontcover&source=gbs_ge_summary_r&cad=0#v=onepage &q&f=false> (accessed 3 September 2010).

Quotations from the Bible marked NASB taken from *The New American Standard Bible®*, © 1960, 1962, 1963, 1968, 1971, 1972, 1973, 1975, 1977, 1995 by The Lockman Foundation. Used by permission. Those marked NIV taken from *The Holy Bible, New International Version®*, © 1973, 1978, 1984 by Biblica. Used by permission of Zondervan.

Wolpe, P. R., K. R. Foster, and D. D. Langleben (2010) 'Emerging Neurotechnologies for Lie-Detection: Promises and Perils,' *American Journal of Bioethics*, 10: 40–48.

Worster, D. (1985) *Nature's Economy: A History of Ecological Ideas*, Cambridge: Cambridge University Press.

Wright, R. (1990) 'The Intelligence Test: Stephen Jay Gould and the Nature of Evolution,' *The New Republic*, 29 January, 28–36.

―――― (1994) *The Moral Animal: Evolutionary Psychology and Everyday Life*, New York: Vintage.

―――― (2006) 'Dignity and Conflicts of Constitutional Values: The Case of Free Speech and Equal Protection,' *San Diego Law Review*, 43: 527–75.

―――― (2006) 'The Uses of Anthropomorphism,' in S. Macedo and J. Ober (eds.), *Primates and Philosophers: How Morality Evolved*, Princeton, NJ: Princeton University Press.

―――― (2009) *The Evolution of God*, New York: Little, Brown.

Wu, J., B. Zhang, Z. Zhou, Z. He, X. Zheng, R. Cressman, and Y. Tao, Y. (2009) 'Costly Punishment Does Not Always Increase Cooperation,' *Proceedings of the National Academy of Sciences*, 106: 17448–51.

Wyman, E., H. Rakoczy, and M. Tomasello (2009) 'Normativity and Context in Young Children's Pretend Play,' *Cognitive Development*, 24: 146–55.

Wynne, C. (2007) 'Aping Language: A Skeptical Analysis of the Evidence for Non-human Primate Language,' *eSkeptic*, 31 October. Available HTTP: <http://www. skeptic.com/eskeptic/07–10–31.html#feature> (accessed 10 January 2008).

Yamagishi, T., S. Tanida, R. Mashima, E. Shimoma, and S. Kanazawa (2003) 'You Can Judge a Book by Its Cover: Evidence That Cheaters May Look Different from Cooperators,' *Evolution and Human Behavior*, 24: 290–301.

Yang, T., and M. N. Shadlen (2007) 'Probabilistic Reasoning by Neurons,' *Nature*, 447: 1075–80.

Yong, A. (2007) *Theology and Down Syndrome: Reimagining Disability in Late Modernity*, Waco, TX: Baylor University Press.

Zagzebski, L. (2001) 'The Uniqueness of Persons,' *Journal of Religious Ethics*, 29: 401–23.

Zaki, J., J. Weber, N. Bolger, and K. Ochsner (2009) 'The Neural Bases of Empathic Accuracy,' *Proceedings of the National Academy of Sciences*, 106: 11382–87.

Zhang, J., K. Berridge, A. Tindell, K. Smith, and J. Aldridge, J. (2009) 'A Neural Computational Model of Incentive Salience,' *PLoS Computational Biology*, 5: e1000437.

Zhang, R., C. Vogler, and D. Metaxis (2007) 'Human Gait Recognition at Sagittal Plane,' *Image and Vision Computing*, 25: 321–30.

Zorich, Z. (2010) 'Should We Clone Neanderthals?' *Archaeology*, 63, no. 2.

Contributors

Roberto Andorno is a senior research fellow at the School of Law, University of Zurich, Switzerland. Originally from Argentina, he holds doctoral degrees in law from the Universities of Buenos Aires (1991) and Paris XII (1994), both on topics related to the ethical and legal aspects of assisted reproductive technologies. Between 2001 and 2005, he conducted various research projects relating to bioethics at the Universities of Göttingen and Tübingen, in Germany. From 1998 to 2005, he served as a member of the International Bioethics Committee of UNESCO and participated in this capacity in the drafting of the Universal Declaration on Bioethics and Human Rights (2005).

David H. Calhoun is associate professor of philosophy at Gonzaga University in Spokane, Washington, where he has been teaching since 1989. Calhoun regularly teaches courses on ancient philosophy, philosophy of human nature, and love and friendship and has recently developed a course on Christianity and science that critically examines the popular view that science and Christianity are at war with one another. He has published articles and book reviews on philosophical theology, ethics, and philosophy of education and spoken at conferences on topics such as the New Atheism, Christianity and science, and philosophy and popular culture. Calhoun directs the Gonzaga Socratic Club, which promotes philosophical inquiry into Christian worldview issues after the model of C. S. Lewis's Oxford Socratic Club.

Audrey R. Chapman is professor of community medicine and health care and holds the Healey Memorial Chair in Medical Ethics and Humanities at the University of Connecticut School of Medicine. She has worked on a wide range of ethical, theological, and human rights issues related to health equity, genetic developments, stem cells, and reproductive technologies. She is the author, coauthor, or editor of 16 books and numerous articles and reports. Her books include *Truth and Reconciliation in South Africa* (with Hugo van der Merwe), *Designing Our Descendants: The Promises and Perils of Genetic Modifications* (with Mark Frankel), and *Unprec-*

edented Choices: Religious Ethics at the Frontiers of Genetic Science. She is currently chair of the University of Connecticut Stem Cell Oversight Committee, is a member of the John Dempsey Hospital Ethics Committee, and serves on various expert committees for the state of Connecticut.

William P. Cheshire, Jr., is professor of neurology at Mayo Clinic in Florida. His clinical expertise has focused on disorders of the autonomic nervous system, which is that part of the brain and peripheral nervous system that governs blood pressure, heart rate, body temperature, sweating, and other functions below the level of conscious awareness. He is director of the Clinical Neurophysiology Laboratory, serves on the editorial board of the journal *Autonomic Neuroscience,* and has chaired the autonomic section of the American Academy of Neurology. He contributes an ongoing series entitled Grey Matters to the journal *Ethics and Medicine.*

Paul Copan is professor and Pledger Family Chair of Philosophy and Ethics at Palm Beach Atlantic University (West Palm Beach, Florida). Copan earned a doctorate in philosophy from Marquette University. He is author of *Loving Wisdom: Christian Philosophy of Religion* (Chalice Press) and coauthor (with William Lane Craig) of *Creation Out of Nothing: A Biblical, Philosophical, and Scientific Exploration* (Baker Academic). He is co-editor (with Paul K. Moser) of *The Rationality of Theism* (Routledge) as well as the co-editor (both with Chad V. Meister) of *The Routledge Companion to Philosophy of Religion* (Routledge) and of *Philosophy of Religion: Classic and Contemporary Issues* (Blackwell). He has authored and edited a number of other books and contributed essays and written reviews for journals such as *The Review of Metaphysics, Faith and Philosophy, Philosophia Christi,* and *Trinity Journal.* He lives with his wife, Jacqueline, and their six children in West Palm Beach, Florida.

Stephen Dilley is associate professor of philosophy at St. Edward's University in Austin, Texas. In addition to ethics, his areas of interest include the history and philosophy of biology. He has published essays in *British Journal for the History of Science, The Journal of the International Society for the History of Philosophy of Science, Philosophia Christi, Journal of Interdisciplinary Studies,* and elsewhere. Dilley is presently at work on a book that examines the epistemic role of theology in *On the Origin of Species.*

Suzanne Holland is Philip M. Phibbs Research Professor in the Department of Religion at University of Puget Sound in Tacoma, Washington. Holland is a bioethicist whose research focuses on stem cell research and ethics, though she also works on the ethics and justice of new genetic technologies, biotechnologies (including assisted reproduction), science and technology, and, more broadly, issues in religion, gender, and culture. With colleagues at the University of Washington she has published a book on translational genomics and ethics, *Achieving Justice in Genomic Translation: Re-thinking the Pathway to Benefit* (Oxford University Press). Holland is also co-editor of the first book published on stem cell ethics, *The Human Embryonic Stem Cell Debate: Science, Ethics, and Policy* (MIT Press). She is the author or coauthor of more than 25 scholarly articles and is presently at work on a monograph, *Technologies of Desire.*

John Loike is the co-director for Graduate Studies in the Department of Physiology Cellular Biophysics and director of Special Programs in the Center for Bioethics at Columbia University College of Physicians and Surgeons. He serves as the editor of the *Columbia University Journal of Bioethics* and is a course instructor for Frontiers in Bioethics, Ethics for Biomedical Engineers, and Stem Cells: Biology, Ethics, and Applications. Each summer, Loike directs a bioethics program in Bangkok, Thailand, called BioCEP, where international undergraduate and graduate students encounter real-life situations in bioethics and medical ethics. His biomedical research focuses on how human white blood cells combat infections and cancer.

Richard McClelland was educated at Reed College (BA, 1970), Princeton Theological Seminary (1970–1971), Oxford University (BA, 1975; MA, 1980), and the University of Cambridge (PhD, 1985). He has taught philosophy at the University of Notre Dame (1981–1985), Seattle Pacific University (1985–1992), and Seattle University (1993–1999) and is currently professor of philosophy at Gonzaga University. His main research interests lie at the junction of contemporary neuroscience, philosophy of mind, and empirical psychology. He has also trained and worked as a psychotherapist and retains interests in contemporary object relations theory and attachment theory. He is married and has three grown children and one grandchild.

Thomas R. McCormick joined the faculty in the University of Washington School of Medicine in 1974 to initiate the first formal program of bioethics for medical students. Since that time he has offered electives and taught a number of hours in the required core curriculum and has received an award for directing an outstanding Continuing Medical Education Bioethics Seminar. In addition to many lectures in the United States and Canada, he has lectured in Japan, Taiwan, Germany, and Italy. He is the author or coauthor of more than 50 articles and numerous videotapes used as teaching resources in bioethics. He has a strong interest in ethical issues at the end of life, the care of dying patients, and strengthening the doctor–patient relationship.

Nathan J. Palpant is a senior research fellow at the University of Washington's Institute for Stem Cell and Regenerative Medicine. His current biomedical research focuses on cardiac stem cell biology and organo-genesis. His work in bioethics has focused on embryonic stem cells. Pal-pant has presented his ethics research at the American Association for the Advancement of Science Conference on Science and Technology in Society and at the annual international conference of the Society for the Social Studies of Science. He is presently co-editing a volume with Ronald Green, *Suffering in Bioethics* (Oxford University Press). His other areas of research include the bioethics of genetic engineering technology (with a focus on advances in zinc finger nucleases for genome editing) and examination of tensions in the goal of medicine between the alleviation of suffering and the preservation of nonhedonic goods.

Scott Rae is professor and chair of philosophy of religion and ethics at Biola University. His primary interests are medical ethics and business ethics, dealing with the application of Christian ethics to medicine and the mar-ketplace. He has authored 10 books in ethics, including *The Ethics of Commercial Surrogate Motherhood*; *Moral Choices: An Introduction to Ethics*; *Brave New Families: Biblical Ethics and Reproductive Tech-nologies*; and *The Virtues of Capitalism: A Moral Case for Free Markets*. His work has appeared in *The Linacre Quarterly, Religion and Liberty, National Catholic Bioethics Quarterly, Ethics and Medicine*, and others. He is a consultant for ethics for four Southern California hospitals. He is a fellow of the Center for Bioethics and Human Dignity and a fellow of the Wilberforce Forum. Recently he has joined the ministry of Chuck Colson's "Doing the Right Thing: An Interactive Exploration of Ethics."

O. Carter Snead is professor of law at the University of Notre Dame. His principal area of research is public bioethics—the governance of science, medicine, and biotechnology in the name of ethical goods. His schol-arly works have appeared in such publications as *New York University Law Review; Harvard Law Review Forum; Yale Journal of Health Policy, Law, and Ethics*; and *Political Science Quarterly*. Snead previously served as general counsel to the President's Council on Bioethics (chaired by Leon R. Kass) and is currently a member of UNESCO's International Bioethics Committee. Snead is also the W. P. and H. B. White Director of the University of Notre Dame Center for Ethics and Culture.

Christopher Tollefsen is professor of philosophy at the University of South Carolina. In 2011–2012, he was visiting fellow in the James Madison Program at Princeton University. Tollefsen has published numerous arti-cles, book chapters, and reviews on bioethics and natural law ethics and is the author, coauthor, or editor of five recent books, including *Biomedi-cal Research and Beyond: Expanding the Ethics of Inquiry; John Paul II's Contribution to Catholic Bioethics*; and, with Robert P. George, *Embryo: A Defense of Human Life*. He is the series editor for the Springer book series *Catholic Studies in Bioethics*. Tollefsen is a regular contributor to *Public Discourse* and a senior fellow of Witherspoon Institute. His PhD is from Emory University.

Mark Dietrich Tschaepe is assistant professor of philosophy at Texas A&M (Prairie View). He specializes in the philosophy of science and biomedical ethics, and his dissertation addressed pragmatic considerations as they pertain to mechanistic explanations, especially with regard to models for behavior. His publications include work on scientific explanation and the pragmatics of explanation. Currently, Tschaepe's focus is primarily the work of John Dewey and Hans Reichenbach as it pertains to scientific explanation and probability. In addition, he is writing an extensive monograph on Dewey's conception of love as it pertains to his ethics.

Index

A

abortion 7, 12–15, 50, 62–4, 132, 142–4, 147–54, 157–64, 204, 206, 210–11, 219–38, 243–4, 259, 323, 326, 328, 331, 335, 347–8, 353, 356, 359, 363
advance directive 272–3
agnosticism 10, 104, 167, 190
Ai, A. 271, 323
Aikman, D. 83, 323
Almeida-Porada, G. 283, 323
altruism 79, 105, 121, 288, 301, 325, 333–4, 363
amygdala 106–8, 113–14, 318, 334, 357, 361, 363
Andorno, R. 5, 11, 86, 95, 127–8, 130, 132, 134, 136, 138, 140, 323, 369
animal rights 352
Annas, G. 128, 134, 315, 323, 345
anthropology 22–3, 26, 30, 33–4, 36, 42, 44–5, 49, 73, 167–9, 181, 188, 327, 329, 331, 341, 345, 362
Aquinas 27, 52, 55, 57, 71, 198, 324, 328–9, 337, 363
Aristocratic 146, 251
Aristotle 21–3, 28, 35–6, 39–41, 43, 57, 76, 110, 137, 172, 180–2, 184, 196, 198, 324, 337, 353
Arnhart, L. 34, 43, 168–9, 172, 177, 192, 194, 197–8, 324
artificial: intelligence 316; nutrition 156–8, 363; wombs 348, 357
Asai, A. 140, 324
Ashcroft, R. 86, 247, 278–9, 324
Asian values 73, 80, 348
assisted reproductive technologies 7, 13, 63, 144, 235, 260, 369
atheism 12, 44, 104, 167–8, 170, 174, 181, 183, 186–7, 190–1, 331, 369

Augustine 26, 41, 193, 324, 339
Aurelius 22, 39, 324
autonomy 4–6, 8, 14, 16–17, 19, 30–1, 35, 37, 42–3, 59, 64, 79, 99, 103–4, 131, 146–9, 155, 157–9, 162–3, 167, 173, 184, 186, 189, 202, 207, 209, 212–14, 217, 219–20, 224, 231–7, 244, 247, 249–50, 253, 255–7, 259, 267, 269–70, 272–3, 277–81, 288–9, 291–2, 304, 306–7, 311, 314–15, 317, 320, 360, 362
Ayton-Shenker, D. 140, 324

B

Bailey, K. 83, 325
Balaban, E. 283, 290, 325
Barilan 227–8, 325
Barth, K. 71
Bauckham, R. 78, 326
Bayertz, K. 134, 326–7
Baylis, F. 95, 136, 285–6, 297, 326, 357
Beauchamp, T. 207, 224, 326
Beckwith, F. 225, 230, 236, 238, 326, 353
behavior 10, 21, 33, 73, 88, 105–11, 113, 116, 118, 120–3, 137, 139, 143, 169, 174–5, 177, 179, 192, 195, 211, 255, 290, 300–2, 304–5, 309, 319, 325, 327, 329, 332, 337, 350–1, 357–8, 363, 367, 373
Bell, R. 122, 326, 329
Belmont Report 16, 352
Benatar, S. 102, 326
beneficence 224, 249, 267
Berger, P. 20, 43, 327
Beyleveld, D. 5, 214, 247, 278, 327
Bible 6, 24, 71, 83, 194, 288, 339, 342–3, 366

biocentrism 172, 194
bioethics 3–5, 9–13, 19–20, 35, 50,
       56, 59–60, 62, 77, 84, 86,
       127–8, 131–2, 135–6, 138–45,
       147, 150, 152, 158–60, 164–5,
       168, 192, 199, 202, 205,
       212, 219–20, 223–4, 235,
       238, 243–4, 251, 261, 278–9,
       287, 289, 300–1, 306–7, 313,
       316–17, 319, 323–36, 338–40,
       343–8, 350–5, 357–60, 362,
       366, 369, 371–3
biological nihilism 169–70, 172–3, 324
biology 11, 94, 104–5, 122–3, 132–3,
       142, 151, 166–71, 174, 177,
       179–82, 192–3, 195–6, 205,
       258, 283, 285–6, 293, 325,
       332, 334–6, 338, 344, 351, 353,
       355–9, 363, 365, 367, 370–2
biotechnology 12, 101, 142–4, 206–7,
       213, 219, 222–3, 235, 238, 289,
       300, 306, 317, 320, 327, 329,
       338, 341, 346, 351, 355, 363,
       372
Birnbacher, D. 4, 251–2, 257, 327
body 11, 15, 23, 42, 53, 57–8, 72, 74–5,
       77, 79, 93, 112–13, 115–17,
       122–3, 132, 151, 159, 164, 170–1,
       180–1, 193–4, 196, 205–8, 212,
       225, 228, 235, 241–2, 245, 252,
       260, 267–8, 275, 282–3, 312,
       316, 324–5, 327, 331, 333, 335,
       338–9, 343, 347, 350–1, 354–5,
       360, 362–4, 370
Boehm, C. 106, 110, 123, 327
Bonhoeffer, D. 9, 35, 72–3, 327
Bontekoe, R. 69, 327
Bostrom, N. 304, 312, 318, 328, 331,
       358
Braaten, C. 71, 328
Braine, D. 194, 328
Brownsword, R. 5, 103, 203, 214, 244,
       249, 278, 319, 328–9
Brugger, E. 57, 328
Bush, G. 150–4, 157–8, 162–4, 219–20,
       252, 254, 329, 360

C
Calhoun, D. 7–8, 12–13, 19–20, 22,
       24, 26, 28, 30, 32, 34, 36, 38,
       40, 42, 44, 122, 166, 168, 170,
       172, 174, 176, 178, 180, 182,
       184, 186, 188, 190, 192, 194,
       196, 198, 369

Calne, D. 318, 329
Cameron, N. 222, 329, 331–2
capabilities approach 215–16, 352
capital punishment 50, 67, 328
Capron, A. 264, 329
*Carey v. Population Services
       International* 231
Carroll, W. 28, 191, 198, 329, 363
Catholic Church 8, 14, 45, 50, 55–7,
       60, 64, 71, 193, 203, 206–7,
       213, 219–20, 239–40, 295, 326,
       351, 364
Caulfield, T. 103, 202–3, 244, 249, 329
central nervous system 302
Chambers, R. 188–9, 246, 329
Chapman, A. 13–14, 201–4, 206, 208,
       210–14, 216, 218, 326, 329–30,
       351, 369
Chasaide, A. 117, 338
Chatterjee, A. 303, 330
cheaters 108, 122, 326, 329, 332, 367
Cheshire, W. 17–18, 300–4, 306,
       308–10, 312–14, 316, 318–20,
       330, 370
Childress, J. 207, 224, 326
chimeras 7, 10, 13, 16–17, 86, 95–8,
       282–7, 289–99, 323, 333–4,
       336, 342, 344, 346, 361
Christianity 8, 19, 20, 25–6, 33, 36–8,
       41–2, 64–5, 72–8, 81–5, 103,
       146, 166, 172–3, 185, 187, 190,
       193, 206, 222, 260, 274, 279,
       288, 323–4, 328–31, 339–41,
       343–8, 352, 355, 357, 361, 366,
       369, 370, 372
Churchland, P. S. 308, 310, 330
Cicero 6, 20, 22–3, 26, 39–42, 329–30
cingulate cortex 107–8, 363
Clark, W. 237–8, 331
cloning 4, 63, 133–6, 142, 144, 150–4,
       161–2, 164, 202–3, 211–14,
       217, 222, 235, 244, 292, 296,
       298–9, 323, 328, 330, 340, 344,
       349, 352–3, 355, 359–63
Coady, C. 297, 331
coalitions 123, 330
Cochrane, T. 313, 331
cognition 13, 96, 106–7, 112, 171,
       177–81, 193, 196, 198, 289,
       302, 316, 324, 326–7, 330, 339,
       348, 351, 358–9
Colson, C. 219–20, 222, 329, 331–2, 372
common descent 12–13, 33, 166–8,
       171–2, 181, 187, 191

complicity 15, 240, 261–3, 328
consequentialism 248
Constitution 104, 135, 143, 145, 155, 161, 164, 253–4, 359
Convention on Human Rights and Biomedicine (Oviedo Convention) 132, 135, 142, 345
Cooper, J. 42, 171, 188, 194, 331
cooperation 11, 54, 63, 104–5, 107, 109, 111, 119–22, 261, 288, 328, 331, 334, 336, 338, 344–5, 352–3, 355–6, 362, 366
Copan, P. 9, 67–8, 70, 72, 74, 76, 78, 80, 82, 84, 331, 370
cosmetic neurology 303, 330
Cosmides, L. 108, 331–2, 361, 363
Council of Europe 135, 142, 212, 278
Craig, W. L. 68, 85, 332, 345, 370
creation 12, 16, 25–6, 28, 30, 41, 51–3, 55, 67, 71–4, 96, 103, 136, 151, 153, 160, 166–8, 170, 183, 188–9, 191, 205–6, 212, 214, 222, 292–3, 296–7, 329, 334, 346, 356, 359, 363, 370
Crick, F. 180, 191, 308–9, 332
Cruzan v. Missouri 16, 155, 161, 165, 268–9, 271–3, 277
cultural diversity 11, 132, 136–8, 141, 324
cultural relativism 11, 84, 138–9, 356
Cunningham, P. 223, 332

D
Damasio, H. 301, 332
Darwin, C. 6, 8, 12, 33–4, 40, 42–4, 69, 115, 122, 166–98, 324, 326–9, 332–6, 339–42, 344–5, 348–51, 353–4, 356–7, 365
Darwinian existentialism 13, 172–3, 186, 196
Darwinism 7, 12–13, 34, 44, 166, 168–70, 173–5, 181, 183–4, 186–8, 191–4, 196, 326, 348, 356
Dawkins, R. 33, 168, 172, 174–5, 180, 188, 191, 193, 333
death 7, 9, 15, 34, 42, 60, 65, 68, 72, 74–5, 77–9, 86, 118, 143–4, 155–6, 160, 162, 166, 171, 174, 188, 244, 253, 263–81, 288, 298, 304, 310, 334, 338, 344–6, 352, 360, 365; of God 174; with dignity 155–6, 162, 264–5, 267–70, 275, 278–81, 334

decisional autonomy 14, 231–7
DeGrazia, D. 96–7, 101, 284, 286, 290, 292–3, 334
deism 30, 183–4
Deleuze 90
Dembski, W. 183, 334
Dennett, D. 6, 34, 36, 44, 166, 168–9, 172–5, 180, 191–2, 194–5, 197, 334, 363
Descartes, R. 8, 28–30, 42–3, 176–7, 179, 181, 188, 196, 332, 334, 340
Dewey, J. 89–90, 92–3, 95, 99–101, 323, 335, 373
dignitas 14, 21–3, 38–9, 41, 221, 246, 277–8, 300, 331, 364
Dignitas Personae 14, 221, 331, 364
dignity 3–28, 30–45, 47, 49–55, 57–101, 103–5, 107, 109–11, 113–23, 125, 127–43, 145–59, 161–3, 165–77, 179, 181–3, 185–7, 189, 191–5, 197, 199, 201–39, 241, 243–61, 263–82, 287–96, 298–301, 303–21, 323–36, 338–9, 342, 344–50, 352–5, 357–63, 366, 372
Dilley, S. 3–4, 6, 8, 10, 12, 14, 16, 18, 84, 167, 187, 190, 197, 335, 370
disability rights 143, 157
Dobzhansky, T. 168, 335
Dolly (cloned sheep) 135, 298
Donnelly, J. 137, 335
Donum Vitae 206, 221–2, 331, 364
double-effect 280
Down syndrome 68, 76–8, 367
dualism 42–3, 53, 170–1, 181, 186, 193–4, 312, 331, 347, 361
Dworkin, R. 137, 335

E
ecocentrism 172
Eddington, A. 310, 336
Edwards, R. 239–40, 295, 326, 336
Eisenstadt v. Baird 231, 233, 336
Elliott, C. 312, 336
embryo 15, 132, 144, 151–4, 164, 202–4, 209–10, 217, 221–2, 225, 227, 229–30, 240–62, 288–9, 295–6, 304, 338–9, 342, 344, 352, 361–2, 372
embryonic stem cells 150, 152, 154, 161, 164, 239–43, 250, 259–60, 262, 284, 287, 346, 354, 372

emergence 21, 44, 69, 83, 172, 180,
    183, 185, 196, 198, 324, 335,
    345
emotions 98, 106, 108, 110, 112, 114–15,
    178–9, 216, 228, 312, 324, 332,
    335, 351, 355–8, 364
empathy 100, 106–8, 118, 122, 310–11,
    314, 325, 334, 338, 341, 346,
    348, 353–5, 360, 364
end of life 15–16, 74, 264–6, 269, 271,
    273–4, 277, 279–81, 300, 304,
    329, 371
enhancement 17, 75, 133, 203, 217,
    219, 235, 287, 295, 303, 311–16,
    328, 330–1, 358
equality 4–5, 10, 12, 22, 55, 63, 65,
    105, 109, 133, 145, 149–50,
    153–4, 157–9, 164, 180, 231–3,
    331, 357, 359
essentialism 90, 94, 182, 252–3; the
    problem of 252
eugenic abortion 224
euthanasia 7, 9–10, 12–13, 50, 55,
    62–3, 95, 98–9, 132, 158, 162,
    264, 276–8, 281, 335, 337–8
evolution 10, 12, 33, 44, 69, 84, 104,
    114, 123, 166–8, 170–1, 175,
    179, 183, 185–94, 197–8, 266–7,
    285, 324, 326–31, 334–8, 341–3,
    345, 350, 352–3, 355, 357, 359,
    362–4, 366–7
existentialism 13, 34, 172–3, 186,
    195–6, 358

F
face 26, 34, 37, 75–6, 79, 101, 112–14,
    121, 134, 166, 170, 175, 229,
    234, 251, 266–7, 276, 280–1,
    337, 345, 348, 357, 363–4, 366
fairness 107–10, 120–1, 328, 332, 340,
    353, 360, 362
Farah, M. 303–4, 314, 336, 339
Fehr, E. 105, 110, 336–7, 341
fertility 153, 201, 257, 260–2, 284,
    335, 342
fetus 14, 147, 150, 153, 160, 162–4,
    203–4, 210–11, 220, 224–37,
    283–4, 291, 296, 339
fideism 172
Finnis, J. 27, 50, 56, 65, 163, 337, 340
First cause 169, 183–4, 186–7, 198
Fischbach R. 295, 337
Foucault M. 70, 90, 337
FOXP2 285, 336–7, 346

Freud, S. 33, 168, 186–7, 196, 337
Fukuyama, F. 289, 302, 304, 318, 338

G
Gage, L. 198, 338
Gage, P. 301, 332
gait 111–17, 325, 327–8, 332, 348,
    351, 367
Galileo 28, 193, 338, 352, 357
Gandhi 80
Gazzaniga, M. 152, 181, 318, 338–9
gene 144, 208, 210, 243, 285, 310,
    331, 333, 336
genealogy 173
George, R. 37, 44, 49, 51, 57, 60, 103,
    195, 225, 243, 248, 252, 258,
    328, 338, 347, 372
germline interventions 133, 135, 208
Gewirth, A. 120, 338
Glannon, W. 313, 338
Glendon, M. 80, 82, 129, 338
Gobl, C. 117, 338
God 6, 8–9, 12–13, 23–8, 30, 33,
    38, 40–2, 44, 50–6, 61–79,
    81–6, 103, 146, 167, 169, 172,
    174–5, 182–5, 187–90, 193–4,
    197–8, 206–7, 221–3, 225, 235,
    238, 246, 253, 274, 276, 279,
    288–9, 296–8, 308, 310, 312,
    326, 328, 331–2, 338–43, 345,
    347–8, 350, 353, 357, 360–1,
    366
God of the gaps 183
Gonzales v. Carhart 14, 148, 161–3,
    220, 231, 233, 338
Gormally, L. 49, 53, 60, 339
Gould, S. 12, 171, 175, 186, 190–1,
    194, 196, 339, 366
great apes 96–7, 101–2, 177, 292–3
great-souled man (Aristotle) 21, 39
Greely, H. 303, 311, 339
Green, J. 42, 171, 188
Green, R. 238, 261–3, 326–7, 339–40,
    342, 350, 372
Gros Espiell, H. 135, 340
Guenin, L. 164, 263, 340
Gustafson, J. 160, 279–80, 340

H
Habermas, J. 82, 134, 210–11, 340
Haidt, J. 110, 120, 123, 340
Hare, J. 68, 340
Harris, J. 59–60, 136, 254, 256, 327,
    337, 339–41

Hauerwas, S. 71, 75–6, 78–9, 327, 340
Healey, S. 70, 82, 361, 369
healing imperative, the 249
health care 50, 63–4, 98, 140, 144, 260, 266–7, 272–4, 277–8, 281, 298, 304–6, 323, 328, 331, 334, 352, 360, 362, 369
Heidegger, M. 34, 41, 341
Heraclitus 22
Hickman, L. 10, 88, 92–5, 97, 99–102, 341
Hippocratic oath 260
Hobbes, T. 29, 32, 42, 341
holism 172–3, 194
Holland S. 14–15, 239–40, 242–4, 246, 248, 250, 252, 254, 256, 258, 260, 262, 342, 354, 370–1
Horton, R. 251, 342
Hospice 266, 271, 274, 280–1
House of Commons Science and Technology Committee 203, 209, 342
human: cloning 63, 134–5, 142, 144, 150–1, 154, 161–2, 164, 202, 211–14, 217, 222, 235, 292, 296, 298–9, 330, 349, 355, 359–60; exceptionalism 8, 12–13, 19–21, 23–5, 27, 29, 31, 33, 35–7, 39, 41, 43–5, 122, 166–70, 172–3, 176, 182, 187, 194–5, 197, 315; goods 62; nature 7, 17–18, 21–2, 24, 26–7, 35–6, 39, 56, 58, 78, 90, 171, 192, 194, 204, 209, 258, 274, 289, 300–1, 307–11, 315, 317–19, 324, 330, 335, 340–3, 351, 366, 369; worth 28, 72, 202, 314, 362
Humana Vitae 221
humanity 9, 24, 31, 40–1, 72–3, 75–7, 79–81, 83–4, 91, 96, 135, 137, 139, 212–13, 216, 223–4, 229–30, 233, 247–9, 251, 253–7, 259, 307, 312, 315, 320–1, 334, 339, 361
Hume, D. 30, 172, 188, 190, 194, 342
Huxley, T. 33, 44, 166, 188, 190–1, 197, 342

I
Ignatieff, M. 80–1, 84, 342
Illes, J. 311, 339
imago Dei 8–9, 12–13, 19–21, 23, 25–31, 33–7, 39, 41, 43, 45, 65, 67, 71–2, 76, 81, 84, 122, 146, 166, 170, 182–3, 185–7, 192–4, 253, 276
infertility 201, 207, 240, 260, 284, 295
instrumentalist 88, 93, 97, 100
insula 108, 113
intelligent design (ID) 197–8, 334, 363
in vitro fertilization 15, 63, 150–1, 201, 204–5, 207, 210, 235, 239–40, 250, 294–5, 359

J
James, W. 32, 86–7, 89–94, 100–1, 342
Jameson, F. 101, 343
Jeeves, M. 310, 319, 343
Jenkins, P. 83, 343
Jesus Christ 25, 41, 71–3, 75–8, 172, 323, 325, 339
Jing-Bao, N. 140, 343
Jonas, H. 133, 344
Jonsen, A. 159–60, 277, 344
Judaism 8, 20, 41, 173, 185, 194, 260, 274, 279, 288, 295, 325
Judeo-Christian 9, 12, 82, 85, 146

K
Kant, I. 6, 8, 15, 20, 31–3, 38–9, 43, 76, 103, 146, 248, 254–6, 288, 307, 344, 353, 359
Karpowicz, P. 288, 344
Kass, L. 5, 8, 19–21, 25, 28, 35, 37–8, 41, 168, 172, 186–7, 196–7, 205–6, 212, 219–20, 223–4, 238, 246, 279, 281, 288–9, 292, 298, 312, 316, 319, 344, 346, 372
Katz, J. 128, 345
kavod (glory) 24–5
Kevorkian, J. 79
Kierkegaard, S. 35, 345
Killmister, S. 103, 122, 345
King, M. L., Jr. 80
Kits Nieuwenkamp, J. 132, 345
Korsgaard, C. 178–9, 196, 345
Kramer, P. 311, 345
Kraynak, R. 6, 50, 64–5, 207, 345
Kriegeskorte, N. 112, 122, 346
Kurzweil, R. 316, 346
Kutukdjian, G. 135, 346

L
Landman, W. 139, 358
Landry, D. 247, 346
Lanza, R. 151, 346
Larson, D. 274, 346

Latour, B. 102, 346
Lauritzen, P. 239–40, 244, 346, 361–2
Lawler, P. 314, 346
Layman, C. S. 75, 346
Lee, P. 37, 44, 49, 51, 57, 60, 103, 195, 224–5, 238, 325, 347
Levinas, E. 89, 101, 347
Lewis, C. S. 176, 192, 196, 319, 347
Lewy, G. 79, 347
liberalism 50, 64–6, 153
liberty 12, 67, 82, 145, 147–9, 152–4, 156–9, 161, 163, 205, 207–8, 231–3, 238, 277, 279, 281, 314, 343–5, 350, 359–60, 372
limbic system 106–8, 115, 318
Lindemann, H. 228–9, 348
Lindsay, R. 95, 348
Linville, M. 68, 348
living will 272, 274
Locke, J. 42, 82, 190, 348
Loike, J. 16–17, 282, 284–90, 292, 294–8, 337, 348, 371
Luther, M. 78
Lyell, C. 34, 167, 171, 188, 191, 193, 348
Lyotard, J.-F. 9–10, 69–70, 87–102, 349

M
Macklin, R. 3–5, 19, 59, 87, 91, 103–4, 131, 146, 191–2, 203, 207, 213–14, 219, 244, 251–2, 278, 288–9, 304, 307, 318, 349
magisterium 54
magnanimity 21, 39
Mandela, N. 80
Mann, J. 128, 349
Manninen, B. 240, 255–8, 349
Marcel, G. 35, 248, 349
Margalit, A. 111, 159, 349
Maritain, J. 8, 35, 38, 50, 84–5, 192, 349
marriage 52, 61–2, 84, 206, 231
materialism 6, 12, 42, 167–8, 170, 174, 181–3, 187, 190–1, 196, 309, 361
Mathieu, B. 134, 349
McClelland, R. 10–11, 103–4, 106, 108, 110, 112, 114, 116, 118, 120, 122–3, 187, 349–50, 371
McCormick, T. 15–16, 264, 266, 268, 270–2, 274, 276, 278, 280, 323, 350, 371
McCrudden, C. 19–20, 35, 38, 103, 129, 141, 191, 350
McCune, J. 283, 350
McMahan, J. 56–8, 350

mechanism 12, 29–30, 42, 113, 188, 192
mechanistic 28–9, 36, 42, 160, 181, 188, 319, 324, 340, 373
medical ethics 37, 86, 128, 168, 191–2, 244, 275, 307, 309, 323–4, 345, 369, 371–2
Meilaender, G. 5, 19, 21, 37, 40–2, 45, 153, 159, 162, 182, 185, 191, 206, 209, 248, 307, 312–13, 319, 350
Melo-Martin, I. de 287, 291, 333
Menschenwürde 4, 15, 245–7, 249, 251–8, 275, 278–9, 327
Midgley, M. 137, 350
Mill, J. S. 32, 44, 79, 207, 277, 281, 350
Milton, J. 28, 350–1
miracles 198, 347
Mirkes, R. 312–13, 351
Mitchell, C. B. 307–8, 319, 330, 351
modernity 28, 32, 38, 49–50, 82, 100, 185–7, 367
monism 70, 331
Montepare, J. 114–15, 351
moral: argument 348; individualism 55–6; luck 228–9, 238, 348, 351
Moreland, J. P. 194, 238, 312, 334, 345, 348, 351
Morris, R. 351, 361
Mother Teresa 79
motion 111–13, 115, 184, 198, 324, 327, 329–30, 342, 353–4, 356
Murray, T. 201, 209, 351

N
Nagel, T. 170, 196, 238, 351
Nagy, A. 283, 351
narrative 28, 32, 35, 86–9, 91–3, 97, 101–2, 171, 174, 188, 193, 317
natural: death 266, 272–3, 288, 352, 365; law 52, 61–2, 65, 71, 327–8, 337, 372; theology 33, 187, 339, 348, 353, 357
Natural Death Act 272–3, 352, 365
naturalism 59, 67–70, 167, 172, 192, 326, 331, 345, 347, 355, 357
naturalistic fallacy 194
neanderthal genomics 352
Neuhaus, R. J. 49, 66, 253, 308, 352
neuroethics 301, 303, 345
neurotransmitters 118, 123, 301–2, 308
Nickel, J. 129, 352
Nietzsche, F. 8, 33–4, 44, 173–4, 195, 352
nihilism 12, 34, 169–70, 172–4, 181, 192, 196, 324

Non-Overlapping Magisteria (NOMA)
    12, 171, 194, 196
noninvasive prenatal diagnosis 326
Noonan, J. 287, 352
Nordenfelt, L. 15, 122, 245–7, 249,
    275–6, 278–9, 352
Nuremberg Code 128, 132, 323, 345
Nussbaum, M. 22–3, 39–41, 45, 197,
    215–16, 278, 337, 352–3

O
Oduncu, F. 247, 253–5, 353
ontology 68, 87, 136, 254, 358

P
Paley, W. 28, 33, 42, 167, 183, 187,
    190, 198, 353
palliative care 7, 264, 266, 271, 280
Palpant, N. 3–4, 6, 8, 10, 12, 14–16,
    18, 239–40, 242, 244, 246, 248,
    250, 252, 254, 256, 258, 260,
    262, 371–2
Paris, P. 80, 353
Parmenides 22
partial-birth abortion 148–50, 160–2,
    220, 232
Patient Self-Determination Act 160,
    273, 353
Pavlischek, K. 236, 353
Payne, C. 296, 353
Peirce, C. S. 99, 101, 354
Pellegrino, E. 306–7, 310, 328, 330,
    344, 350–2, 354, 359
Penn, D. 178–9, 198, 327, 354
Persaud, R. 309, 354
personalism 50
personhood 15, 27, 31, 52–3, 78, 102,
    129, 131, 144, 148, 150, 153,
    162–3, 165, 203, 228, 230, 233,
    248–9, 252–6, 259, 266, 312,
    326–7, 336, 340
Peters, T. 250, 257, 354
Phillips, R. S. 363
philosophical Darwinism 168, 170,
    173–4, 181, 183, 186, 191, 193
philosophical reductionism 306
physician-assisted suicide 144, 154–5,
    158, 160, 268
Pico della Mirandola 27, 354
Pinker, S. 19, 36, 59–60, 191, 196, 219,
    307, 318, 350, 354
Planned Parenthood v. Casey 14,
    148–9, 161–2, 220, 230–4, 236,
    355, 359

Plantinga, A. 170, 190, 196, 326, 355
Plato 22–3, 39–40, 54, 137, 172, 181–2,
    194, 331, 355
Pollard, K. 315, 355
Pontifical Academy for Life 206–7, 355
Pope John Paul II 8, 35, 49–56, 61–2,
    193, 339, 341, 343
Porges, S. 118–19, 123, 324–5, 334–5,
    341, 355
post-postmodernism 10, 88, 97, 100–1,
    341
postmodernism 10, 70, 87–8, 92, 94,
    96–7, 100–2, 341, 343
posture 111, 114–15, 117, 123, 356
potentiality 15, 39, 203, 231, 241, 248,
    256–8
pragmatics 88–9, 91–3, 96, 99, 373
prefrontal cortex 107–8, 112, 336
pregnancy 147, 150, 163, 201, 204,
    210–11, 220, 227–9, 231–3,
    236, 241
preimplantation genetic diagnosis
    (PGD) 86, 132, 164, 204, 210–11,
    217, 240, 247, 251, 293
prenatal screening 204, 208
President's Council on Bioethics 3–4,
    19, 152, 160, 164–5, 168, 192,
    202, 205, 220, 238, 307, 313,
    316–17, 319, 328, 330, 333–4,
    338, 344–7, 350, 352–5, 359,
    362, 372
Prince v. Massachusetts 231
privacy 12, 145, 149, 158–9, 162, 208,
    231, 272, 310–11
pro-choice 14, 226, 229–30, 366
pro-life 14, 222–3, 226, 332, 347, 361
progress 7, 133, 175, 243, 250, 257, 305,
    309, 320, 333, 339, 354–5, 365
prosody 111, 113–19, 327, 339, 346,
    354, 356, 364–5
Protestant 7, 9, 14, 28, 67–9, 71–3, 75,
    77, 79, 81, 83, 85, 203, 219–20,
    222, 351
psychotropic drugs 7, 13, 17, 300–7,
    309–11, 313–15, 317–21
Pufendorf, S. 31, 43, 356–7
punishment 10, 50, 56–7, 67, 103,
    108–11, 119, 121–2, 128, 143,
    328–9, 331–2, 335–7, 341,
    356–7, 366

Q
Quinlan 16, 162, 267–9, 271–3, 277,
    342

R

Rachels, J. 34, 56, 138, 168–9, 173, 177, 194–5, 356

Rae, S. 14, 194, 219–20, 222, 224, 226, 228, 230, 232, 234, 236, 238, 312, 351, 372

rationality 13, 17, 22–3, 27, 31, 40, 42, 58, 68, 70, 78, 91, 103, 119–20, 174, 177, 182, 185–6, 195, 249–50, 256, 258, 329, 357, 370

Rawls, J. 68, 130, 356

Reagan, R. 223, 225, 237, 331, 356

reason 5, 21–5, 27–30, 37, 42–3, 45, 51–4, 58, 60–1, 64, 68, 73, 84, 86, 101, 116–18, 130–2, 138–9, 141, 162, 170, 176, 180–1, 188, 196–7, 202, 213, 216–17, 222, 227, 240–1, 251–3, 255, 258, 262, 292, 301–2, 309, 315, 318, 329, 356

reciprocity 10, 105–11, 118–19, 246, 328, 335, 338, 344, 353, 356–7

reductionism 4, 36, 181, 192, 306–8, 361

Renaissance humanism 20, 39

reproduction 63, 133–4, 150, 160–1, 201–2, 206–7, 209, 221, 260, 262, 295, 298, 353, 355, 370

reproductive: cloning 4, 133, 135–6, 164, 203, 212–14, 296, 349, 363; freedom 75, 208–9; technologies 7, 13–14, 50, 63, 144, 201–11, 213–17, 235, 260, 342, 355, 357, 362, 369, 372

respect for persons 4–5, 16, 91, 131, 244, 264, 267, 279–81, 289, 307, 333

retribution 110, 329

rhetoric 4, 7, 13–15, 17, 19, 129, 187, 199, 220, 223, 226–8, 236–7, 239, 244–5, 248, 252, 259–60, 304, 317–20, 342, 366

rights 4–5, 8–11, 14–15, 19, 31, 35, 37, 43, 45, 50, 57, 64–5, 67–70, 72–5, 79–84, 86, 91, 103, 109, 120, 127–43, 147–8, 152, 157, 159, 161–2, 185, 192, 195, 206–9, 212–14, 216, 219, 223–9, 235–6, 245–9, 251–2, 254, 257–8, 272–3, 278–9, 281, 286–7, 290, 293, 306–8, 323–6, 328–31, 335–9, 342–3, 345–52, 356–9, 361–6, 369

right to die 155

Robertson, J. 201, 204, 208, 357, 360

Roe v. Wade 144, 161–3, 223, 233, 236, 238, 326, 353

Rorty, R. 10, 87, 90–3, 95–8, 100, 102, 336–7, 353, 357

Rose, S. 315, 357

Rosen, C. 284, 357

Rue, L. 34, 36, 168–9, 192, 194, 198, 357

Russell, B. 9, 68–9, 357

S

sanctity of life 74, 157, 222–3, 234, 238, 326–7, 348

Sartre, J.-P. 6, 34, 173–4, 358

Savulescu, J. 154, 315, 331, 358

Scherer, K. 117, 325, 339, 358

Schopenhauer, A. 8, 32–3, 44, 358

Schroeder, D. 104, 111, 288, 358

Schuklenk, U. 139, 358

Schulkin, J. 107–8, 123, 351, 358

Schulman, A. 20, 142, 146–7, 162, 359

science 7, 11–13, 17, 28–9, 34, 36, 42, 65, 87, 95, 116, 120, 125, 142–4, 151, 153, 159–60, 164, 168, 171–2, 174, 183–91, 193–4, 196, 203, 205, 207, 209, 224, 239, 241, 244, 260, 298, 302–4, 309, 311, 324–36, 339–42, 344–50, 352, 354, 357–63, 365–6, 369–73

Second Vatican Council 49, 51–5, 61, 63, 65, 359

secularization 30, 36, 185

selection technologies 203–4, 210–11

Sen, A. 139–40, 278, 359

Sententia, W. 314–15, 359

sentimental 91, 93, 96, 98, 100

Shell, S. 20, 244, 359

Siegel, R. 231–4, 359

signal 11, 44, 111, 114–17, 119, 123, 341

Silveira, M. 273–4, 360

Singer, P. 76, 78, 107, 122, 194–5, 315, 354, 360

Sinnott-Armstrong, W. 69, 360

Skinner, B. F. 33, 169, 179, 192, 360

slavery 5, 30, 33, 81, 128, 130, 185, 191, 198, 255, 335, 361, 365

Snead, O. C. 12, 142, 144, 146, 148, 150, 152, 154, 156, 158–60, 162, 164, 360, 372

social rank 21, 24, 32, 39, 41, 275

Soloveitchik, J. 288, 360

soul 23–4, 39–40, 42–3, 53, 55, 57–9, 62, 65, 170–1, 180–2, 193–4, 196, 207, 221, 253, 260, 292,

331–2, 337, 339, 342–3, 345, 350–1, 362–3
Spaemann, R. 137, 361
special creation 12, 166–8, 183, 188, 191
Stackhouse, M. 9, 70, 73, 80–2, 353, 361
status of the embryo 15, 153, 202–3, 210, 243–4
Steinbock, B. 204, 228, 361
stem cell 7, 12–13, 15–16, 142, 150–4, 157–61, 164–5, 239–44, 247, 249, 252, 259, 261–3, 283–4, 294, 297, 323, 327–31, 339, 342, 344, 346, 349, 351–5, 357–8, 360, 363, 370–2
stewardship of life 279
Stith, R. 225–6, 238, 361
Stoicism 39, 173
Stolberg, S. 151, 361
striatum 107, 113, 285
subaltern 351, 361
suffering 75–9, 86, 146, 152, 156–7, 187, 201, 250–1, 257, 259–60, 264, 266, 270–2, 275, 278–80, 288, 298, 304, 310, 372
Sulmasy, D. 19, 24, 37–9, 44–5, 49, 60, 131, 146, 182, 191, 362

T
Taylor, C. 30–1, 185, 362–3
teleology 42, 324
Tendler, M. 285–90, 292, 294, 297, 348
Teno, J. 273, 363
theism 9–10, 30, 36, 67–9, 71, 76–7, 86, 104, 120, 167, 174, 183, 187, 189–91, 193, 196, 331, 370
theistic evolution 183, 189
Thomson, J. J. 230, 236, 363
Tollefsen, C. 8–9, 49–50, 52, 54, 56, 58, 60, 62, 64, 66, 243, 248, 252, 258, 338–9, 363, 372–3
Tomasello, M. 105–6, 109, 122, 356, 363, 366
Tooby, J. 123, 331–2, 361, 363
Trivers, R. 105, 363
Tschaepe, M. 9–10, 86, 88, 90, 92, 94, 96, 98, 100, 102, 373
Turgenev, I. 87
Tutu, D. 80

U
UNESCO 4, 132, 134–6, 138–40, 142, 212, 278, 323–4, 340, 343, 358, 363, 365, 369, 372

United Nations 3, 19, 45, 80, 84, 86, 103, 127, 130, 136, 142, 192, 252, 308, 324, 331, 342, 364
United States Conference of Catholic Bishops 204, 207, 364
Universal Declaration of Human Rights 4, 11, 19, 86, 127–31, 136, 142, 147, 192, 208, 252, 308, 335, 338, 349, 352, 364
Universal Declaration on Bioethics and Human Rights 132, 138, 140, 142, 323, 369
Universal Declaration on the Human Genome and Human Rights 4, 346, 363
universalism 55, 79, 82, 139, 348
Unmoved Mover 41, 198
utilitarianism 68, 195, 248–9

V
vagus nerve 118
Vatican Congregation for the Doctrine of the Faith 204, 364
Volodin, V. 140, 365
vulnerability 36, 39, 146, 153, 216, 248, 250, 255, 321

W
Waal, F. de 106–7, 122, 174, 177, 195, 333, 345, 355
Wallace, A. R. 193, 365
Warren, M. 228, 365
Washington v. Glucksberg 16, 155–6, 161, 165, 268–9, 277, 280, 365
Wegner, W. 309, 365
Western culture 11, 20, 36, 41, 45, 67, 80–4, 127, 133, 139–40
Wilson, E. O. 30, 169–70, 174–5, 177–8, 180, 184, 188, 191–2, 194, 196, 366
Wiredu, K. 67, 366
Witte, J. 84, 361, 366
Wolf, N. 229–30, 236, 366
Wollstonecraft, M. 43–4, 366
women's rights 31, 228
World War II 11, 19, 45, 84, 103, 127–8, 136, 192, 244, 277, 351

Y
Yamanaka, S. 247, 362
Yong, A. 77–8, 367

Z
Zorich, Z. 287, 367